RESIDENTIAL AND COMMERCIAL ELECTRICAL WIRING

RESIDENTIAL AND COMMERCIAL ELECTRICAL WIRING

William J. Whitney

Albert Lea Area Vocational Technical Institute
Albert Lea, Minnesota

John Wiley & Sons

New York Chichester Brisbane Toronto Singapore

Library of Congress Cataloging in Publication Data:

Whitney, William J., 1918–
Residential and commercial electrical wiring.

Bibliography: p.
Includes index.
1. Electric wiring. I. Title.

TK3201.W54 621.319'24 82-2030
ISBN 0-471-09319-X AACR2

Printed in the United States of America

10 9 8 7 6 5 4 3 2

To my wife Elizabeth
My active partner
for 40 years

PREFACE

The purpose of this book is to aid you in your study of residential and commercial wiring. The text presents in nontechnical terms the mathematical theory behind the electrical devices and equipment used in residential and commercial construction. I draw on my experience as a journeyman electrician, electrical contractor, apprentice, adult educator, and vocational technical institute instructor, to explain, as simply as possible, electrical wiring as it applies to single- and multiple-family dwellings, farm operations, store buildings, mobile and recreational vehicle parks, and commercial industry. My aim has been to downplay the use of complex technical terms and to emphasize the reasons for electrical installations and practices. Much of the text and many of the illustrations and examples have been reviewed by manufacturers and others to ensure that the content is up to date and corresponds with the 1984 National Electrical Code.

Because of our changing times, I place emphasis on intermediate metal conduit, PVC conduit, and other new materials and equipment now available to the electrical industry.

It is my hope that high school students, vocational students, Vocational Technical Institute students, apprentices, tradespeople, adult classes, and newcomers to the industry will find this book a useful tool for understanding residential and commercial wiring.

I think you will enjoy the book's format. It includes instructional objectives and several self-evaluation questions at the beginning of each chapter, text discussion with a summary, and problems at the end of each chapter. This book is designed to focus your attention on what's important, point you in the right direction, and make you sure you understand the material in each chapter.

As you read the book, you will notice that the National Electrical Code is included in the text material; however, it is suggested that a copy of the 1984 National Electrical Code be referred to while studying this material.

The material covered in this text has been classroom tested by second-year construction electrician students at the Albert Lea, Minnesota Area Vocational Technical Institute.

I gratefully acknowledge the support of the Wiley staff and the tireless work of Mrs. Irene Holz, who typed the manuscript and its many revisions.

William J. Whitney

ACKNOWLEDGMENTS

Many manufacturers of equipment have contributed illustrations, photos, written materials, and suggestions and their efforts, time, and assistance have been of great value. I would like to thank each of them for their contributions.

B and K Lumber Company. Northwood, Iowa
Underwriters' Laboratories, Inc. Chicago, Illinois
Raco, Inc. South Bend, Indiana
Lennox Industries, Inc. Marshalltown, Iowa
Brown and Sharpe Mfg. Co. Providence, Rhode Island
Martin Industries. Florence, Alabama
Ideal Industries. Sycamore, Illinois
General Electric Company. Plainville, Connecticut
Pass and Seymour. Syracuse, New York
The Singer Company. Carteret, New Jersey
Blackhawk Industries. Dubuque, Iowa
Midland-Ross Corporation. Pittsburgh, Pennsylvania
Square D. Company. Lexington, Kentucky
Nutone Division. Scovill Mfg. Co. Cincinnati, Ohio
Cutler-Hammer, Inc. Milwaukee, Wisconsin
Essex International. Fort Wayne, Indiana
Albert Lea Vocational Technical Institute. Albert Lea, Minnesota
Elektra Systems, Inc. Farmingdale, New York
Progress Lighting. Philadelphia, Pennsylvania
ERICO Products, Inc. Cleveland, Ohio
Jefferson Electric. Bellwood, Illinois
Crouse-Hinds-Arrow Hart Div. Hartford Connecticut
Carlon Products Corporation. Cleveland, Ohio
ITT Hobub Industries. Sycamore, Illinois
Lightcraft of California. Philadelphia, Pennsylvania
Teledyne Big Beam. Crystal Lake, Illinois
Edwards Company, Inc. Norwalk, Connecticut
Midwest Electric Products. Mankato, Minnesota
The Aluminum Association. Washington, D.C.
The L.S. Starrett Co. Athol, Massachussetts
The Wiremold Company. West Hartford, Connecticut

Burndy Corporation. Norwalk, Connecticut
ITE Imperial Corporation. Farmingdale, New York
Ronk Electrical Industries, Inc. Nokomis, Illinois
General Electric Lamp Market. Cleveland, Ohio
Onan Div., Onan Corp. Minneapolis, Minnesota
Republic Steel Corporation. Cleveland, Ohio
Allied Tube & Conduit Corporation. Harvey, Illinois
Touch-Plate Electro-Systems, Inc. Paramount, California
Thomas & Betts Corporation. Raritan, New Jersey
City of Albert Lea, Minnesota
State of Minnesota

W.J.W.

CONTENTS

PART 1 INTRODUCTION

PART 2 METHODS AND MATERIALS

RESIDENTIAL AND COMMERCIAL ELECTRICAL WIRING

PART 1

INTRODUCTION

chapter 1

BASIC ELECTRICITY

Instructional Objectives

1. To learn how Ohm's law applies to a circuit.
2. To solve simple Ohm's law problems.
3. To understand the formula for making Ohm's law.
4. To name and define the units of current, electromotive force (emf), and resistance.
5. To distinguish between emf and potential difference (pd).

Self-Evaluation Questions

Test your prior knowledge of the information in this chapter by answering the following questions. As you read the chapter, watch for the answers. When you have completed the chapter return to this section and answer the questions again.

1. State Ohm's law when voltage and resistance are known.
2. State Ohm's law when voltage and current are known.
3. State Ohm's law when current and resistance are known.
4. What effect does resistance have on voltage delivered to the load?
5. What is meant by the term "drop of potential"?
6. Name the three primary circuit elements.

Electricity alone is a subject that would require several chapters to cover the basics. However, in this chapter we discuss basic electricity as it relates to electrical wiring for the construction industry.

1-1 OHM'S LAW

Experience with direct-current circuits shows that the current established in closed metallic paths is directly proportional to the emf of the source of electricity in the circuit and inversely proportional to the resistance of the path. These facts were first exposed by Ohm, and the relation between the three factors involved is known as Ohm's law. Since its first phrasing in 1827, this law has had outstanding importance in electrical calculations.

Ohm's law states that the current in a metallic circuit is equal to the emf available in the circuit divided by the resistance of that circuit. In order to unite the formula for this law, let

I = current maintained in the circuit
E = emf of the source of electricity included in the circuit
R = total resistance of the circuit, including the internal resistance of the source

Then, by the foregoing statement of Ohm's law,

$$\text{Current} = \frac{\text{emf}}{\text{resistance}}$$

The value of any one of the three factors can be calculated when the value of the other two are known.

To find the current in a circuit when the emf and resistance are known, divide the emf by the resistance

$$I = \frac{E}{R}$$

Introducing the units in Ohm's law, ampere for current, volt for emf, and ohm for resistance, it follows that in a metallic circuit

$$\text{Amperes} = \frac{\text{volts}}{\text{ohms}}$$

EXAMPLE 1

A particular incandescent lamp is connected to an electric generator which develops an emf of 110 volts; under these conditions the lamp has a resistance of 275 ohms. What current will the lamp take?

The current is given by Ohm's law as

$$I = \frac{E}{R} = \frac{110}{275} = 0.40 \text{ ampere}$$

EXAMPLE 2

An electric heater has a resistance of 20 ohms and another has a resistance of 40 ohms. Two sources of emf are available, one of 240 volts and the other of 120 volts. Find how much current each heater will take when connected to the source.

The formula $I = E/R$ is used to complete the results desired; first take $R = 20$ ohms with $E = 240$ volts and then $E = 120$ volts.

(a) For the 20-ohm heater operating on 240 volts

$$I = \frac{240}{20} = 12 \text{ amperes}$$

(b) on 120 volts

$$I = \frac{120}{20} = 6 \text{ amperes}$$

(c) For the 40-ohm heater operating on 240 volts

$$I = \frac{240}{40} = 6 \text{ amperes}$$

(d) On 120 volts

$$I = \frac{120}{40} = 3 \text{ amperes}$$

This second example shows the effect of changing the available emf or changing the resistance. Halving the emf reduces the current to one-half, as indicated by results (a) and (b), or by (c) and (d). Doubling the resistance also reduces the current to one-half, as indicated by results (a) and (c), or by (b) and (d). These numerical results illustrate the statement of Ohm's law that the current varies directly with the emf and inversely with the resistance.

To find the resistance of a circuit when the emf and current are known, divide the emf by the current.

$$R = \frac{E}{I} \qquad \text{ohms} = \frac{\text{volts}}{\text{amperes}}$$

EXAMPLE 3

What is the resistance of the heater in an electric iron that takes 5.7 amperes when operated at 115 volts?

Ohm's law gives the resistance of the heater as

$$R = \frac{E}{I} = \frac{115}{5.7} = 20.2 \text{ ohms}$$

EXAMPLE 4

A 240-volt rheostat has two resistors and each is equipped with a switch so that one or both of them may be placed in the circuit. When one switch is closed the current is 5 amperes, and when the other switch is closed the current is 10 amperes. Find the resistance of the rheostat under each of these conditions.

$$R = \frac{E}{I} = \frac{240}{5} = 48 \text{ ohms}$$

With the second switch closed the resistance is

$$\frac{240}{10} = 24 \text{ ohms}$$

With both switches closed the resistance is

$$\frac{240}{15} = 16 \text{ ohms}$$

To find the emf needed to maintain a certain current in a circuit of known resistance, multiply the current by the resistance (remember formula $I = E/R$ this time for emf).

$$E = I \times R \text{ volts} = \text{amperes} \times \text{ohms}$$

EXAMPLE 5

A telegraph sounder requires a current of 0.2 ampere to operate it satisfactorily. If the sounder circuit has a total resistance of 40 ohms, what emf must be supplied to the circuit?

The answer is given by Ohm's law as

$$E = I \times R = 0.2 \times 40 = 8 \text{ volts}$$

All three forms of the equation that represents Ohm's law can be visualized by the fraction

$$\frac{E}{I \times R}$$

If a problem seeks the value of the current I, cover I in this fraction with a pencil and the uncovered parts show that I is given by E/R ($I = E/R$). Again, if a problem calls for the resistance R, cover R and the visible part of the fraction is E/I ($R = E/I$). Lastly, if the emf E is sought, cover E and the result is $I \times R$ ($E = I \times R$).

1-2 APPLICATION OF OHM'S LAW TO PARTS OF CIRCUITS

Ohm's law applies equally well to a part of a circuit as it does to an entire circuit, however, care must be exercised to have all quantities apply to the part involved. Figure 1.1 shows a circuit with several resistors, one of them marked R, connected to a battery marked B. The usual representation of a battery is a group of parallel lines alternately long and short, as indicated.

Let

R_1 = resistance of one part of a circuit
I = current in that part of the circuit
E_1 = potential difference across the same part, the circuit

Then by Ohm's law,

$$I = \frac{E}{R}$$

It should be noted particularly that the numerator in this equation is the difference of potential across a part of a circuit, whereas in $I = E/R$, the

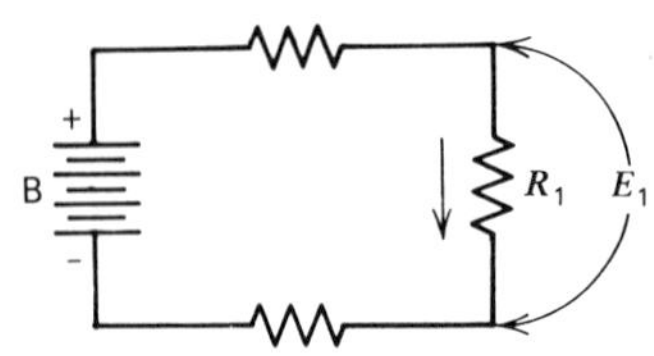

Figure 1.1 Application of Ohm's law to part of a circuit.

numerator is the electromotive force supplied by the source of electricity to the entire circuit.

Formula $I = E/R$ can be transposed in order to solve problems where it is desired to compute the resistance or the potential difference. The equations will be like formulas $R = E/I$ and $E = I/R$, except that the emf E will be replaced by the potential difference E. For example, to determine the potential difference (pd) necessary to establish a current through any part of a circuit, the current through it is multiplied by the resistance of that part of the circuit, that is,

$$E = I \times R$$

To find the drop in potential across a resistor carrying current, multiply the current by the resistance of the resistor.

EXAMPLE 6

Among the devices connected to a source of electricity in the manner indicated in Figure 1.1 is a resistor having a resistance of 3 ohms. What pd will there be at the terminals of this resistor when the current in it is 8 amperes?

$$E = I \times R = 8 \times 3 = 24 \text{ volts}$$

If the emf available for the circuit is changed, or if the resistance of the other devices is altered, the current through the resistor R will be different and consequently the potential difference across it will change accordingly.

1-3 EFFECT OF INTERNAL RESISTANCE OF SOURCE

All sources of electricity possess some resistance, even though they are designed to have very little, and as a result some difference of potential must be involved in establishing current through these sources themselves. Because of this, not all of the emf of a source is available for use in its external circuit, and the loss is spoken of as the drop in potential due to internal resistance.

Let

R = total external resistance of the circuit
r = internal resistance of the battery or generator
E_d = drop in potential across either source of electricity

Then the total resistance of the circuit (external + internal) is $R + r$, and Ohm's law becomes

$$I = \frac{E}{R + r}$$

This current multiplied by the internal resistance gives the potential drop in the source, or

$$E_d = I \times r$$

This result is obtained from formula $E_1 = I \times R$ by using the internal resistance r instead of the total resistance R.

To find the drop in potential due to the current through the source of electricity, multiply the current by the internal resistance of the source.

EXAMPLE 7

A storage battery, having an emf of 2.0 volts and an external resistance of 0.05 ohm, is connected to an electromagnet that has a resistance of 0.35 ohm. How much of the emf is used in establishing current through the battery and what is the potential difference available for supplying current to the electromagnet?

The total resistance of the circuit is the sum of the internal resistance of the battery and the resistance external to it, that is, $R + r = 0.35 + 0.05 = 0.40$ ohm.

The current in the circuit is

$$I = \frac{E}{R + r} = \frac{2.0}{0.35 + 0.05} = 5 \text{ amperes}$$

This current through the internal resistance of the battery produces a drop of potential amounting to

$$E_d = I \times r = 5 \times 0.05 = 0.25 \text{ volt}$$

Therefore the potential difference available at the electromagnet is $2.00 - 0.25 = 1.75$ volts.

1-4 WIRING CALCULATIONS

The determination of the size of wire required to conduct current from generators or service mains to lamps, motors, or other electrical devices is of frequent occurrence in electrical calculations. It is, of course, desirable to keep such line wires as small as possible for the sake of economy; however, if they are too small, their resistance will be unduly high and the potential drop along the line will be excessive. As a result, the potential difference available at the devices may be too low to operate them satisfactorily, or when lamps are turned on and off the light from the lamps still in circuit will fluctuate. It is common practice to restrict the potential drop along the line wires to less than 3 percent on lighting circuits and 5 percent on power circuits.

A second objection to undersized line conductors is the excessive heating that the current would produce in them, as well as the damage that might occur to the insulation surrounding the wires as a result of heating. For safe operation the current should not exceed the values given for allowable ampacities of insulated conductors in Table 310-16 of the National Electrical Code (Table 1.1).

SUMMARY

1. Ohm's law states that the current in a metallic circuit is equal to the emf available in the circuit divided by the resistance of that circuit.
2. Ohm's law applies equally well to part of a circuit as it does to an entire circuit.

TABLE 1.1 Allowable Ampacities of Insulated Conductors Rated 0-2000 Volts, 60° to 90°C (NEC Table 310-16)

Not More Than Three Conductors in Raceway or Cable or Earth (Directly Buried), Based on Ambient Temperature of 30°C (86°F)

Size	Temperature Rating of Conductor, See Table 310-13								Size
	60°C (140°F)	75°C (167°F)	85°C (185°F)	90°C (194°F)	60°C (140°F)	75°C (167°F)	85°C (185°F)	90°C (194°F)	
AWG MCM	Types †RUW, †T, †TW, †UF	Types †FEPW, †RH, †RHW, †RUH, †THW, †THWN, †XHHW, †USE, †ZW	Types V, MI	Types TA, TBS, SA, AVD, SIS, †FEP, †FEPB, †RHH, †THHN, †XHHW*	Types †RUW, †T, †TW, †UF	Types †RH, †RHW, †RUH, †THW, †THWN, †XHHW, †USE	Types V, MI	Types TA, TBS, SA, AVB SIS, †RHH, †THHN, †XHHW*	AWG MCM
	Copper				Aluminum or Copper-Clad Aluminum				
18	....	....	....	14	....	....	....	....	....
16	....	....	18	18	....	....	....	....	....
14	20†	20†	25	25†	....	....	....	....	....
12	25†	25†	30	30†	20†	20†	25	25†	12
10	30†	35†	40	40†	25†	30†	30	35†	10
8	40	50	55	55	30	40	40	45	8
6	55	65	70	75	40	50	55	60	6
4	70	85	95	95	55	65	75	75	4
3	85	100	110	110	65	75	85	85	3
2	95	115	125	130	75	90	100	100	2
1	110	130	145	150	85	100	110	115	1
0	125	150	165	170	100	120	130	135	0
00	145	175	190	195	115	135	145	150	00
000	165	200	215	225	130	155	170	175	000
0000	195	230	250	260	150	180	195	205	0000
250	215	255	275	290	170	205	220	230	250
300	240	285	310	320	190	230	250	255	300
350	260	310	340	350	210	250	270	280	350
400	280	335	365	380	225	270	295	305	400
500	320	380	415	430	260	310	335	350	500
600	355	420	460	475	285	340	370	385	600
700	385	460	500	520	310	375	405	420	700
750	400	475	515	535	320	385	420	435	750
800	410	490	535	555	330	395	430	450	800
900	435	520	565	585	355	425	465	480	900
1000	455	545	590	615	375	445	485	500	1000
1250	495	590	640	665	405	485	525	545	1250
1500	520	625	680	705	435	520	565	585	1500
1750	545	650	705	735	455	545	595	615	1750
2000	560	665	725	750	470	560	610	630	2000

3. All sources of electricity possess some resistance, even though they are designed to have very little.
4. It is desirable to keep line wires as small as possible for the sake of economy.
5. If conductors are too small, resistance will be unduly high and a potential drop along the line will be excessive.
6. It is common practice to restrict the potential drop to less than 3 percent.

PROBLEMS

1-1 A voltage of 20 volts is required to force a current of 5 amperes though a coil. What is the resistance of the coil?

1-2 How much current will a 30-ohm electric iron take when operated on 115-volt service mains?

1-3 What is the resistance of a heating coil that draws a current of 5.8 amperes when connected across 118 volts?

1-4 A sound system requires a current of 3 amperes to operate satisfactorily. If the system circuit has a total resistance of 40 ohms, what emf must be supplied to the circuit?

1-5 The voltage across the leads of a resistor measures 18 volts and its resistance is 6 ohms. What current will flow through it?

1-6 What is the potential difference across a bank of lamps that has an equivalent resistance of 8.2 ohms and in which the current is 15 amperes?

1-7 To find emf needed to maintain a certain current in a circuit of known resistance, ______________ the current by the ______________.

1-8 Ohm's law states that the current of a circuit is ______________ ______ the emf available in the circuit divided by the resistance of the circuit.

1-9 To find the current in a circuit when the emf and resistance are known, ______________ the emf by the ______________.

1-10 To find the resistance of a circuit when the emf and current are known, ______________ the emf by the ______________.

chapter 2

GENERAL INFORMATION

Instructional Objectives

The purpose of the book is to explain the how and why of the electrical installation so that it will be safe, economical, and convenient to the builder.

1. To learn how the electrical wiring information is conveyed to the electrician.
2. To develop familiarity with working drawings, specifications, electrical symbols, and drawing notations used in architectural plans.
3. To become aware of electrical standards.
4. To be aware of the need of building permits and inspections.
5. To learn about listing and labeling of products.

Self-Evaluation Questions

Test your prior knowledge of the information in this chapter by answering the following questions. As you read the chapter, watch for the answers. When you have completed the chapter, return to this section and answer the questions again.

1. What is a working drawing?
2. How does the architect convey instructions to the electrician on a job site?
3. Why is cooperation between the crafts essential?
4. What name is given to the reproduction of the working drawing?
5. How would you define a floor plan?
6. How does the floor plan differ from a plot plan?
7. Why is it important for the electrician to understand the specifications before starting the job?
8. What is the purpose of using standard electrical symbols?
9. Why do we need electrical standards to protect the builder?
10. What is the National Electrical Code?

2-1 THE WORKING DRAWING

Generally a prospective owner or builder will meet with an architect to discuss the planning of the new house. The architect will ask such important questions as:

1. How large a family do you have?
2. What is the size and shape of your lot?
3. Where is your lot located?
4. What quality materials do you want used?
5. What name brand of equipment do you prefer?
6. Do you intend to select and purchase the fixtures?
7. In what price range do you consider building?
8. What is your preference in layout and design?

When these questions are answered by the owner or builder, the architect is ready to create a set of drawings containing all of the information and dimensions necessary to successfully complete the project. These drawings are referred to as working drawings. Reproduction of the drawings are called a set of blueprints. The architect uses this set of blueprints to convey instructions to all of the crafts who are to plan, erect, and complete the structure.

The newcomer, apprentice, or electrician finds it an advantage to be able to read a set of blueprints early in the job experience. When this ability is acquired, that person becomes a part of a team representing many skills working out a series of construction problems. "Each must know how to 'take off' dimensions accurately so that all of the outlets and equipment are located according to the blueprint."

All crafts must follow the print carefully if the work is to progress smoothly. Cooperation between crafts is essential because the work of one person hinges on that of another in many instances. The carpenter sets the windows and door frames for the brick mason, and also must provide floor and wall framing for the plumber before the piping and fixtures for the bathroom can be installed. Sheet metal workers must provide flashing and gutters in cooperation with roofers, carpenters, and brick masons. A fine sense of good will is essential and generally found among workers as they work toward the common objective of creating a building in the most efficient manner.

A Set of Working Drawings

When the owner/builder, developer, or contractor decides to construct a commercial building, an architect is usually commissioned to create a set of blueprints called working drawings and to write the specifications for the project.

In nearly all cases, regardless of the project size, the working drawings will consist of the following (see Figure 2.1):

1. *Site Plan.* This plan shows the location of the building on the property, and also shows the location of the existing or new utilities such as electricity, telephone (overhead or underground), water, and sewer.

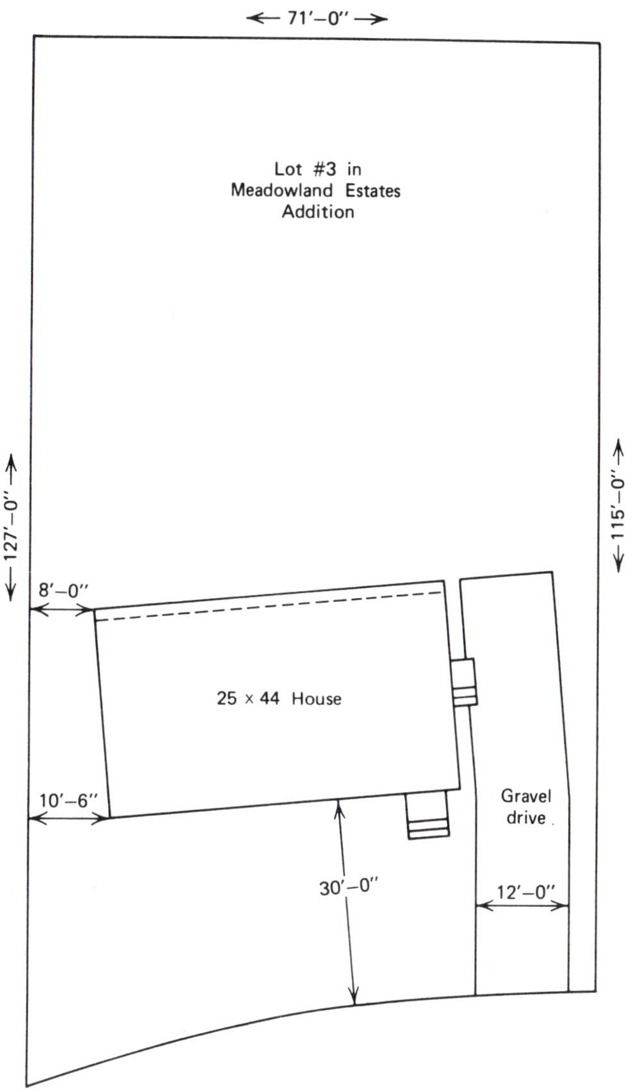

Plot Plan
1" = 12'–0"

(a)

Front Elevation

(b)

Figure 2.1 The working drawing. *(a)* Plot plan. *(b)* Front elevation. *(c)* Right elevation. *(d)* Left elevation. *(e)* Detail drawing: of *(b)* and *(c)*. *(f)* Sectional drawing. *(g)* Floor plan. (Courtesy of B & K Lumber Company)

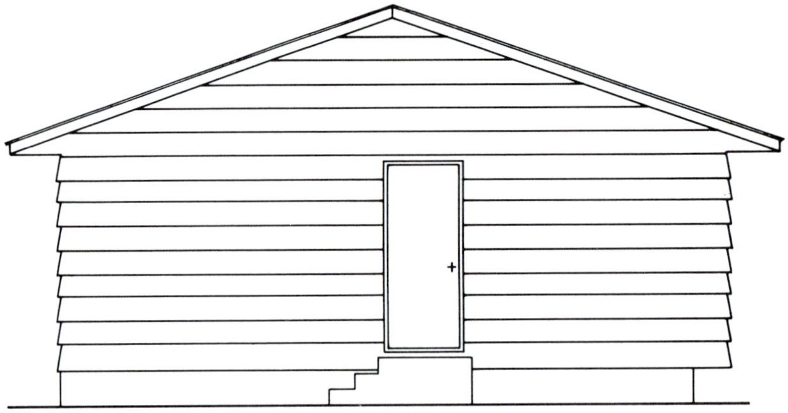

Right Elevation

(*c*)

Left Elevation

(*d*)

2. *Architectural Plans*
   (a) *Floor Plans.* Shows walls, partitions, doors, and windows for each floor or level of the building.
   (b) *Elevations Drawing.* Shows all exterior faces of the building.
   (c) *Sectional Drawing.* Shows the various details of the building such as the roof structure, ceilings, stairways, and similar details.
   (d) *Detail Drawings.* Shows interior trim and cabinet details.
   (e) *Electrical Drawings.* These drawings cover the complete design and layout of the wiring for lighting, power, signal and communication equipment, and special electrical systems. Floor plans show the crafts on the project the location of panelboards, power risers, power outlets, receptacles, lighting, and lighting fixtures, and also show listing of symbols, schedules, and schematic diagrams.
   (f) *Mechanical Drawings.* These drawings cover the complete layout and design of the heating, ventilating, air conditioning, plumbing, and related mechanical construction of the building. These drawings include floor plans of all levels which show air-ducts, water and sanitary pipes, and any other mechanical equipment.

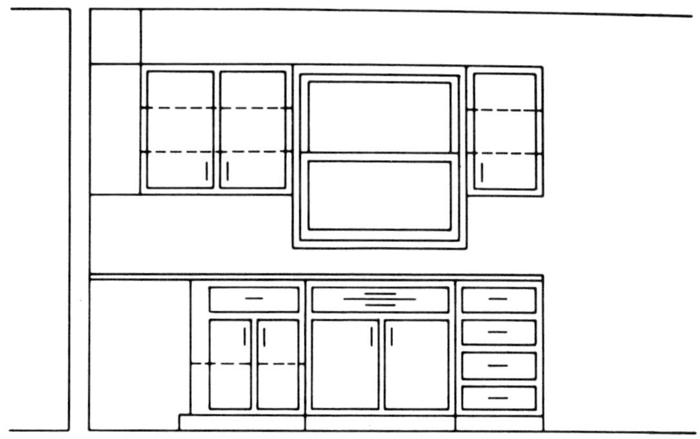

Cabinet View "A"

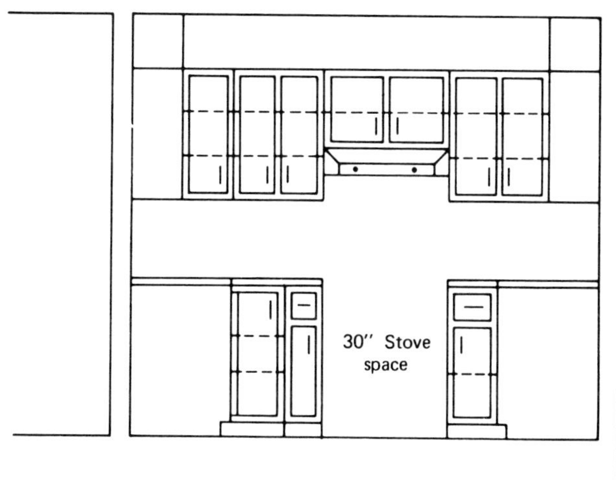

Cabinet View "B"

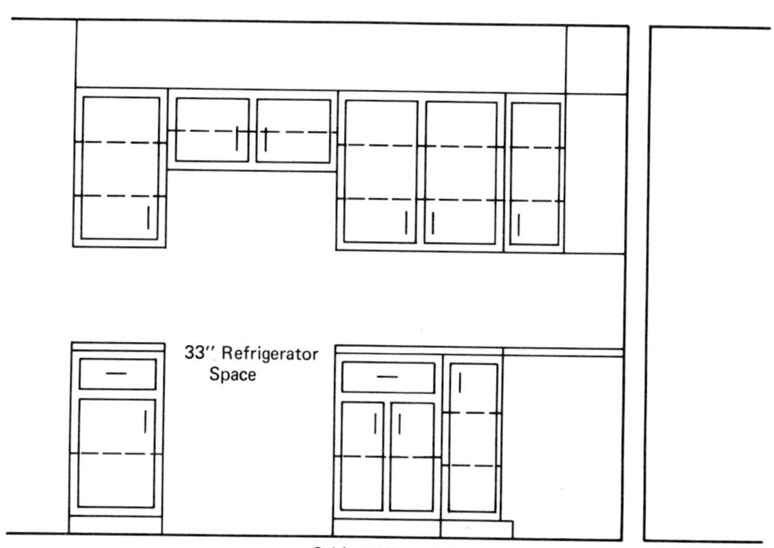

Cabinet View "C"

(*e*)

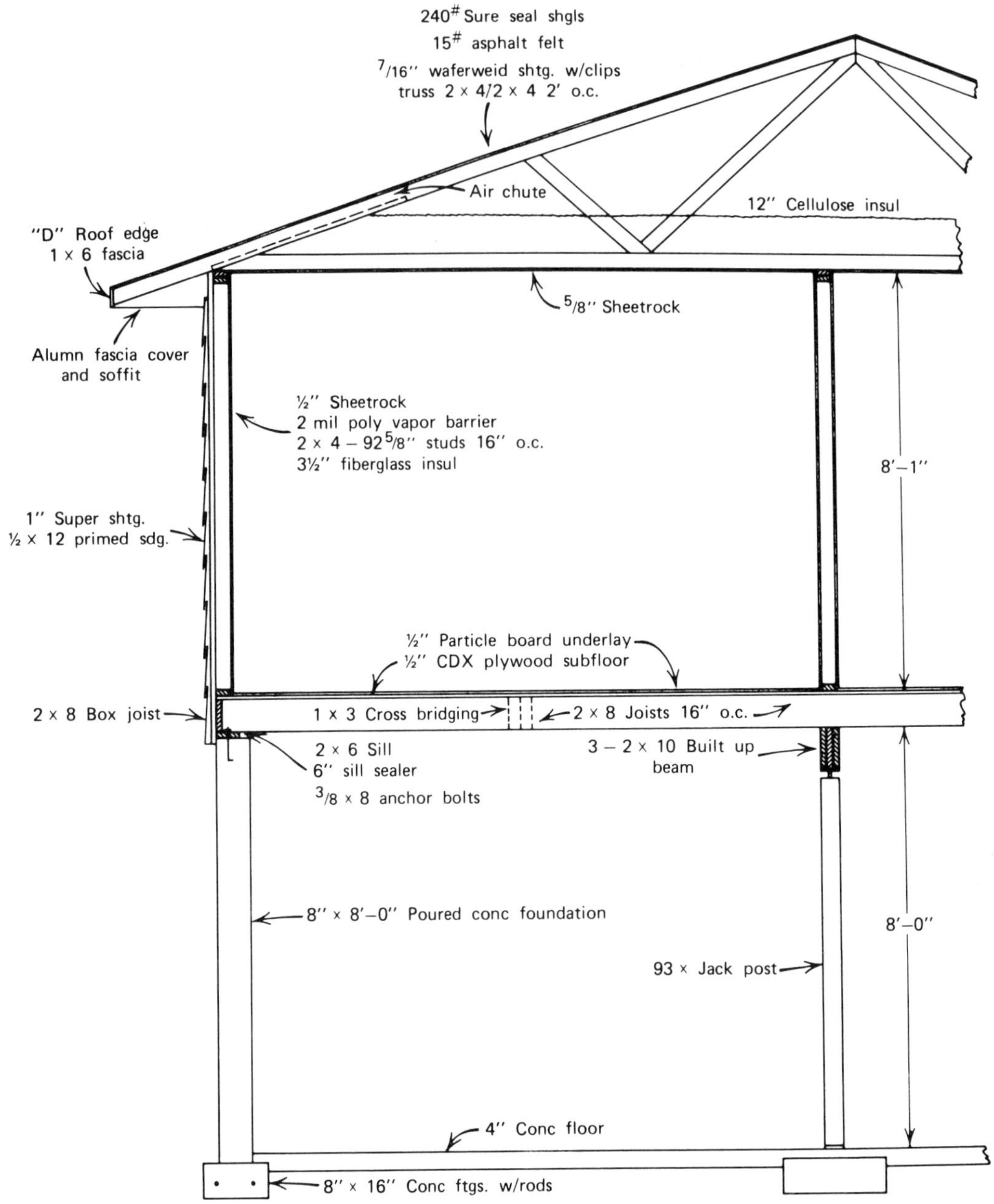

(f)

2-2 SPECIFICATIONS

Specifications are a vital part of the building plans. They are the rules governing the type, kinds, quality, and colors of the materials used and the work to be performed by all of the crafts on the construction project. When the details are clear in the specifications, there is little room for misunderstanding or misinterpreting the plans. The specifications enable the electrical contractor to estimate the bid in terms of the legal responsibilities, guarantees, permits, inspections, brand names, and quality of the materials required. Lighting fixtures are usually chosen by the owner. However, they are usually purchased through the contractor, so the architect specifies a certain amount of money in the specs for a fixture allowance. Specifications for a job are written in sections. Each section pertains to a different craft.

It is most important for the electrician to review all of the drawings and

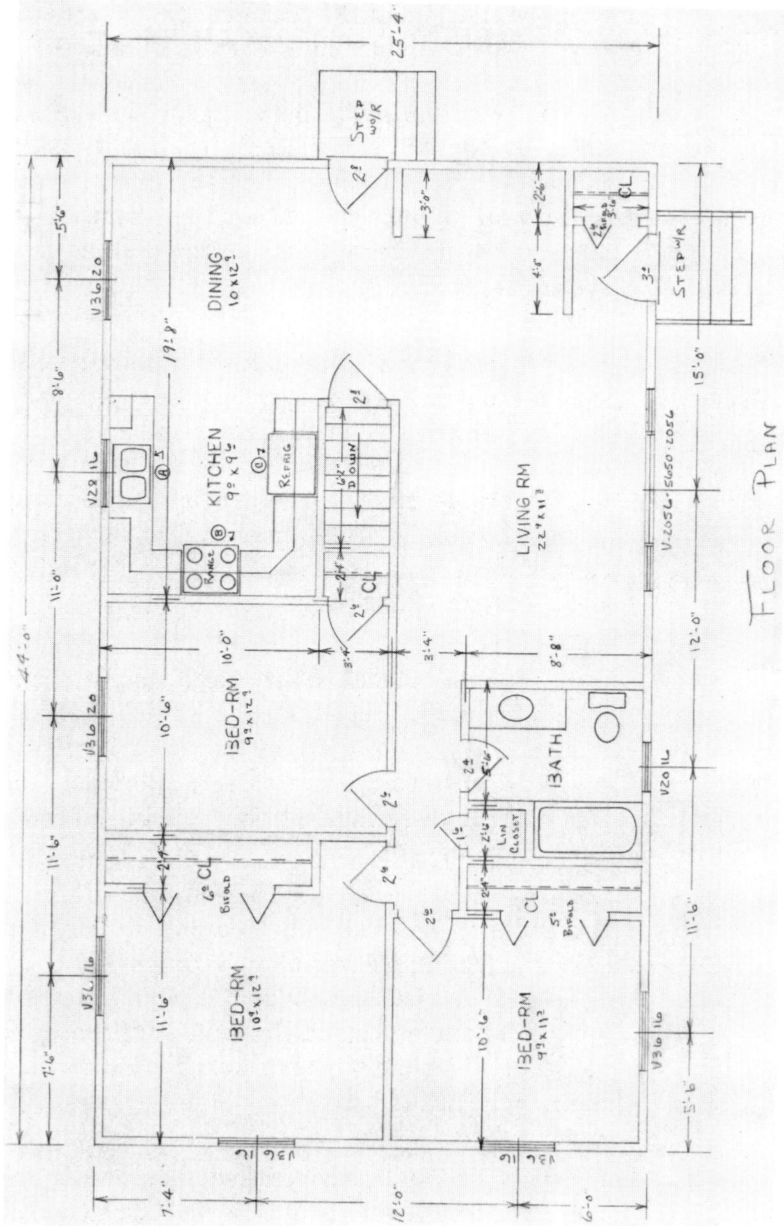
FLOOR PLAN
DINING
KITCHEN
LIVING RM
BED-RM
BATH
REFRIG
STEP W/R

(g)

thoroughly understand the specifications before starting the job. The specification is usually divided into several sections.

1. *General Provisions.* The general provisions of the electrical specifications normally consist of a selected group of considerations and regulations that apply to all sections of the trade. Items covered may include the scope of the work in the electrical contract, electrical reference symbols, codes, inspections and fees, tests, demonstration of the completed electrical system, and identification of the equipment and panels used in the installation.
2. *Basic Materials and Methods.* This portion of the specifications should establish the means of identifying the type, quality, and color of materials and equipment installed. This section should establish the accepted method of installing various materials such as conduits, raceways, under floor ducts, floor boxes, cabinets, panelboards, switches and receptacles, wires and cables, wire connections and devices, motor disconnects, overcurrent protection, and supporting devices.
3. *Power Generation.* This section covers items of equipment used for emergency or standby power facilities, the type equipment used to take over essential electrical service during a normal power outage. This section cites requirements for a complete installation of all emergency circuits on a given project, including emergency service or standby power in the form of a generator set, automatic control facilities, feeders, panelboards, disconnects, branch circuits, raceways, and outlets.
4. *Service and Distribution.* Power distribution (under 600 volts) for the service entrance, metering, distribution switchboards, branch circuit feeders, and branch circuit panelboards is usually described in this section by clauses covering selected related equipment.

   Clearly defined should be the voltage, phase, size, and number of service conductors, installation and supporting methods, and location and rating of all main circuit breakers and other disconnecting means. Other items may include overhead or underground service, grounding, and transformers.
5. *Lighting.* This section covers general conditions relating to lighting equipment, so that all equipment is furnished and installed exactly as designated by the architect. It usually establishes the quality and type of interior lighting fixtures, luminous ceilings, lamps, ballasts, and related accessories, poles, and standards. Methods of installation are included in this clause.
6. *Communications.* Equipment items that are interconnected to permit audio or visual contact between two or more stations or to monitor activity and operations at remote points are covered in this section. Items include radio-intercom-alarm and detection systems, smoke-heat detectors, clock and program equipment, telephone equipment, and public address equipment.
7. *Controls and Instruments.* As the name implies, this section covers all types of controls and instruments used on a given project. Examples include recording and indicating devices, lighting control equipment, control of electric heating and cooling, motor control centers, and numerous other such devices and systems.

It is essential that the architect and all of the crafts on the project understand each other. The architect will place all the important information possible on the drawing; the electricians must know how to interpret the drawing and apply it to the construction project.

The architect is working at a small scale, usually ¼ inch = 1 foot 0 inches, so many lines on the drawing must be omitted so that the blueprint may be readable. To do this, standard symbols are used to represent locations and types of materials required for the installation (Figure 2.2).

Most symbols and notations have a standard interpretation in the United States. When the plans are drawn, architects do their best to follow the accepted standards in representing materials and equipment. New materials being developed require new symbols and designations. The American National Standards Institute (ANSI) and many other trade groups are working constantly to standardize procedures.

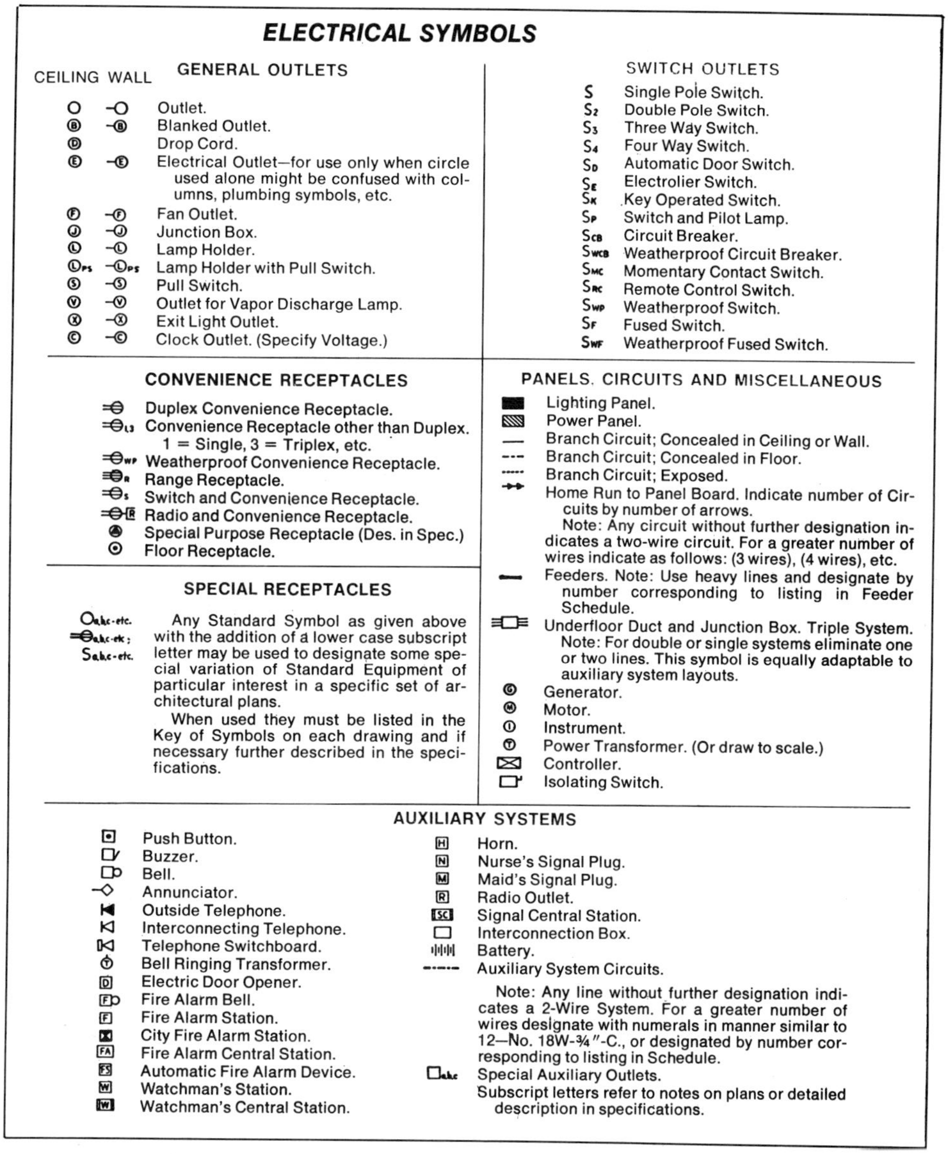

Figure 2.2 Standard electrical symbols. (Courtesy of Crouse-Hinds Company, Arrow Hart Div.)

The notations that the architect places on the drawing next to a specific symbol provide information on the type, size, or quality of the device required (Figure 2.3). The newcomer or apprentice to the electrical construction industry should make every effort to learn material identification in order to associate symbols with wiring devices and other electrical materials.

The drawing most often used by the electrician is the floor plan because most of the wiring requirements are found on this plan. A typical electrical floor plan will show electrical outlets, fixtures, switches, and light-power panels, all designated by standard electrical symbols. The plans will show curved broken lines to indicate which lighting outlets the switch controls (Figure 2.4).

The National Electrical Code defines an outlet as "a point of the wiring system at which current is taken to supply utilization equipment." The term "outlet" is used very broadly by experienced electricians to include duplex receptacles, lighting outlets, switches, and other similar control devices in the wiring system. However, each type of outlet is represented on the plans as a symbol.

A study of a floor plan for a commercial building will show that there are many different electrical symbols used to represent the various electrical devices and equipment used in commercial construction.

2-4 ELECTRICAL STANDARDS

There has been a tremendous growth in the use of electrical energy for light, power, communications, and heating, and because there is the ever present danger of fire and electrical shock through some failure of the electrical system, the electrician has a real responsibility to perform all work in accordance with recognized standards.

The National Electrical Code

The National Electrical Code is the only nationally accepted code recognized in the United States as the safety standard for the electrical construction industry. This code is a set of rules and regulations. It is the basic standard

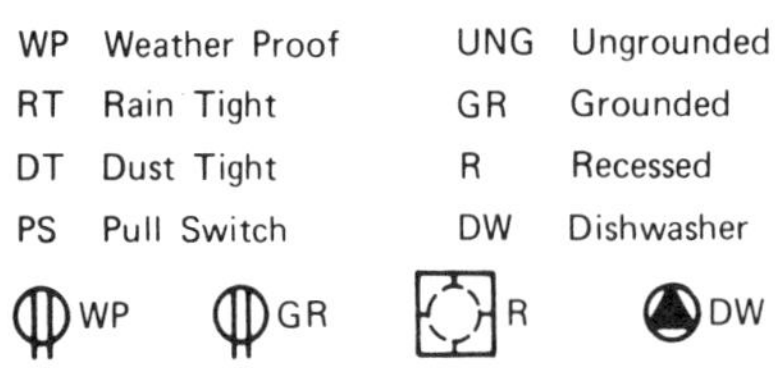

Figure 2.3 Building plan notations.

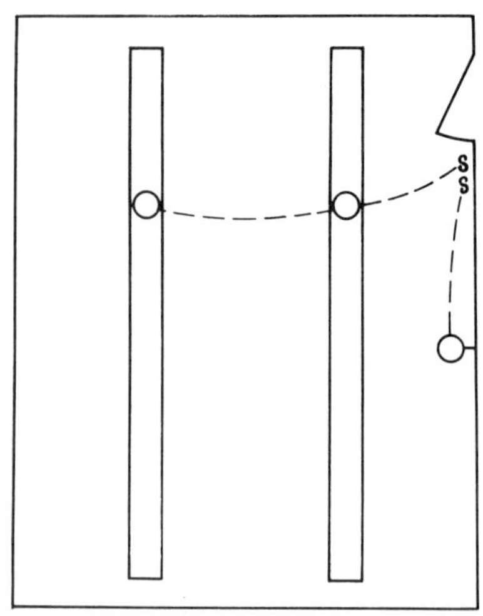

Figure 2.4 Electrical floor plan showing notations and switch control.

that governs the proper method for installing electrical materials so that the completed job may be adequate to protect both life and property.

The National Electrical Code sets forth the minimum requirements for electrical safety; however, it is not intended that the code book be a design manual. It is the National Electrical Code that the electrical inspector refers to when inspecting an electrical installation or alteration. It is obvious that this code is among the most important reference books in the hand of the electrician. For convenience, the National Electrical Code will be designated as NEC throughout this book.

When the NEC is legally adopted by a town, city, or state as the official electrical code of that government body, it becomes law. Compliance with the code in that locality then becomes mandatory when it is officially adopted or accepted.

Since the NEC is revised every three years the current edition must always be consulted for new and remodel work. The purpose of the NEC is discussed in Article 90.1.

> 90-1. Purpose (a). The purpose of this code is the practical safeguarding of persons and property from hazards arising from the use of electricity.

Local Electrical Codes

Some large cities, counties, states, and municipalities adopt their own electrical code or electrical ordinances usually based on the requirements of the National Electrical Code. Electrical installations in any locality must at least meet the requirements of the electrical code of that locality. In some instances, these local codes include special provisions, so it is important for electricians to acquaint themselves with such local regulations that may affect them.

The responsibility of the interpretation of local codes and ordinances rests with the local electrical inspection authority having jurisdiction in that area. In NEC Article 90-4 under enforcements, it is clear that the local electrical inspector has the final word in any situation.

> NEC Article 90-4. Enforcement. The authority having jurisdiction of enforcement of the codes will have the responsibility of making interpretations of the rules, for deciding upon the approval of equipment and materials, and for granting the special permission contemplated in a number of the rules.

2-5 PERMITS AND INSPECTIONS

In many areas regardless of population, if the locality has a building code, it is usually necessary to obtain a permit from the city or county building inspection office before a wiring installation may be started (Figure 2.5).

After the rough-in of the electrical system has been completed, the electrician will call for a rough-in inspection. Concrete cannot be poured in forms containing conduits, and walls and ceilings cannot be covered until the inspector has made the inspection and signed the building permit.

The utility company usually will not install the electrical meter in the service or supply power to the project until an inspection has been made and an inspection certificate has been turned in (Figure 2.6).

CITY OF ALBERT LEA DEPARTMENT OF FIRE AND INSPECTION DIVISION
BUILDING PERMIT APPLICATION

IMPORTANT
BEFORE STARTING WORK CONTACT
N. W. Bell Telephone Co.____373-6441
City of Albert Lea_________373-2393
Interstate Power Co.________373-2371
Williams Bros. Pipeline__612-633-1555
For buried cables, pipes, etc.

Approved By ____________________

Disapproved By ____________________

Hold ____________________

C-No. ____________

Date ____________, 19____ Zoning Dist. ____________ Fire Zone ____________

Owner ____________________ Address ____________________

Contractor ____________________ Address ____________________

Architect or Eng. ____________________ Address ____________________

Elect. Contractor ____________________ Address ____________________

House Number	Street or

Lot	Block	Addition or Sub.

DESCRIPTION OF BUILDING

Front or Width Feet	Side or Length Feet	Height Feet	Number of Stories	Constructed of	Contents Cubical or Sq. Ft.	Cost of Work Covered by this Permit
						$

Structure Used As	Number of Rooms	Number of Units	Interior Finish	Kind of Heat	Basement	Garage

REMARKS

The foregoing is a true and correct description of the improvement contemplated by the undersigned applicant, and the applicant states that he will have full authority over the construction of same, and hereby agrees to comply with all ordinances of the City applicable to building and zoning and assumes all responsibility for such compliance. It is understood that the improvement shall not be used until Certificate of Occupancy and Compliance has been issued by the Building Inspector.

Owner or Authorized Agent

PERMIT AND RECEIPT

X ____________________

Fee________ State Tax________ Total________

In consideration of the above application and the payment of $____________permit fee a permit is hereby granted for the above described improvement conditioned upon the terms and specifications set forth above, and the faithful observance of all the provisions of the City building code, zoning ordinance and all other ordinances applicable to same. All permits issued are subject to all property restrictions.

Date Issued ____________________, 19____ By ____________________
Permit Clerk

This permit shall expire by limitation and become null and void, if the building or work authorized by this permit is not commenced within six months from the date of this permit, or if the building or work authorized by this permit is suspended or abandoned at any time after the work is commenced for a period of six months.

A Certificate Of Occupancy Is Required Before Occupancy Of All Buildings Or Change in Use Thereof

CALL FOR ALL INSPECTIONS 373-6429

TRADES

Figure 2.5 Building permit application. (Courtesy City of Albert Lea, Minnesota)

It is best to check with the building inspection authority in your area for local rules and ordinances. In some localities, electrical installations can only be legally made by a licensed electrician.

2-6 TESTING LABORATORIES

Underwriters' Laboratories

Even though electrical wiring devices and equipment are carefully installed, the devices and equipment themselves can be hazardous unless properly designed for a specific purpose.

Minnesota State Board of Electricity
1954 University Ave., St. Paul, Minn. 55104—Phone 645-7703

REQUEST FOR ELECTRICAL INSPECTION **P 85343**

CHECK BELOW WORK COVERED BY THIS REQUEST

Type of Building	New	Add.	Rep.	Check Appliances Wired For		Check Equipment Wired For	
Home	☐	☐	☐	Range	☐	Temporary Wiring	☐
Duplex	☐	☐	☐	Water Heater	☐	Lighting Fixtures	☐
Apt. Bldg.	☐	☐	☐	Dryer	☐	Electric Heating	☐
Commercial Bldg.	☐	☐	☐	Furnace	☐	Silo Unloader	☐
Industrial Bldg.	☐	☐	☐	Air Conditioner	☐	Bulk Milk Tank	☐
Farm	☐	☐	☐	List Others Here ________		List Others Here ________	
Other________	☐	☐	☐				

COMPUTE INSPECTION FEE BELOW

Service Entrance Size:	#	Fee	Feeders& Subfeeders:	#	Fee	Circuits:	#	Fee
0 to 100 Amps.			0 to 30 Amperes			0 to 30 Amperes		
101 to 200 Amps.			31 to 100 Amperes			31 to 100 Amperes		
Above 200___Amps.			Above 100___Amps.			Above 100___Amps.		
Transformers			Remote Control Circ.			Partial or other fee		
Signs			Special Inspection			Minimum fee $5.00		
Remarks						**TOTAL FEE**		

I, the Electrical Inspector, hereby certify that the above inspection has been made.

(Rough-in)________________________________ Date ______________

(Final) ________________________________ Date ______________

This request void 18 months from

This request void 18 months from

P 85343

Date of this Request______________________.

I, as ☐ Licensed Electrical Contractor ☐ Owner, do hereby request inspection of the above electrical wiring installed at:

Street Address or Route No. ________________________________City__________

Section_______ Township ____________________ Range_______County ______________

Which is occupied by ____
(Name of Occupant)

Is a roughin inspection required on this job? No ☐ Yes ☐ Ready Now ☐ Will Call ☐

Power Supplier ____________________________ Address ______________________

Electrical Contractor ____________________________ Contractor's License No. _____
(Company Name)

Mailing Address ____
(Electrical Contractor or Owner Making This Installation)

Authorized Signature ____________________________ Phone No. ___________
(Electrical Contractor or Owner Making This Installation)

STATE BOARD COPY

This inspection request will not be accepted by the State Board unless proper inspection fee is enclosed.

Figure 2.6 Application for electrical inspection (state of Minnesota).

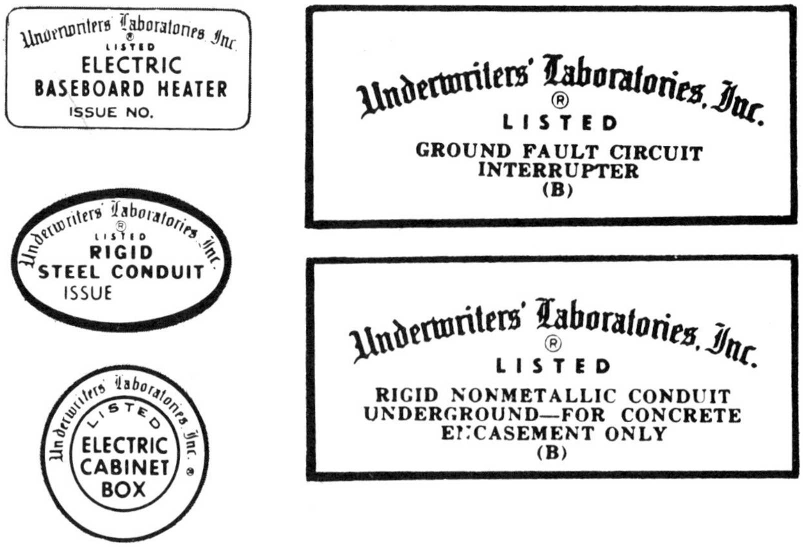

Figure 2.7 Types of labels that appear on products covered by Underwriters Laboratories, Inc.

The National Board of Fire Underwriters Laboratories, Inc., is a nonprofit testing laboratory established in many areas of the United States to investigate materials, devices, products, equipment and construction methods, and systems in order to define any hazards that may affect life and property. Manufacturers submit samples of their products, devices, or equipment to these laboratories for testing. When the product meets the minimum underwriters' standards of safety the UL Label is attached.

Tested products that meet the UL safety standards are listed under various categories. The tested, listed products that are of interest to you are contained in the Underwriters' Laboratories "Electrical Construction and Material List."

When purchasing and installing devices or equipment for your commercial electrical installation, it is essential that you look for the UL listing label (Figure 2.7).

Canadian Standards Association

The parallel organization to Underwriters' Laboratories in Canada is the Canadian Standards Association, or CSA. However, in comparison to the UL, the CSA has more authority to remove products from the market which do not meet CSA standards. The UL program is strictly voluntary. If an electrical product used in Canada is connected in any way with the use of energy from the electrical power source owned by the provinces, that product must have CSA approval.

The objective of the Canadian association is to provide voluntary certification services for national standards.

CSA is the testing authority concerned with safety standards acceptable to inspection authority in Canada.

A manufacturer, who may be from any part of the world, files an application with the CSA, and engineers and technicians test and inspect the product for compliance with an applicable standard. If the product meets the standard, the manufacturer will apply a CSA mark to the product indicating certification.

CANADIAN
STANDARDS
ASSOCIATION

Figure 2.8 Symbol of Canadian Standard Association.

Products certified by CSA are eligible to bear the CSA certification mark. However, misuse of the mark may result in suspension or cancellation of product certification. CSA may resort to legal action to protect its trademark in the event of abuse.

The CSA symbol on electrical equipment is an assurance that the equipment has passed rigid inspection and is safe to use (Figure 2.8).

2-7 NATIONAL ELECTRICAL MANUFACTURERS ASSOCIATION

The National Electrical Manufacturers Association (NEMA) is a nonprofit trade association of manufacturers of electrical equipment and supplies. Standards are in the public interest and are designed to eliminate misunderstandings between the manufacturer and the purchaser and to assist the purchaser in selecting and obtaining the proper product for its particular need.

NEMA is divided into numerous subdivisions, each covering a group of related apparatus or supplies. Each subdivision has a definite program of developing standards along with its numerous other activities.

2-8 OSHA

The Occupational Safety and Health Act of 1970—better known as OSHA—has as its objectives to assure, so far as possible, safe and healthful working conditions for every working man and woman in the nation, and to preserve our human resources.

The U.S. Department of Labor has been responsible for setting the standards, that is, safety requirements, which all businesses must meet. The Department of Labor currently has a field force of approximately 1000 compliance officers who conduct safety inspections throughout the country.

The compliance officers are empowered to issue citations to the businesses inspected if conditions or practices observed are unsafe. The officers may also recommend that specific penalties be imposed for citations.

The inspections are conducted using the applicable OSHA or state safety and health standards, whichever of the two codes is more severe. The Department of Labor foresees that the state and federal standards will be merged into one uniform code in the next few years.

OSHA administrators have adopted the 1971 National Electrical Code as the standard of electrical safety for every new electrical installation and all new equipment installed, replaced, modified, repaired, or rehabilitated after March 15, 1972. However, the electrician must remember that all new installations and all rewiring projects must comply with the 1984 National Electrical Code.

SUMMARY

1. With a working drawing, the electrician has several guidelines that tell how to proceed with the installation.
2. Specifications are a part of the building plans.
3. Symbols and notations have a standard interpretation.
4. The National Electrical Code is a set of rules and regulations that governs the methods of the installation for protection.
5. Because of the ever-present danger of fire and electrical shock, there is a need for recognized standards.
6. The electrical inspector has the final word as the inspection authority.
7. Materials approved for a specific purpose will be tested, listed, and labeled by Underwriters' Laboratories.
8. It is usually necessary to obtain an electrical permit before a wiring installation may be started.

PROBLEMS

2-1 Explain how the architect uses a set of blueprints to convey instructions to the electrician on the job. ______________

____

____

____

2-2 The electrician can find out how a house is constructed and what materials will be used from (circle one):

a. A picture of the house.
b. A set of blueprints/specifications.
c. The National Electrical Code.

2-3 What is the purpose of specifications? ______________

____

____

2-4 What is the purpose of an electrical symbol? ______________

____

____

____

2-5 The National Electrical Code is a standard for the electrical industry. It is used by electricians to determine (circle one):

a. How to wire a house.
b. How to design an electrical system.
c. Minimum safety standards for installing and maintaining electrical wiring and equipment.

2-6 List six items that might be covered in the specifications.

1. ______________________ 4. ______________________
2. ______________________ 5. ______________________
3. ______________________ 6. ______________________

2-7 What do the letters UL and CSA signify? ______________________

____

2-8 What authority enforces the standards set by the NEC? ______________________

____

2-9 Where are notations found? ______________________

____

2-10 What is the purpose of the National Electrical Code? ______________________

____

____

2-11 By name, what type of plan would you expect to show the location of the building on the property? ______________________

____

2-12 When you need information about the outside finish of a building you look on the ________ views.

chapter 3

ELECTRICAL SYMBOLS AND OUTLETS

Instructional Objectives

1. To become more familiar with electrical symbols.
2. To learn how to identify the different types of outlets, boxes, and switches used in a dwelling.
3. To understand the need for calculating the number of conductors in a box.
4. To make it possible to know how to use the NEC book to find the correct information relative to the location of outlets.
5. To become familiar with the methods of mounting the various electrical devices used in a residence.

Self-Evaluation Questions

Test your prior knowledge of the information in this chapter by answering the following questions. Watch for the answers as you read the chapter. Your final evaluation of whether you understand the material is measured by your ability to answer these questions.

1. What is an outlet?
2. What symbols are used to show different types of outlets?
3. How are locations of outlets usually determined?
4. What symbol represents a ceiling outlet?
5. What symbol represents a three-way switch?
6. How can you tell, from looking at the plan, which switches control the various outlets?
7. At what height do you install convenient outlets?
8. How do you mount lighting fixtures?
9. Can you identify a recessed fixture?
10. What electrical symbol represents a wall bracket?

Electrical symbols drawn on an architectural plan are used to show the location and type of electrical devices required for the installation.

When plans are drawn, architects do their best to follow the accepted standards in representing materials and equipment. New materials are always being developed, requiring new symbols and designations. The American National Standards Institute (ANSI) and many other trade groups are working constantly to standardize procedures. Figure 2.2 shows electrical wiring symbols and their meanings.

The newcomer or apprentice to the electrical construction industry should make every effort to learn material identification in order to associate symbols with wiring devices and other electrical materials.

As stated in Chapter 2, the drawing most often used by the electrician is the floor plan, because most of the wiring requirements are found on this plan. Figure 3.1 shows a simplified electrical floor plan with typical electrical outlets and switches designated by standard electrical symbols. The curved broken lines indicate which outlet the switch controls.

The National Electrical Code defines an outlet as "a point on the wiring system at which current is taken to supply utilization equipment." The term "outlet" is used very broadly by experienced electricians to include duplex receptacles, lighting outlets, switches, and similar control devices in a wiring system. However, each type of outlet is represented on the plans as a symbol (Figure 2.2).

A study of the floor plan for a single family dwelling, Figure 3.1, shows

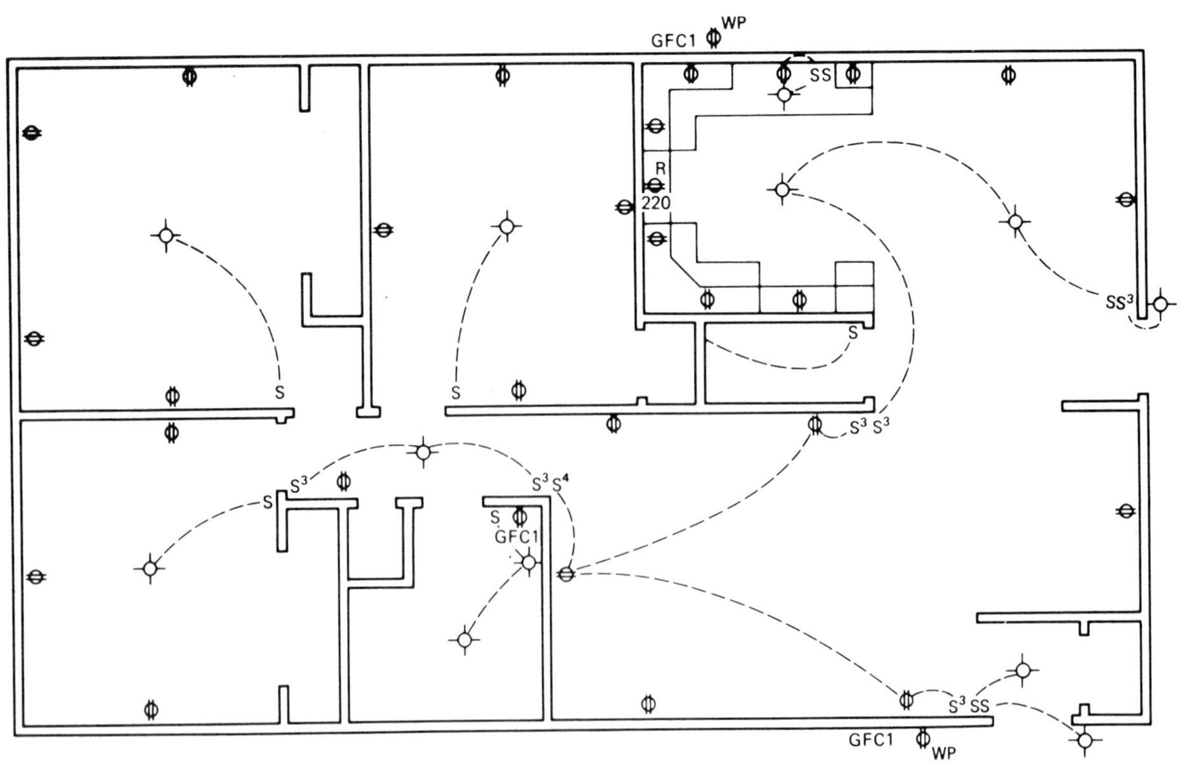

Electrical Plan

Figure 3.1 Electrical floor plan.

that many different electrical symbols are used to represent the various electrical devices and equipment used in residential construction.

3-2 FIXTURES AND OUTLETS

It is common practice among architects in residential construction to include in the specifications a certain amount of money as a ''fixture allowance'' for the purchase of the electrical light fixtures. The selection of fixtures is left up to the owner or builder. However, the electrical contractor includes this allowance in the bid. Should the fixture cost exceed the fixture allowance in the bid and specifications for an exclusive custom-built home, the owner is expected to pay the difference. For a smaller conventional house, the contractor usually purchases and hangs competitive fixtures as a part of the contract.

If the builder has made a selection of the fixtures prior to the drawing of the plans, the architect can specify these fixtures in the plans and specifications. This will provide the electrician with advance information relative to the roughing in or mounting that may be required for the fixtures. This is especially true when recessed fixtures are specified. If the fixtures are not selected in advance to construction, the electrician will usually install outlet boxes that have standard fixture-mounting studs where heavy fixtures will be anticipated. Other lightweight fixtures can be mounted either to an outlet box or box plaster ring using a bar strap and two 8-32 machine screws. Figure 3.2 illustrates several types of ceiling fixture outlet boxes.

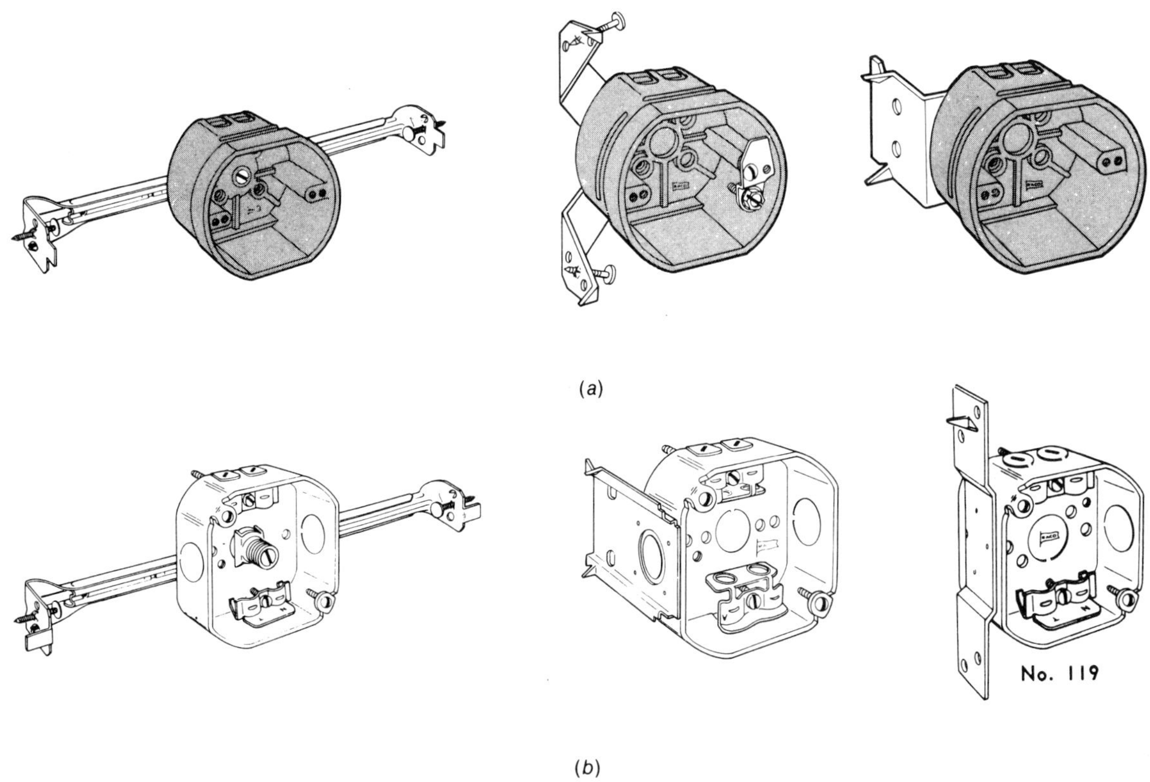

Figure 3.2 Ceiling fixture outlet boxes. (*a*) Four-inch ceiling boxes (nonmetallic). (*b*) Four-inch setup boxes (metal). (Courtesy of Raco, Inc.)

3-3 SWITCH OUTLETS

The symbol for a residential-type single pole switch is shown on the floor plan by the letter "S," a three-way switch "S_3," and a four-way switch "S_4." Figure 3.3 illustrates several switch outlet boxes. A flush-type switch will fit into a standard 1¾ × 2¾ inch sectional box (either steel or plastic). If and when the plans call for two or more switches or a switch and a

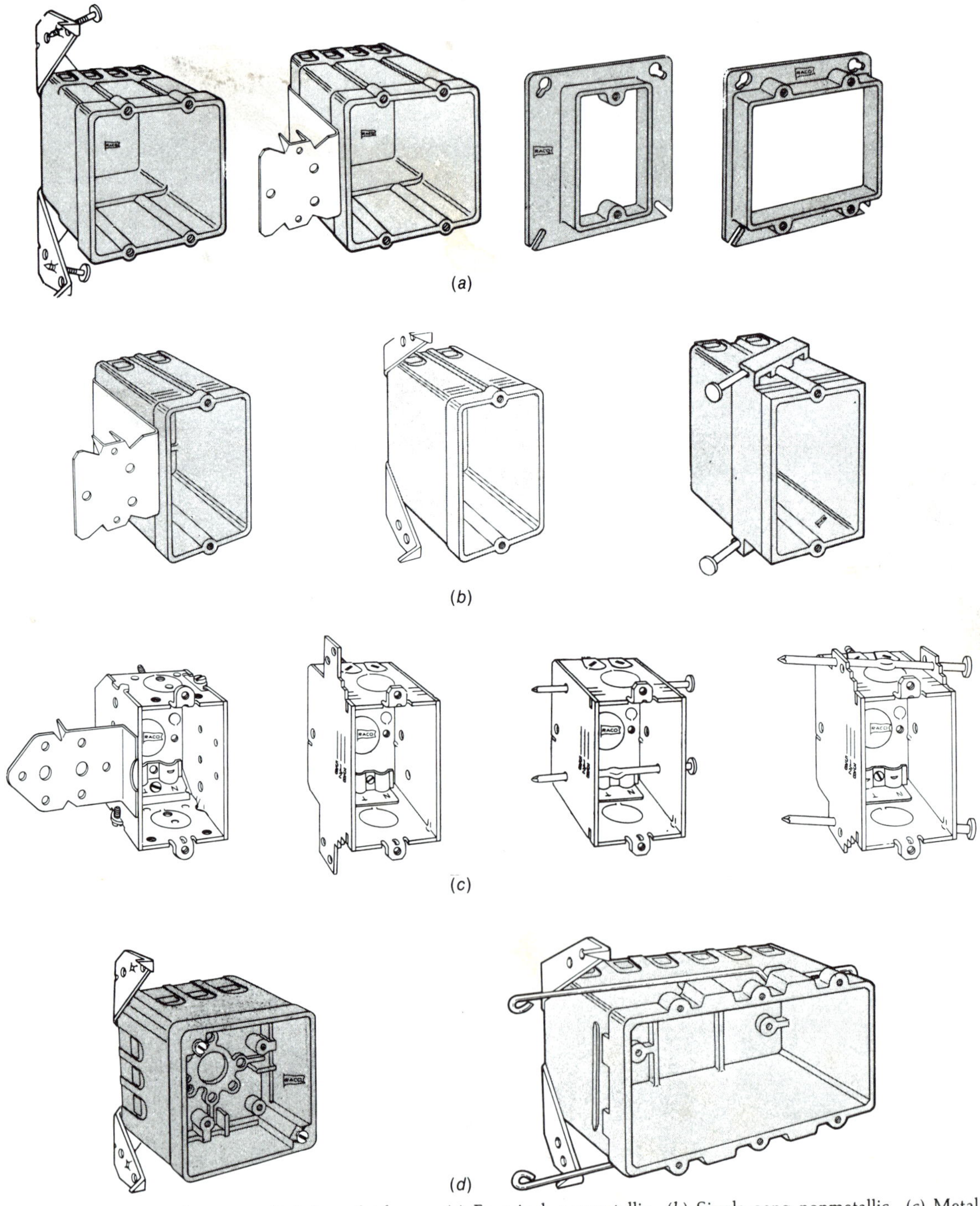

Figure 3.3 Switch outlet boxes. (*a*) Four-inch nonmetallic. (*b*) Single gang nonmetallic. (*c*) Metal. (*d*) Three gang nonmetallic. (Courtesy of Raco, Inc.)

convenience outlet at the same location, a two gang box is used. Single outlet boxes can be ganged together by removing one side of each of the boxes and fastening them together.

NEC Section 370-10. In wall or ceiling. In walls or ceilings of concrete, tile, or other noncombustible material, boxes and fittings shall be so installed that the front edge of the box or fitting will not sit back of the finished surface more than ¼ inch. In walls and ceilings constructed of wood or other combustible material, outlet boxes and fittings shall be flush with the finished surface or project therefrom (Figure 3.4).

3-4 JUNCTION BOXES

Junction boxes are at times placed in the circuit for convenience. A 4-inch square box serves a twofold purpose: (1) as a convenient outlet box (when a single-gang plaster ring is used) and (2) as a box to make up connections. At times, junction boxes are installed in the attic for the door chime transformer or ventilation fan thermostat (Figure 3.5).

NEC Section 300-15. Boxes and fittings—where required: (b) Box only. A box shall be installed at each conductor splice connection point, outlet, switch point, junction point, or pull point for the connection of metal-clad cable, mineral-insulated metal sheathed cable, type AC cable, nonmetallic-sheathed cable or other cables, at the connection point between any such cable system and a raceway system and at each outlet and switch point for concealed knob-and-tube wiring.

3-5 SPECIAL PURPOSE OUTLETS

It is the responsibility of the electrician to check when a special-purpose outlet is indicated on the plans or in the specifications. It may indicate a special procedure such as a need for a polarized or grounded receptacle or a special 240-volt circuit.

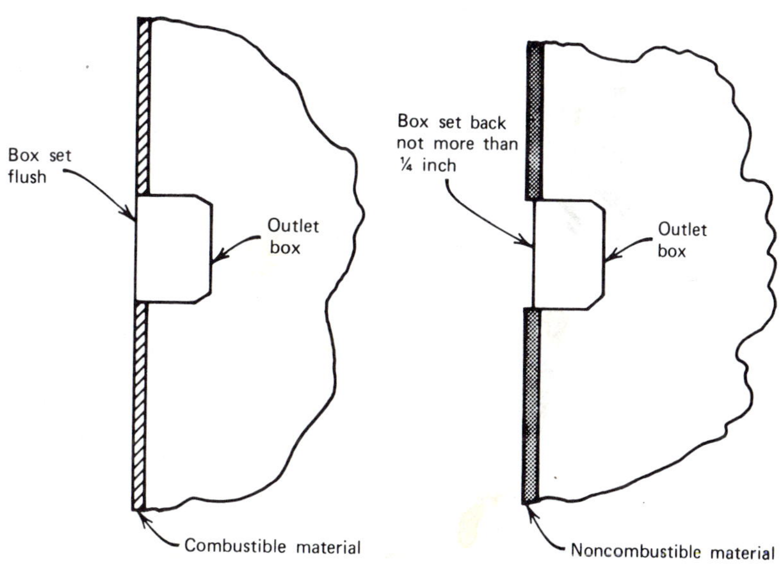

Figure 3.4 Switch outlet boxes installed in wall.

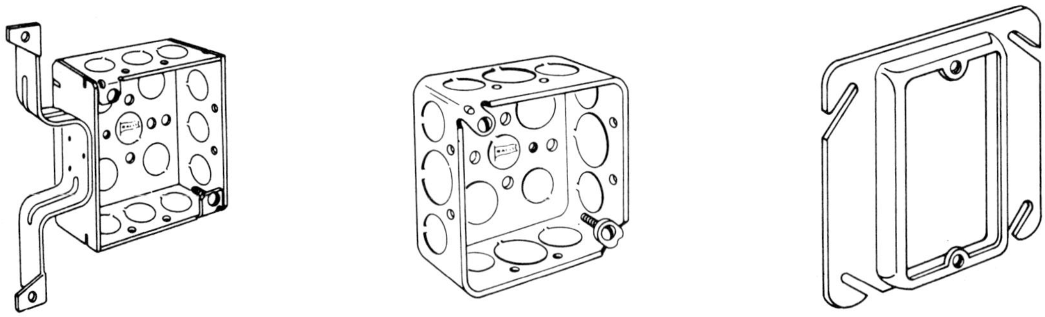

(a)

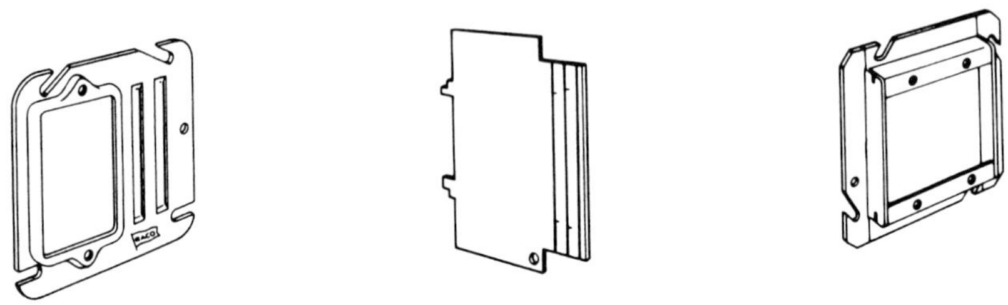

(b)

Figure 3.5 Junction boxes, plaster rings, dividers. (*a*) Four-inch square junction boxes. (*b*) Plaster rings and dividers. (Courtesy of Raco, Inc.)

Special-purpose outlets are installed in the rough-in for the dishwasher, clothes dryer, built-in cook-top, oven, and others. These outlets are usually indicated on the floor plans and are described by a notation by the symbol.

3-6 NUMBER OF CONDUCTORS IN A BOX

The number of conductors allowed in a standard-size outlet or switch box is specified in the National Electrical Code (Table 3.1, Table 370-6(a), NEC).

> Section 370-6. Number of conductors in switch, outlet, receptacle, device and junction boxes. Boxes shall be of sufficient size to provide free space for all conductors in the box.
>
> (a) Standard boxes. The maximum number of conductors, not counting fixture wires permitted in standard boxes, shall be listed in Table 370-6(a).
>
> (1) Table 370-6(a) shall apply where no fittings or devices, such as fixture studs, cable clamps, hickeys, switches, or receptacles are contained in the box and where no grounding conductors are part of the wiring within the box. Where one or more fixture studs, cable clamps, or hickeys are contained in the box, the number of conductors shall be one less than shown in the Tables; an additional deduction of one conductor shall be made for each strap containing one or more devices; and a further deduction of one conductor shall be made for one or more grounding conductors entering the box. A conductor running through the box shall be counted as one conductor, and each conductor originating outside of the box and terminating inside the box is counted as one conductor. Conductors, no part of which leaves the box, shall not be counted. The volume of a wiring enclosure (box) shall be the total volume of the assembled sections, and where used, the space provided by plaster rings, domed covers extension rings, etc. that are marked with their volume in cubic inches.

TABLE 3-1 Number of Conductors in a Box [NEC Table 370-6(a)]: Metal Boxes

Box Dimension, Inches, Trade Size or Type	Minimum Cubic Inch Capacity	Maximum Number of Conductors				
		No. 14	No. 12	No. 10	No. 8	No. 6
4 × 1¼ round or octagonal	12.5	6	5	5	4	0
4 × 1½ round or octagonal	15.5	7	6	6	5	0
4 × 2⅛ round or octagonal	21.5	10	9	8	7	0
4 × 1¼ square	18.0	9	8	7	6	0
4 × 1½ square	21.0	10	9	8	7	0
4 × 2⅛ square	30.3	15	13	12	10	0
4¹¹⁄₁₆ × 1¼ square	25.5	12	11	10	8	0
4¹¹⁄₁₆ × 1½ square	29.5	14	13	11	9	0
4¹¹⁄₁₆ × 2⅛ square	42.0	21	18	16	14	6
3 × 2 × 1½ device	7.5	3	3	3	2	0
3 × 2 × 2 device	10.0	5	4	4	3	0
3 × 2 × 2¼ device	10.5	5	4	4	3	0
3 × 2 × 2½ device	12.5	6	5	5	4	0
3 × 3 × 2¾ device	14.0	7	6	5	4	0
3 × 2 × 3½ device	18.0	9	8	7	6	0
4 × 2⅛ × 1½ device	10.3	5	4	4	3	0
4 × 2⅛ × 1⅞ device	13.0	6	5	5	4	0
4 × 2⅛ × 2⅛ device	14.5	7	6	5	4	0
3¾ × 2 × 2½ masonry box/gang	14.0	7	6	5	4	0
3¾ × 2 × 3½ masonry box/gang	21.0	10	9	8	7	0
FS—minimum internal depth 1¾ single cover/gang	13.5	6	6	5	4	0
FD—minimum internal depth 2⅜ single cover/gang	18.0	9	8	7	6	3
FS—minimum internal depth 1¾ multiple cover/gang	18.0	9	8	7	6	0
FD—minimum internal depth 2⅜ multiple cover/gang	24.0	12	10	9	8	4

The code requires that boxes other than those described in Table 370-6(a) must be "durably and legibly marked."

Section 370-6(b). Other boxes. Boxes 100 cubic inches or less, other than those described in Table 370-6(a), nonmetallic boxes and conduit bodies having provisions for more than two conduit entries shall be durably and legibly marked by the manufacturer with their cubic inch capacity, and the maximum number of conductors permitted shall be computed using the volume per conductor listed in Table 370-6(b) and the reductions provided for in Section 370-6(a)(1).

3-7 LOCATION OF OUTLETS

There are no hard and fast rules for locating most outlets; however, the electrician should verify all dimensions before starting the rough-in either by the notes on the floor plan or by the directions written in the specifications. Care should be taken in checking the kitchen and bathroom areas. Usually the kitchen and bath will have a detail drawing showing cabinets and space for equipment, sink, range, oven, and cook-top. Also check for tile, if used. Switch and duplex receptacles should be installed so that when finished they are in the tile. Check for the swing of doors and make sure the switch outlet isn't installed behind the door.

Unless otherwise noted, heights for convenience outlets and wall switches are given in the specifications and are usually from the finished floor to the center of the outlet. Proper allowances must be made when roughing-in for the kind and thickness of sheet-rock or panel used on the walls and ceiling.

Commonly specified heights for wall duplex receptacles outlets are 14 inches and for switch outlets 48 inches. Switches and receptacles over counter-tops in kitchen and bathrooms are mounted 42 inches to bottom of outlet.

Materials vary in thickness, so check plans and spaces.

SUMMARY

1. The American National Standards Institute and trade groups work for standardized procedures.
2. Learn symbols and associate them with wiring devices.
3. The National Electrical Code defines an "outlet" as a point on the wiring system where current is taken to supply a circuit.
4. Understand why it is important to check the plans and specifications before starting the job.`
5. The National Electrical Code has specific information relative to the location of outlets.
6. It is your responsibility to check special purpose outlets before starting the job.

PROBLEMS

3-1 Draw a one-room floor plan. Locate two switch outlets and two light outlets (one ceiling and one wall bracket). Show switch to light relationship with dashed lines.

3-2 Draw a sketch of a concrete wall. Show how an outlet box would be installed to the finished wall.

3-3 Draw a sketch of a combustible material wall showing how an outlet box would be installed to the finished wall.

3-4 Explain the purpose of a floor plan with reference to electrical outlets.

3-5 Draw a special outlet for a dishwasher, water heater, and clothes dryer showing proper symbols and notations.

3-6 Draw the proper symbols for the following outlets: a four-way switch, a duplex grounding type receptable, a single pole switch, and a duplex split-circuit receptacle.

3-7 Electrical symbols drawn on an architectural plan are used to show ______________ and ______________ of electrical devices.

3-8 It is the responsibility of the ______________ to check when a special purpose outlet is indicated on the plans.

3-9 The number of conductors allowed in an outlet or switch box is specified in ______________ ______________ ______________

3-10 The electrician should ______________ all dimensions before starting the rough-in either by notes on the floor plans or by directions in the specifications.

PART 2

METHODS AND MATERIALS

chapter 4

CONDUCTORS—MATERIALS, SIZES, AND TYPES

Instructional Objectives

1. To become familiar with the terms used to size and rate conductors.
2. To identify two commonly used conductors.
3. To learn some disadvantages of aluminum conductors.
4. To learn how to splice aluminum to copper conductors.
5. To become familiar with wire types used in commercial buildings.
6. To learn why insulation is an important factor when determining ampacity.

Self-Evaluation Questions

Test your prior knowledge of the information in this chapter by answering the following questions. Watch for the answers as you read the chapter. Your final evaluation of whether you understand the material is measured by your ability to answer the questions. When you have completed the chapter, return to this section and answer the questions again.

1. What two materials are most commonly used for electrical conductors?
2. What effect does oxidation have on metal conductors?
3. What is the name of the gauge used to measure solid conductors?
4. What is a mil?
5. List the type of wire used in commercial buildings.
6. Define ampacity and list the three factors that affect it.
7. Is copper the best conductor?
8. How do you splice copper to aluminum?
9. What is the correct method of terminating aluminum wire at a wire-binding screw?
10. What is a pressure connector?

4-1 TYPES OF CONDUCTORS

Although silver is the best conductor, its use is limited because of high cost. Two commonly used conductors are aluminum and copper. Each has advantages and disadvantages that will be discussed in this chapter.

Aluminum Conductors

Aluminum is no newcomer to the electrical conductor field. It was first used on an overhead transmission line more than 75 years ago. Today virtually all overhead transmission lines have conductors of aluminum or aluminum reinforced with steel (ACSR).

Its economy and performance record on overhead transmission lines logically led to its use in conductors of other types so that today some 90 percent of overhead distribution, 95 percent of service drop, and 80 percent of service entrance cables are aluminum. More recent developments are in underground distribution and in building wire where the use of aluminum is growing rapidly.

The reason for this growth is largely economics; aluminum building wire, for example, costs substantially less than copper for the same ampacitv rating. Other factors influencing this trend are ready availability and lighter weight, which permit fewer supports and longer distances between junction boxes.

Aluminum building wire actually is just as easy to install as copper, and installation procedures are basically similar. Because aluminum is a different metal with somewhat different properties, however, a few differences in

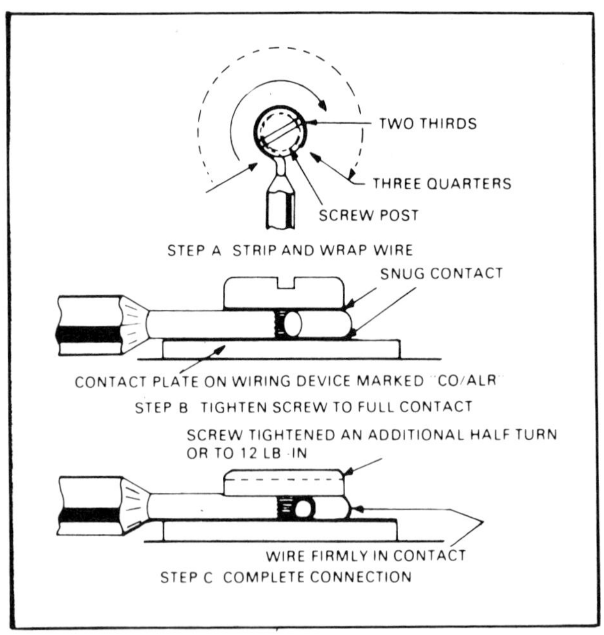

(a)

Figure 4.1 (*a*) Correct method of terminating aluminum wire at wire-binding screw terminals of receptacles and snap switches. (*b*) Incorrect method of terminating aluminum wire. (*c*) Pigtailing is an approved procedure under NEC; however, direct connection of aluminum to CO/ALR devices is preferred. (Courtesy of Aluminum Association)

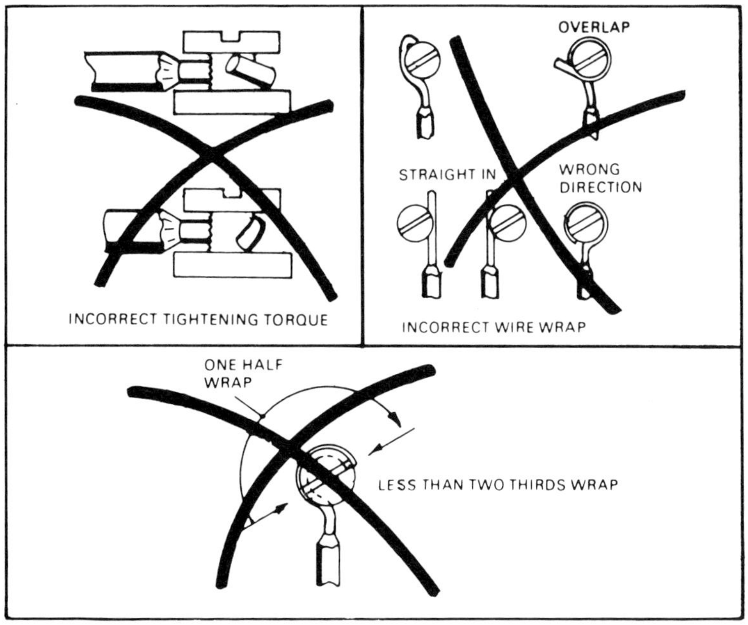

(b)

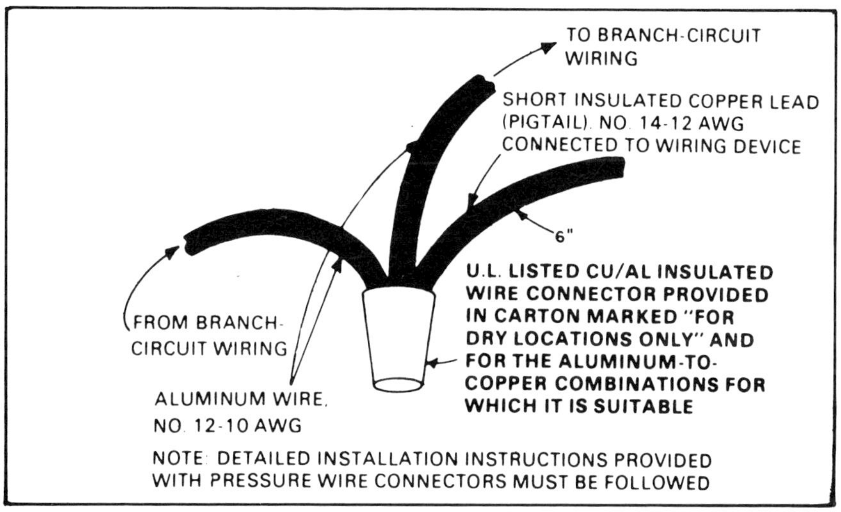

(c)

installation practices must be followed for its successful use in building wiring. Connectors designed for aluminum conductors must be employed and equipment to which aluminum conductors are to be connected must have terminals designed for their use (Figure 4.1).

Aluminum wire and cables are available in sizes to meet all needs and with the same types of insulation as copper. In addition, aluminum sheathed cable with insulated aluminum conductors is available and is recognized by the National Electrical Code for installation without conduit. Likewise, connectors for all types and sizes of aluminum conductors and equipment with suitable terminals are readily available. Such equipment is UL listed and designated "AL/CU" for use with either aluminum or copper conductors (Figure 4.2).

Figure 4.2 (*a*) Basic techniques for installing aluminum wire. (1) Select the right connector. Always use an aluminum connector that's marked for the size you're using. Never use a copper connector on aluminum conductor. (Courtesy of Burndy Corporation)

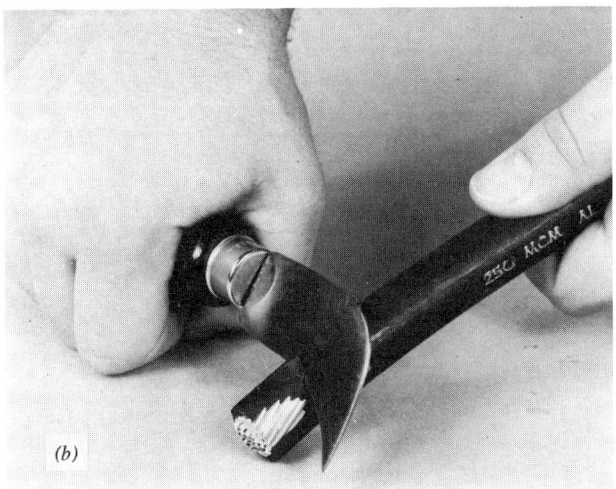

Figure 4.2 (*b*) Basic techniques for installing aluminum wire. (2) Strip carefully. Remove the insulation without nicking the wire. Never ring a conductor when stripping insulation. One way to avoid this is to pencil or whittle the insulation. (Courtesy of Burndy Corporation)

Figure 4.2 (*c*) Basic techniques for installing aluminum wire. (3) Brush wire thoroughly. Always wire brush the stripped portion of the wire. An unplated terminal pad and the surface to which the terminal will be attached should also be brushed. (Courtesy of Burndy Corporation)

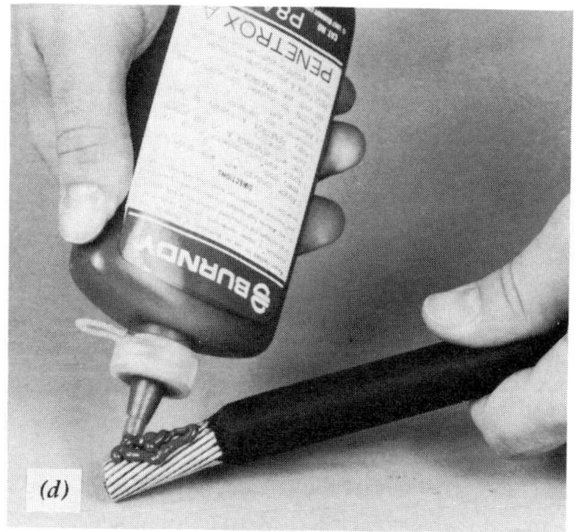

Figure 4.2 (*d*) Basic techniques for installing aluminum wire. (4) Apply Penetrox. For mechanical connectors, apply Penetrox joint compound liberally to the conductor to prevent the formation of surface oxides once the connection is made. Also apply to any terminal pad. (Courtesy of Burndy Corporation)

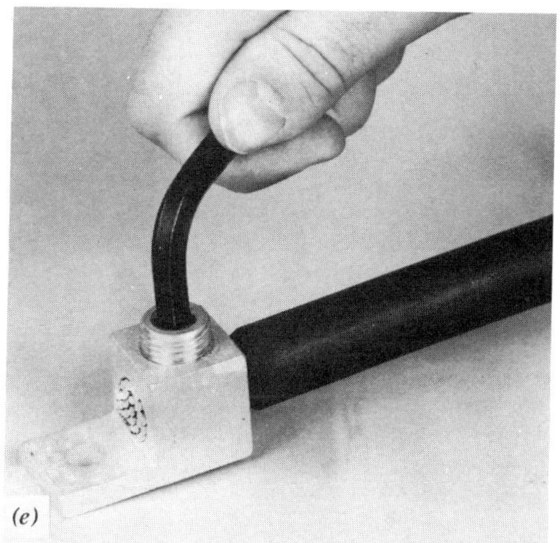

Figure 4.2 (*e*) Basic techniques for installing aluminum wire. (5) Tighten completely. For mechanical connectors, use an allen wrench or screwdriver to securely tighten the aluminum connectors. (Courtesy of Burndy Corporation)

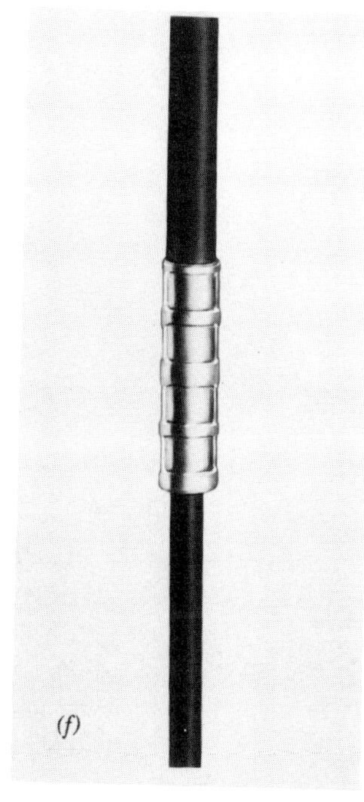

Figure 4.2 (*f*) Basic techniques for installing aluminum wire. (6) Crimp. For compression connectors, choose the right die and right tool. Insert the wire completely. Apply enough pressure to fully distort all strands and make the recommended number of crimps. (Courtesy of Burndy Corporation)

Receptacles and snap switches for 15 to 20 ampere service have a special marking CO/ALR.

Junction boxes are used where abrupt changes in direction of conduit runs are made, to provide for expansion and contraction of the conductors upon heating and cooling, and where tap connectors are to be made. See Figure 4.3.

Connections in aluminum conductors of the larger sizes are made with the aid of joint compound. This is applied to the bare conductor after it has been wire brushed. A coating of joint compound remains on the surface of the conductors, preventing the oxide from reforming when it is broken in the installation operation. Most of the compression type connectors supplied by the manufacturer are filled with joint compound to seal the connection from moisture and other contaminants. However, the electrician must add compound to mechanical connectors. It is preferred practice to use it on all connections.

SERVIT Types KS
For Copper, Copperweld,
& Type SCP Amerductor

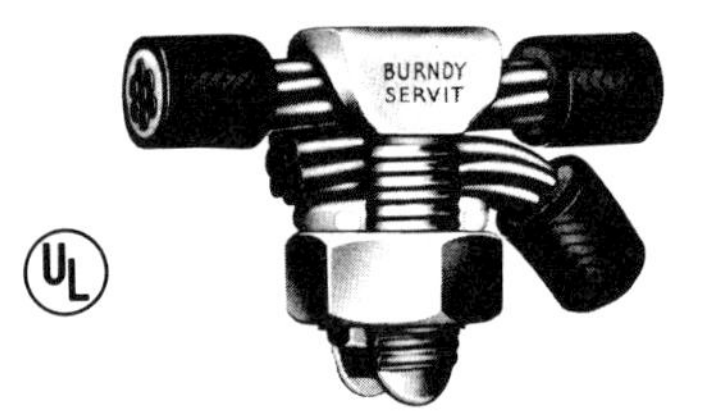

Compact, high strength, high copper alloy SERVIT split-bolt has free-running threads, and easy to grip wrench flats. Highly resistant to season cracking and corrosion, the SERVIT provides maximum pressure and assures secure connection on all combinations of run and tap conductors.

Catalog Number	Copper Conductor		Recommended Torque (inch — lbs.)
	Range For Equal Run And Tap	Min. Tap With Max. Run	
KS90*	12 Str.–10 Str.	16 Str.	75
KS15	10 Str.–8 Str.	14 Str.	
KS17	8 Str.–6 Sol.	14 Str.	150
KS20	8 Str.–4 Sol.	14 Str.	
KS22	6 Str.–2 Sol.	14 Str.	250
KS23	6 Str.–2 Str.	14 Str.	
KS25	4 Str.–1/0 Str.	14 Str.	350
KS26	2 Str.–2/0 Str.	14 Str.	
KS27	1 Str.–3/0 Str.	8 Sol.	450
KS29	1 Str.–250	8 Str.	600
KS31	1/0 Str.–350	1/0 Str.	
KS34	2/0 Str.–500	2/0 Str.	750
KS39	4/0 Str.–750	4/0 Str.	900
KS44	300–1000	4/0 Str.	1000

*Not U/L Listed

(a)

Universal SERVIT Type KSU
For All Combinations of Copper, Alum.†, ACSR†, AAAC, 5005, & Steel

Conductor		Catalog Number	Recommended Torque (inch — lbs.)
Run	Tap		
Copper & Aluminum†	Copper & Aluminum†		
12 Sol.–6 Sol.	12 Sol.–6 Sol.	KSU17	150
10 Sol.–4 Sol.	10 Sol.–4 Sol.	KSU20	
10 Sol.–2 Sol.	10 Sol.–2 Sol.	KSU22	250
8 Str.–2 Str.	8 Sol.–2 Str.	KSU23	
2 Str.–1/0 Str.	10 Str.–1/0 Str.	KSU25	350
2 Str.–2/0 Str.	8 Str.–2/0 Str.	KSU26	
1 Str.–250	8 Str.–250	KSU29	600
1/0 Str.–350	4 Str.–350	KSU31	

†Accommodates compressed conductors within diameter range.

Tin-plated, high strength copper alloy SERVIT with spacer. Spacer separates dissimilar conductors and provides long contact length that prevents high pressure point contacts between run and tap conductors.

Use of PENETROX joint compound recommended with aluminum or ACSR.

(b)

Figure 4.3 Tap connectors. (*a*) Copper to copper. (*b*) Combination, copper to copper or copper to aluminum. (*c*) Junction box, shows connection made with crimp connector and crimping tool. (Courtesy of Burndy Corporation)

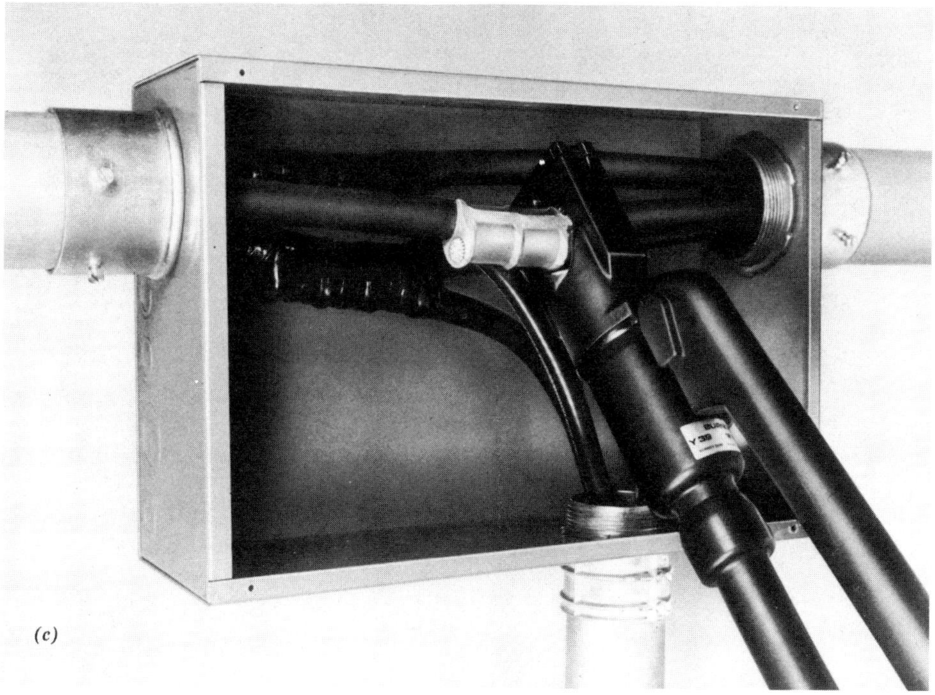

(c)

The student is cautioned that two classes of joint compounds are available and each has its own application. The type used with connectors contains small metallic particles that help penetrate aluminum's oxide. The second type of compound does not contain these particles and should be used only for lug-to-bus and bus-to-bus connections.

Connectors

Only all-aluminum pressure type connectors marked AL-CU to indicate they have been tested and are listed by UL for aluminum, copper, or aluminum to copper connections interchangeably, should be used. The connectors are usually plated to avoid the formation of oxide and to resist corrosion.

Pressure connectors are of two basic types—mechanical screw type and compression type applied with a tool and die (Figure 4.4). The mechanical screw type requires only a wrench to install and is reversible. Except for the cost of the tool, the compression is less expensive but is not reusable.

Both types are designed to apply sufficient pressure to shatter the brittle aluminum oxide from the strand surfaces and provide low-resistance metal to metal contact. However, whichever type you use, follow the manufacturer's instructions very carefully.

In making connections, first strip the insulation. Then apply joint compound if it is not already contained in the connector.

If the connector is a mechanical screw type, apply the manufacturer's recommended torque (Figure 4.5). If it is a compression type, crimp it as recommended by the manufacturer. Be sure to select the correct size die and close the tool completely for full compression. Wipe off any excess compound. Then tape the joint or apply the insulation enclosures that come with some types of connectors.

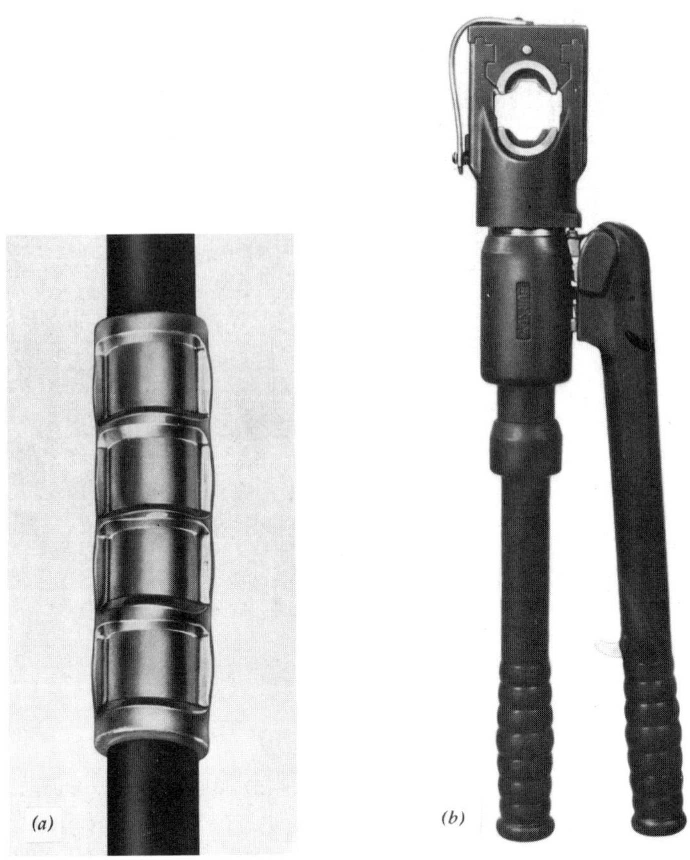

Figure 4.4 (*a*) Compression type connector. (*b*) Crimping tool must be fully closed. Failure to close crimping tool will lead to an unsatisfactory and weak joint. (Courtesy of Burndy Corporation)

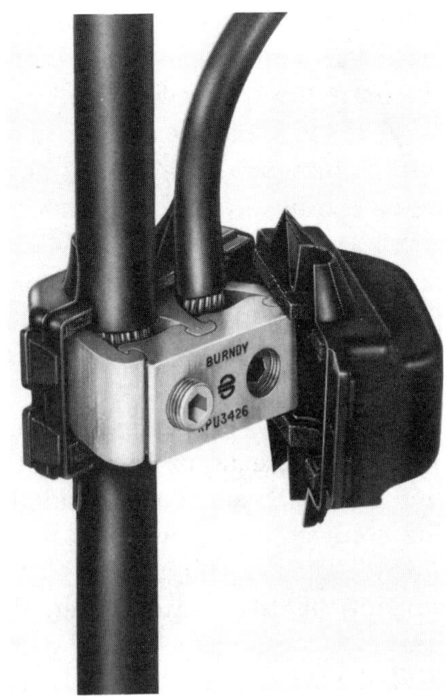

Figure 4.5 Proper torque is important; overtightening may lead to overheating and failure. (Courtesy of Burndy Corporation)

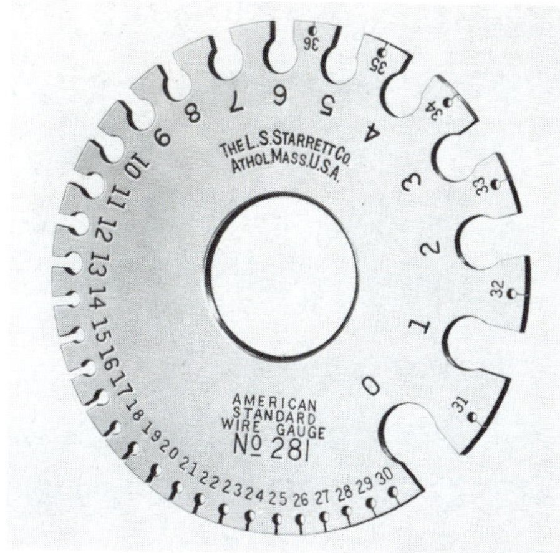

Figure 4.6 American wire gauge used to measure the size of solid wire. (Courtesy of L. S. Starrett Co.)

Copper Conductors

Copper is an excellent conductor, has high conductivity, and is more ductile (can be drawn out thinner). It is easy to work with and handle and does not oxidize as much as aluminum. It is relatively high in tensile strength and can be soldered easily. Copper has become very expensive.

4-2 CONDUCTOR SIZE

Conductor sizes are measured and listed in two ways. One is based on the American Wire Gauge (AWG), a gauge to measure solid wire only. The outer edge of the gauge has slots, which are numbered. The smallest slot into which the wire will fit is the gauge number of the wire (Figure 4.6).

Circular Mil Area

The size of large conductors is determined by calculating the circular mil area of the material. A mil is equal to one-thousandth of an inch (that is, 0.001 inch). A circular mil is equal to a diameter of one-thousandth of an inch. Therefore, a circular mil conductor with a diameter of 1 mil has a cross-sectional area of 1 circular mil. The cross-sectional area of a conductor, stated in circular mils, is equal to the number of 1-mil wires that will fit into it. The cross-sectional area of any round wire in circular mils is the area of the metal only, and is found by squaring the diameter in mils or thousandths of an inch (multiplying the diameter by itself).

The circular mil area is usually stated in an abbreviated form. The NEC abbreviates these numbers by using the Roman numeral "M" for thousands of circular mils. Example, a 500 MCM conductor has a circular mil area of 500,000 circular mils.

Circular mils = diameter squared
Square mils = 3.1416 times radius squared

To find the sectional area of a round wire in circular mils when the diameter is known, express the diameter in mils and square it. Let

d = diameter of wire in mils
CM = sectional area in circular mils

Then the circular mil area is

$$CM = d^2$$

To find the diameter of a wire when its sectional area is known, exact the square root of the circular mil area; the result will be in mils. This rule is derived by extracting the square root of both sides; thus

$$d = \sqrt{CM}$$

4-3 CONDUCTOR INSULATION

Most kinds of wire used in the wiring of commercial buildings have thermoplastic insulation, the thickness of which depends on the size of the conductor. There are several types.

PARASYN® TW

60°C. BUILDING WIRE

UL FILE NUMBER E 14656

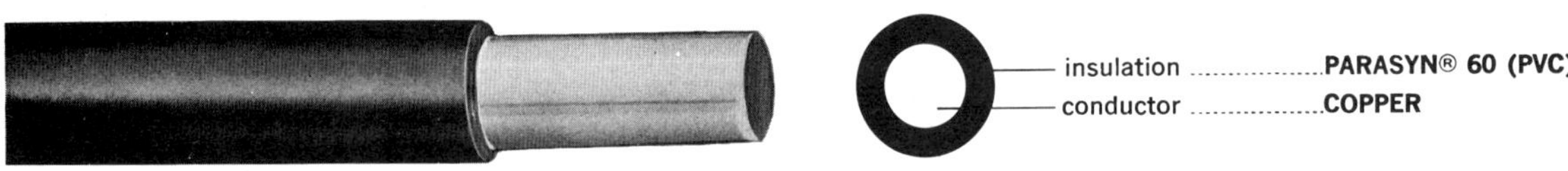

Paranite's PARASYN® type TW is an Underwriter's Laboratories listed type TW cable suitable for use at conductor temperatures up to 60°C in either wet or dry locations. The PARASYN® 60 (PVC) insulation has been specifically designed for high insulation resistance and outstanding moisture stability up to 60°C temperatures.

In addition to Underwriter's Laboratories type TW, PARASYN® TW exceeds the requirements of IPCEA-S-61-402 (3.7), ASTM-D-2219, U.L. 60°C in oil service and Federal Specification JC-30A. The solid bare copper conductors are made to ASTM-B-3 and stranded bare copper conductors to ASTM-B-8 and B-3.

Paranite's PARASYN® TW is available in all standard NEMA colors and each finished package is shipped with U.L. Insulated Wire Label attached.

Size	Number of Strands	Insulation Thick.		Approx. OD Inches	Approx. Weight Per 1000 Ft.	Ampacity**	Package*
		64th	Mils				
14	1	2	.030	.13	20	15	500 Ft. Ctn.
12	1	2	.030	.15	28	20	500 Ft. Ctn.
10	1	2	.030	.17	42	30	500 Ft. Ctn.
14	7	2	.030	.14	20	15	500 Ft. Ctn.
12	7	2	.030	.16	29	20	500 Ft. Ctn.
10	7	2	.030	.18	43	30	500 Ft. Ctn.
8	7	3	.045	.24	72	40	500 Ft. Ctn.

* 14-10 also available in 2500 Ft. Reels.
8 also available in 1000 Ft. Reels.
** Ampacity based on NOT more than three conductors in raceway or conduit with an ambient temperature of (30°C) (86°F) and maximum conductor temperature of 60°C.

Figure 4.7 Type TW building wire is suitable for use at conductor temperatures up to 60°C in either wet or dry locations. (Courtesy of ESSEX INTERNATIONAL)

Types TW, THW. The most commonly used types of plastic-insulated wires are TW and THW. TW may be used in wet or dry locations. Type THW is similar to TW but the insulation withstands a greater degree of heat. Figure 4.7 shows TW, Figure 4.8 shows THW.

PARASYN® THW

75°C. BUILDING WIRE

UL FILE NUMBER E 14656

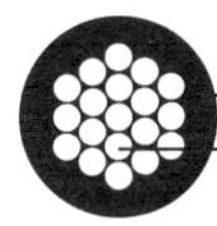

insulation .... **PARASYN® 75**
conductor .... **COPPER OR ALUMINUM**

Paranite's PARASYN® THW is an Underwriter's Laboratories listed type THW cable suitable for use as a power cable at conductor temperatures up to 75°C in either wet or dry locations. The PARASYN® 75 insulation is specifically designed for high insulation resistance and outstanding moisture stability up to 75"C temperatures.

In addition to the Underwriter's Laboratories Type THW, Paranite's PARASYN® THW exceeds the requirements of IPCEA S-61-402 (3.8), ASTM-D-2220, Underwirter's Laboratories 90°C Appliance Wiring Material and Federal Specification JC-30A. All stranded aluminum conductors comply with ASTM-B-231. The solid bare copper conductors are made to ASTM-B-3 and the stranded bare copper conductors to ASTM-B-8 and B-3.

Paranite's PARASYN® THW is available in all standard NEMA colors and each finished package is shipped with U.L. Insulated Wire Labels attached.

Size	Number of Strands	Insulation Thick.		Approx. OD Inches	Approx. Weight Per 1000 Ft.		Ampacity**		Packaging*
		64th	Mils		Copper	Alum.	Copper	Alum.	
14	1	3	.045	.16	22		15		500 Ft. Ctn.
12	1	3	.045	.18	31		20		500 Ft. Ctn.
10	1	3	.045	.20	45		30		500 Ft. Ctn.
14	7	3	.045	.17	24		15		500 Ft. Ctn.
12	7	3	.045	.19	33		20		500 Ft. Ctn.
10	7	3	.045	.21	48		30		500 Ft. Ctn.
8	7	4	.060	.27	77	42	45	40	500 Ft. Ctn.
6	7	4	.060	.31	112	57	65	50	500 Ft. Coil
4	7	4	.060	.36	167	79	85	65	500 Ft. Coil
3	7	4	.060	.39	205	92	100	75	500 Ft. Coil
2	7	4	.060	.42	252	112	115	90	500 Ft. Coil
1	19	5	.080	.49	324	146	130	100	1000 Ft. Reel
1/0	19	5	.080	.53	397	174	150	120	1000 Ft. Reel
2/0	19	5	.080	.58	491	209	175	135	1000 Ft. Reel
3/0	19	5	.080	.63	608	252	200	155	1000 Ft. Reel
4/0	19	5	.080	.69	753	304	230	180	1000 Ft. Reel
250	37	6	.095	.77	899	369	255	205	1000 Ft. Reel
300	37	6	.095	.82	1068	429	285	230	1000 Ft. Reel
350	37	6	.095	.87	1233	488	310	250	1000 Ft. Reel
400	37	6	.095	.92	1396	546	335	270	1000 Ft. Reel
500	37	6	.095	1.01	1722	660	380	310	1000 Ft. Reel
600	61	7	.110	1.12	2080	798	420	340	1000 Ft. Reel
750	61	7	.110	1.23	2570	967	475	385	1000 Ft. Reel
1000	61	7	.110	1.37	3375	1244	545	445	1000 Ft. Reel

* 14-10 also available in 2500 Ft. Reels.
8-2 also available in 1000 Ft. Reels.
1-1000 also available in 500 Ft. Reels.

** Ampacity based on NOT more than three conductors in raceway or conduit with an ambient temperature of (30°C) (86°F) maximum conductor temperature of 75°C.

Figure 4.8 Type THW cable is suitable for use as a power cable at conductor temperatures up to 75°C in either wet or dry locations. (Courtesy of ESSEX INTERNATIONAL)

Type THHN. THHN may be used only in dry locations and has a high temperature rating of 90°C. It is expensive wire; however, the use is often justified because for a given ampacity, smaller conduit may be used than when installing wires with thicker insulation (Figure 4.9).

Type XHHN. XHHN has combined properties of both rubber and polyethylene to provide a thermo-setting material with excellent thermal, electrical, and physical properties. This cable can be used in aerial and raceway applications and is dual rated, 70°C in dry locations and 75°C in wet locations. The reduced insulation thickness permits a greater number of conductors per conduit size (Figure 4.10). Figure 4.11 shows bare copper.

4-4 AMPACITY

The purpose of the conductor is to carry current from one place in a circuit to another. Ampacity refers to the ability of a conductor to carry current.

The ampacity rate of a conductor is determined by three things.

PARASYN® THHN - THWN

60°-105° BUILDING WIRE

UL FILE NUMBER E 14656

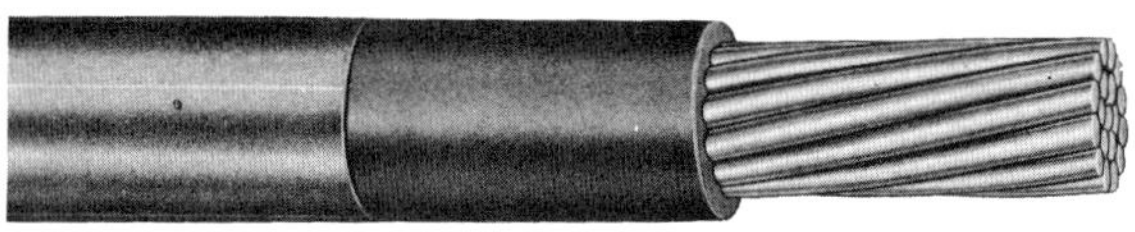

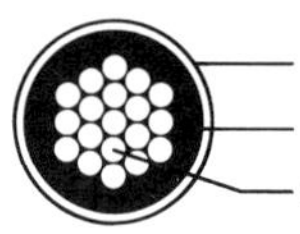

Paranite's Type THHN or THWN is a small diameter general purpose 600 volt building wire for use as power, lighting and control wiring. The thin, high grade PVC insulation and smooth, tough Nylon jacket permits the easy use of more wires in a given size conduit. The wide range of listings by Underwriter's Laboratories Standard 83 include: THWN - 75°C wet or dry, THHN - 90°C dry, machine tool wire 90°C*, appliance wire 105°C, and sizes 14 thru 6 are gasoline and oil resistant 75°C.

The copper conductors meet the requirements of ASTM-B-3 if solid, and ASTM-B-3 and B-8 if stranded. As Type THHN or THWN, this wire also complies with Federal specification JC-30A.

Size	Number of Strands	Thickness In Inches		Approx. O.D. Inches	Ampacity**		Approx. Weight Per 1000 Ft.	Package***
		Insul.	Jacket		THHN 90°	THWN 75°C		
14	Solid	.015	.004	.11	15	15	16	500 Ft. Ctn.
12	Solid	.015	.004	.13	20	20	25	500 Ft. Ctn.
10	Solid	.020	.004	.16	30	30	38	500 Ft. Ctn.
14	7	.015	.004	.12	15	15	17	500 Ft. Spool
12	7	.015	.004	.14	20	20	25	500 Ft. Spool
10	7	.020	.004	.17	30	30	40	500 Ft. Spool
8	7	.030	.005	.22	50	45	65	500 Ft. Spool
6	7	.030	.005	.26	70	65	97	500 Ft. Reel
4	19	.040	.006	.33	90	85	155	500 Ft. Reel
2	19	.040	.006	.39	120	115	192	500 Ft. Reel

* Solid and 7 strand not marked or labeled as machine tool wire.

** Based on 3 conductors in conduit, with ambient temperature of (30°C) (86°F)

*** 14, 12 and 10 also available in 2500 Ft. Reels.

Figure 4.9 Type THHN cable is a small-diameter general purpose 600-volt building wire for use as power, lighting, and control wiring. (Courtesy of ESSEX INTERNATIONAL)

1. *Material.* The material of which the conductor is made determines how easily it will carry current. For example, copper is a better conductor than aluminum and will carry more current.
2. *Size.* The larger the conductor, the more current it will carry. Since conductors are enclosed in conduits and boxes, care must be taken to use

PARATHENE®-XL XHHW

TYPE XHHW

U.L. FILE NUMBER E 1139

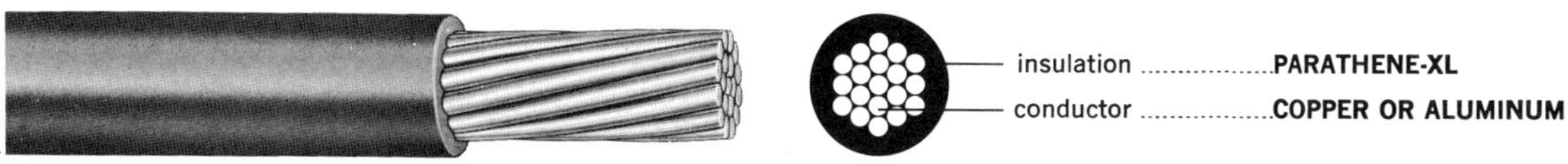

Paranite's PARATHENE-XL is a general purpose chemically cross-linked Polyethylene combining the best properties of both rubber and Polyethylene to provide a thermosetting material with excellent thermal, electrical, and physical properties. This cable can be used in aerial and raceway applications and is dual rated, 90°C in dry locations and 75°C in wet locations. The reduced insulation thickness permits a greater number of conductors per conduit in sizes 4 and larger than the Type USE. This cable meets the requirements of Underwriter's Laboratories Standard 44. In addition it also exceeds the requirements of Interim Standard #2 to IPCEA S-66-524. The construction and components of this cable also conform to Federal Specification J-C-30A. The solid copper conductors are made to ASTM-B-3 and the stranded conductors to ASTM-B-3 and B-8. All aluminum conductors comply with ASTM-B-231.

Size AWG or MCM	Number of Strands	Insul. Thick. in Mils	Approx. O.D. in inches	Net Wt. per M Ft. in Lbs.		Ampacity*				Packaging #
						75°C Wet		90°C Dry		
				Copper	Alum.	Copper	Alum.	Copper	Alum.	
14	1	.030	.14	17	—	15	—	15	—	2500 Ft. Reel
12	1	.030	.15	25	—	20	—	20	—	2500 Ft. Reel
10	1	.030	.17	38	—	30	—	30	—	2500 Ft. Reel
14	7	.030	.15	18	—	15	—	15	—	2500 Ft. Reel
12	7	.030	.17	27	—	20	—	20	—	2500 Ft. Reel
10	7	.030	.19	40	—	30	—	30	—	2500 Ft. Reel
8	7	.045	.25	67	—	45	—	50	—	2500 Ft. Reel
6	7	.045	.29	99	43	65	50	70	55	1000 Ft. Reel
4	7	.045	.34	152	63	85	65	90	70	1000 Ft. Reel
2	7	.045	.40	234	92	115	90	120	95	1000 Ft. Reel
1	19	.055	.46	297	117	130	100	140	110	1000 Ft. Reel
1/0	19	.055	.50	368	141	150	120	155	125	1000 Ft. Reel
2/0	19	.055	.54	458	172	175	135	185	145	1000 Ft. Reel
3/0	19	.055	.60	572	211	200	155	210	165	1000 Ft. Reel
4/0	19	.055	.65	712	258	230	180	235	185	1000 Ft. Reel
250	37	.065	.72	842	304	255	205	290	215	1000 Ft. Reel
300	37	.065	.77	1001	357	285	230	300	240	1000 Ft. Reel
350	37	.065	.82	1164	411	310	250	325	260	1000 Ft. Reel
400	37	.065	.87	1320	459	335	270	360	290	1000 Ft. Reel
500	37	.065	.95	1638	564	380	310	405	330	1000 Ft. Reel
600	61	.080	1.06	1980	689	420	340	455	370	1000 Ft. Reel
750	61	.080	1.17	2460	845	475	385	500	405	1000 Ft. Reel
1000	61	.080	1.23	3250	1100	545	445	585	480	500 Ft. Reel

* Ampacity based on not more than three conductors in raceway or conduit with an ambient temperature of 30°C and maximum conductor temperature of 75°C in wet locations and 90°C in dry locations per table 310-12 or 14 of N.E. Code.

#14-8 also available in 500 Ft. Cartons.

6-2 also available in 500 Ft. Coils.

1-1000 also available in 500 Ft. Reels.

Figure 4.10 Type XHHW cable can be used in aerial and raceway applications, is dual rated, 90°C in dry locations and 75°C in wet locations. The reduced insulation thickness permits a greater number of conductors per conduit in sizes 4 and larger than some other types of cable. (Courtesy of ESSEX INTERNATIONAL) conductors of a large enough size in order to prevent damage to the insulation from overheating.

3. *Insulation.* A conductor with insulation capable of withstanding heat will have a higher ampacity rating than a conductor of the same size with a lower insulator temperature rating. Wires with heat-resistant covering can be contained in a more confined area than normal conductors without heat damage to the insulation. Conductors must be handled carefully to prevent damage to the insulation. Figure 4.12 shows types of insulation-temperature rating and uses of wires and cables.

BARE COPPER

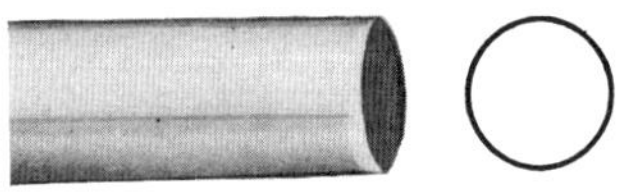

SOLID CONDUCTOR

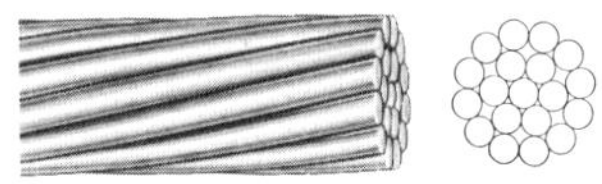

STRANDED CONDUCTOR

Paranite's bare copper conductors are made from pure electrolytic-grade copper. Stranded bare copper is a concentric lay conductor composed of a central core surrounded by one or more layers of helically laid wires. Each of these succeeding layers is applied with opposite direction of twist. In the manufacture of this type conductor the number of wires laid up around the center wire is six (6) and each succeeding layer consists of six (6) additional wires, so that the number of wires in the strands is 7, 19, 37, 61, etc. Our copper conductors are manufactured in compliance with ASTM standards B-1, B-2, B-3, and B-8. In addition, Paranite's copper conductors comply with the applicable types and classes of Federal specifications QQ-H-343.

Size	Number of Strands	Diameter Inches	Approx. Weight Per 1000 Ft.	Standard Package
14	Solid	.064	12.4	1000 Ft. Spool
12	Solid	.081	19.8	1000 Ft. Spool
10	Solid	.102	31.4	500 Ft. Spool
8	Solid	.129	50	500 Ft. Spool
6	Solid	.162	79	250 Ft. Spool
4	Solid	.204	126	250 Ft. Spool
2	Solid	.258	201	250 Ft. Spool
8	7	.146	51	250 Ft. Spool
6	7	.184	81	250 Ft. Spool
4	7	.232	129	250 Ft. Spool
2	7	.292	205	250 Ft. Spool
1	19	.332	259	1000 Ft. Reel
1/0	19	.373	326	1000 Ft. Reel
2/0	19	.418	411	1000 Ft. Reel
3/0	19	.470	518	1000 Ft. Reel
4/0	19	.528	653	1000 Ft. Reel
250 MCM	37	.575	772	1000 Ft. Reel
300 MCM	37	.630	925	1000 Ft. Reel
350 MCM	37	.681	1080	1000 Ft. Reel
400 MCM	37	.728	1236	1000 Ft. Reel
500 MCM	37	.813	1544	1000 Ft. Reel
600 MCM	61	.893	1850	1000 Ft. Reel
750 MCM	61	.998	2316	1000 Ft. Reel
•1000 MCM	61	1.152	3086	1000 Ft. Reel

Sizes #1 thru 1000 MCM also available on 500 Ft. reels.

Above stranded conductors are class B stranding.

Figure 4.11 Bare copper, solid conductor. (Courtesy of ESSEX INTERNATIONAL)

TYPES OF INSULATION — TEMP. RATING — USE OF WIRES AND CABLES

Trade Name	Type Letter	Max. Oper. Temp.	Insulation	Outer Cover	Special Provisions and Use
Rubber-Covered Fixture Wire Solid or 7-Strand	RF-1	60°C 140°F	Code Rubber	Non-Metallic Covering	Fixture wiring limited to 300 volts
	RF-2	60°C 140°F	Code Rubber	Non-Metallic Covering	Fixture wiring and as permitted
Rubber-Covered Fixture Wire Flexible Stranding	FF-1	60°C 140°F	Code Rubber	Non-Metallic Covering	Fixture wiring limited to 300 volts in Section 725-14
	FF-2	60°C 140°F	Code Rubber	Non-Metallic Covering	Fixture wiring and as permitted in Section 725-14
Heat-Resistant Silicone Solid or 7-Strand	SF-1	200°C 392°F	Silicone Rubber	Lacquered Glass	High Temperature Fixture wiring limited to 300 volts
	SF-2	200°C 392°F	Silicone Rubber	Lacquered Glass	High Temperature Fixture wiring and as permitted in section 725-14
Heat-Resistant Rubber-Covered Fixture Wire—Solid or 7-Strand	RFH-1	75°C 167°F	Heat-Resistant Rubber	Non-Metallic Covering	Fixture wiring limited to 300 volts
	RFH-2	75°C 167°F	Heat-Resistant Rubber	Non-Metallic Covering	Fixture wiring and as permitted in Section 725-14
Heat-Resistant Rubber-Covered Fixture Wire Flexible Stranding	FFH-1	75°C 167°F	Heat-Resistant Rubber	Non-Metallic Covering	Fixture wiring limited to 300 volts
	FFH-2	75°C 167°F	Heat-Resistant Rubber	Non-Metallic Covering	Fixture wiring and as permitted in Section 725-14
Thermoplastic-Covered Fixture Wire—Solid or Stranded	TF	60°C 140°F	Thermoplastic	None	Fixture wiring and as permitted in Section 725-14
Thermoplastic-Covered Fixture Wire—Flexible Stranding	TFF	60°C 140°F	Thermoplastic	None	Fixture wiring and as permitted in Section 725-14
Cotton-Covered Heat-Resistant Fixture Wire	CF	90°C 194°F	Impregnated Cotton	None	Fixture wiring limited to 300 volts
Asbestos-Covered Heat-Resistant Fixture Wire	AF	150°C 302°F	Impregnated Asbestos	None	Fixture wiring limited to 300 volts Indoor Dry Location
Heat-Resistant Rubber	RH	75°C 167°F	Heat-Resistant Rubber	Moisture-Resistant Flame-Retardant Non-Metallic Covering	Dry Locations
Heat-Resistant Rubber	RHH	90°C 194°F	Heat-Resistant Rubber	None	Dry Locations
Moisture-Resistant Rubber	RW	60°C 140°F	Moisture-Resistant Rubber	Moisture-Resistant Flame-Retardant Non-Metallic Covering	Dry and Wet Locations
Heat and Moisture-Resistant Rubber	RHW	75°C 167°F	Heat and Moisture-Resistant Rubber	None	Dry and Wet Locations
Heat-Resistant Latex Rubber	RUH	75°C 167°F	90% Unmilled Grainless Rubber	Moisture-Resistant Flame-Retardant Non-Metallic Covering	Dry Locations
Moisture-Resistant Latex Rubber	RUW	60°C 140°F	90% Unmilled Grainless Rubber	Moisture-Resistant Flame-Retardant Non-Metallic Covering	Dry and Wet Locations
Thermoplastic	T	60°C 140°F	Flame-Retardant Thermoplastic Compound	None	Dry Locations
Moisture-Resistant Thermoplastic	TW	60°C 140°F	Flame-Retardant Moisture-Resistant Thermoplastic	None	Dry and Wet Locations
Heat-Resistant Thermoplastic	THHN	90°C 194°F	Flame-Retardant Heat-Resistant Thermoplastic	Nylon Jacket	Dry Locations
Moisture and Heat-Resistant Thermoplastic	THW	75°C 167°F	Flame-Retardant, Moisture and Heat-Resistant Thermoplastic	None	Dry and Wet Locations
Moisture and Heat-Resistant Thermoplastic	THWN	75°C 167°F	Flame-Retardant, Moisture and Heat-Resistant Thermoplastic	Nylon Jacket	Dry and Wet Locations
Mineral Insulation (Metal Sheathed)	MI	85°C 185°F	Magnesium Oxide	Copper	Dry-Wet Locations with Type O termination fittings. Max. oper. temp. for special appl. 250°C
Thermoplastic and Asbestos	TA	90°C 194°F	Thermoplastic and Asbestos	Flame-Retardant Cotton Braid	Switchboard wiring only
Varnished Cambric	V	85°C 185°F	Varnished Cambric	Non-Metallic Covering or Lead Sheath	Dry Locations Only
Asbestos and Varnished Cambric	AVA	110°C 230°F	Impregnated Asbestos and Varnished Cambric	Asbestos Braid	Dry Locations Only
Asbestos and Varnished Cambric	AVL	110°C 230°F	Impregnated Asbestos and Varnished Cambric	Lead Sheath	General Use and Wet Locations
Asbestos and Varnished Cambric	AVB	90°C 194°F	Impregnated Asbestos and Varnished Cambric	Flame-Retardant Cotton Braid	Dry Locations Only
Asbestos	A	200°C 392°F	Asbestos	Without Asbestos Braid	Dry locations only. In raceways only for leads to or within apparatus limited to 300 volts
Asbestos	AA	200°C 392°F	Asbestos	With Asbestos Braid	Dry locations. Open wiring. In raceways for leads to or within apparatus limited to 300 volts
Asbestos	AI	125°C 257°F	Impregnated Asbestos	Without Asbestos Braid	Dry locations only. In raceways only for leads to or within apparatus limited to 300 volts
Asbestos	AIA	125°C 257°F	Impregnated Asbestos	With Asbestos Braid	Dry locations. Open wiring. Raceways for leads to or within apparatus
Heat and Moisture-Resistant Cross-Linked Polyethylene	XHHW	75°C 167°F	Cross-Linked Polyethylene	None	Wet and Dry Locations
		90°C 194°F			Dry Locations
Moisture, Heat and Oil-Resistant Thermoplastic	THW-MTW	60°C 140°F	Thermoplastic	None	Wet and Dry Locations
		90°C 194°F			Special applications within electric discharge lighting equipment (size 14-8 only).

Figure 4.12 Shows types of insulation-temperature rating and uses of wire and cables. (Courtesy of ESSEX INTERNATIONAL)

SUMMARY

1. The material of which the conductor is made determines how easily it will carry current.
2. Conductors must be handled carefully to prevent damage to the insulation.
3. Aluminum building wire costs less than copper for the same ampacity rating.
4. Two commonly used conductors are aluminum and copper.
5. Only all-aluminum pressure type connectors marked AL-CU to indicate they have been tested and listed by UL for aluminum and copper connections should be used.

PROBLEMS

4-1 Check Figure 4.11. Give the diameter of copper wire with the folowing AWG.

a. No. 14 ____________

b. No. 12 ____________

c. No. 10 ____________

4-2 Check Figure 4.8. Give the ampacity of copper wire with the following AWG.

a. No. 2 ____________

b. No. 6 ____________

c. No. 10 ____________

d. No. 14 ____________

4-3 Check Figure 4.10. Give the ampacity of 90°C dry aluminum wire with the following AWG.

a. 1/0 MCM ____________

b. 2/0 MCM ____________

c. 4/0 MCM ____________

d. 250 MCM ____________

e. 500 MCM ____________

4-4 What is the sectional area of a metal rod ¼ inch in diameter?

4-5 What is the diameter of a wire having a sectional area of 26,250 circular mils?

4-6 Two types of conductors most commonly used in the trade are ______________ and ______________________________________ .

4-7 Pressure connectors are of two basic types. Name them.

1. ______________

2. ______________

4-8 ______________ is a conductor easy to work and handle and does not oxidize as much as ______________________________________ .

4-9 The purpose of the conductor is to carry current. Ampacity refers to:

____

____

4-10 The ampacity rate of a conductor is determined by: ______________

__________________ ______________________________________ .

chapter 5

METHODS OF CONDUIT WIRING

Instructional Objectives

1. To identify different types of raceway systems.
2. To compare steel conduits to polyvinyl chloride (PVC).
3. To develop an ability to make bends in conduit.
4. To learn why intermediate metal conduit was introduced to the trade.
5. To understand the need for raceway systems.
6. To learn how to use an electrical metallic tubing (EMT) bender.
7. To understand the NEC relating to conduit bending.
8. To become familiar with conduit fittings.
9. To learn how to bend PVC.
10. To learn some advantages of using rigid nonmetallic conduit.

Self-Evaluation Questions

Test your prior knowledge of the information in this chapter by answering the following questions. Watch for the answers as you read the chapter. Your final evaluation of whether you understand the material is measured by your ability to answer the questions. When you have completed the chapter, return to this section and answer the questions again.

1. What are some construction specifications for rigid metal conduit?
2. For what locations and conditions of use is electrical metallic tubing permitted?
3. How are bends made with EMT?
4. What is the limitation as to the number of bends in one run of conduit?
5. Is PVC different than EMT?
6. How do you make a joint in PVC?
7. How do you bend PVC?
8. Is intermediate metal conduit (IMC) heavier or lighter than rigid metal conduit?

9. Can you thread intermediate metal conduit?
10. What is a hot box?

5-1 TYPES OF RACEWAY SYSTEMS

There are several types of approved raceway systems used in commercial buildings. Provisions for each type are in a separate article of the National Electrical Code. These include rigid metal conduit, intermediate metal conduit, electrical metallic tubing, flexible metal conduit, liquidtight flexible metal conduit, surface metal raceways, underfloor raceways, and a rigid nonmetallic conduit called PVC (polyvinylchloride).

Rigid Metal Conduit—NEC Article 346

Since the early nineteen hundreds the National Electrical Code has stated that "the function of an electrical raceway system is to facilitate the insertion and extraction of the conductors, and to protect them from mechanical injury." The raceway systems in the late 1970s combine this same function with a high degree of safety, distribution, and flexibility.

Rigid metal conduit with its strength and durability offers a permanent protection against physical damage to circuit conductors. Rigid conduit looks like water pipe; however, it differs in several ways. (1) Conduit pipe is annealed and heat treated to permit easy bending. (2) The inside of the conduit is carefully prepared and treated with a smooth and uniform coating of zinc. (3) After the zinc has been applied, an additional protective treatment is applied to both inside and outside surfaces, leaving a smooth raceway so that the conduit conductors can be pulled in with a minimum of effort and without damage to the insulation of the conductors. Some manufacturers coat galvanized rigid metal conduit with a tough, corrosion resistant coating of polyvinylchloride to the outside surface of the conduit. This conduit is designed for use in any area where corrosion is severe enough to attack metal conduit whether exposed or underground.

Conduit is manufactured in 10-foot lengths. Each length is threaded on both ends, one threaded coupling is furnished with each 10-foot length, and each length must bear a UL label (Figure 5.1).

Elbows and nipples, as well as a full line of standard fittings, outlet boxes, and accessories, are available in trade sizes from ½ to 6 inches (Figure 5.2).

Conduit is measured by its internal diameter; however, the actual size is larger than the indicated diameter. A ½-inch conduit has an internal diameter of nearly ⅝ inch and external diameter of about ⅞ inch.

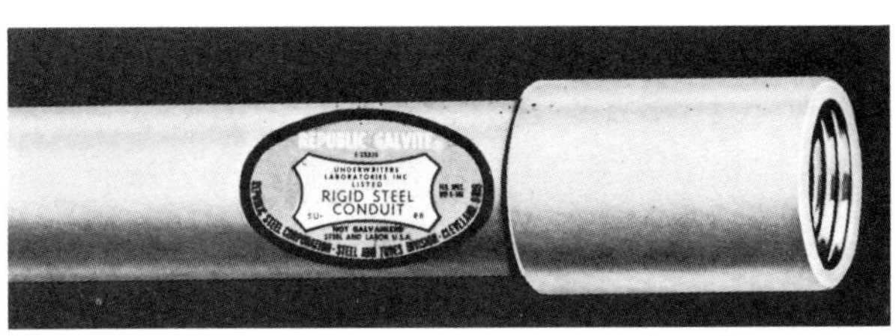

Figure 5.1 Rigid steel conduit, showing UL label. (Courtesy of Republic Steel Corporation)

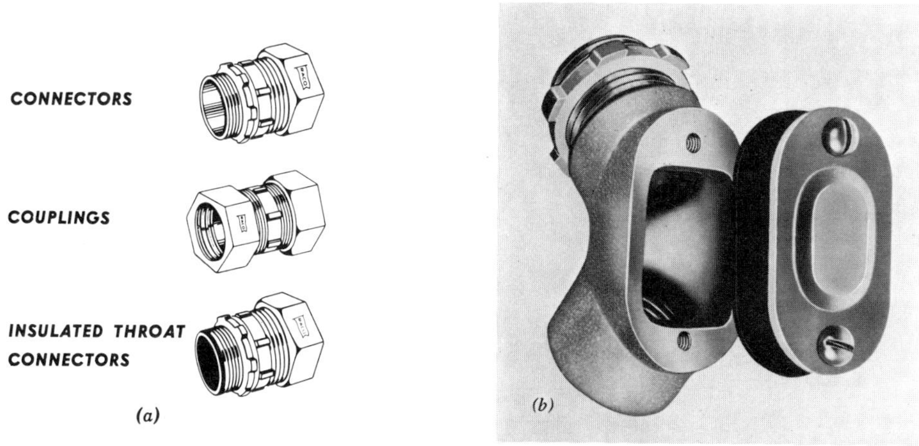

Figure 5.2 Rigid conduit fittings. (*a*) Compression type connector and coupling. (Courtesy of Raco, Inc.) (*b*) Rigid conduit pulling elbow. (Courtesy of Ideal Industries)

Hand Bending Rigid Conduit

The smaller sizes of rigid conduit are bent with a hickey, which is sometimes called a bending tool. The bend is made by inching the hickey through the space to be bent, approximately 10° to 20° bends with each move. Exact measurements and calculations for bending rigid conduit with a hickey are extremely hard to determine.

To bend, start by determining the necessary data. For example, to bend a ½-inch electrical trade-size rigid conduit at 90°, to be stubbed up 12 inches from the base, the NEC states that the radius of the inner edge of any field bend shall not be less than shown in Table 346-10, which would be 4 inches for a ½-inch conduit having conductors, other than lead covered, pulled in (Table 5.1). This limits the inner radius of the ½-inch conduit to 4 inches; however, it does *not* say it could not be more. For all practical applications, the outside radius of the conduit is used for determining the bend. The only additional information that is necessary is the outside diameter (O.D.) of the conduit. This value is added to the radius given in Table 346-10. See Table 5.1.

To determine the point where the bend starts so the rise will be exactly 12 inches, subtract the outside radius of the bend from the distance of the rise, 12 inches − 4⅞ inches = start of bend. The developed end of the conduit is

TABLE 5.1 Radius of Conduit Bends (Table 346-10 Exception)

Size of Conduit, Inches	Radius to Center of Conduit, Inches
½	4
¾	4½
1	5¾
1¼	7¼
1½	8¼
2	9½
2½	10½
3	13
3½	15
4	16
4½	20
5	24
6	30

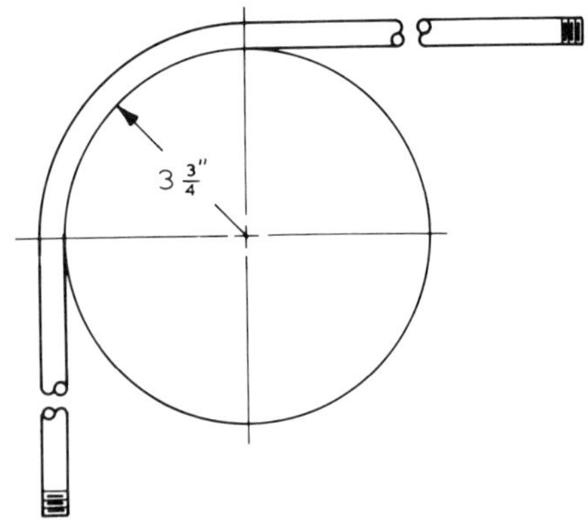

Figure 5.3 Illustrating the minimum radius of ½-inch rigid conduit. (Albert Lea Voc Tech)

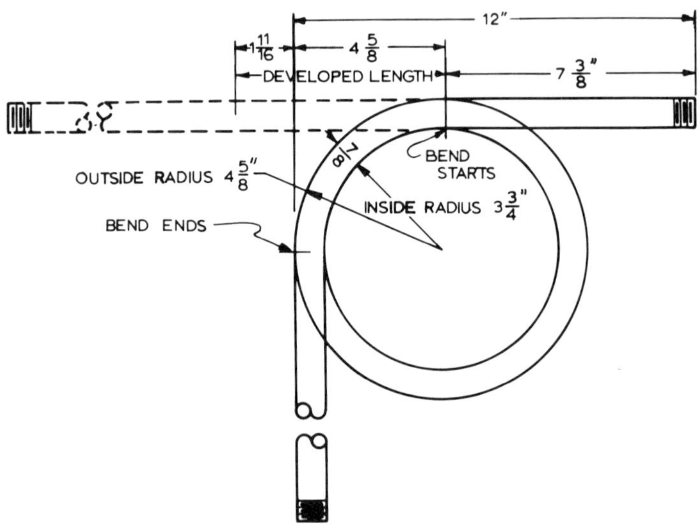

Figure 5.4 Factors involved in determining the measurement for making a 90° bend in a ½-inch rigid conduit. (Albert Lea Voc Tech)

that part which is included in making the bend. It is called a developed end because it takes more than 12 inches of straight run to make a 90° bend and have a 12-inch rise. Actually it takes 1.5 times the outside radius of the bend. Following through with Figures 5.3 and 5.4, mark off a piece of conduit for bending. The length of the rise is to be 12 inches. The radius of the bend to the outside edge is 4 inches + ⅞ = 4⅞.

To determine the place to start the bend, subtract the radius from the distance of the rise, 12 inches − 4⅞ inches = 7⅛ inches. To determine the entire length of the bend, the multiplying factor is 1.5 times the radius of the bend:

1.5 × 4⅞ inches = 7⁵⁄₁₆ inches

The overall dimensions would be

7⁵⁄₁₆ inches + 7⅛ inches or 14⁷⁄₁₆ inches

Place the heel of the hickey at 14⁷⁄₁₆ inches, and start the bend, inching the hickey to the rise end until a 90° bend is made; however, do not complete the

Dimension comparison (nominal) of 3/4" trade sizes

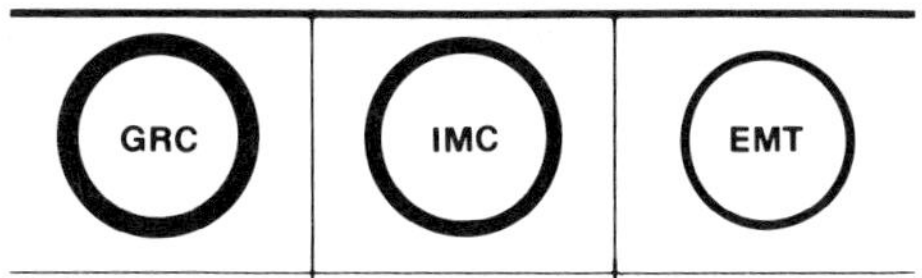

GRC	IMC	EMT
O.D. 1.050"	O.D. 1.029"	O.D. 0.922"
I.D. 0.824"	I.D. 0.863"	I.D. 0.824"
Wall 0.113"	Wall 0.083"	Wall 0.049"

Weight comparison of trade sizes
Approximate Lbs. per 100 Ft.

Trade Size	GRC W/Coupling	IMC W/Coupling*	E.M.T.
1/2"	79	60	28
3/4"	105	82	43
1"	153	116	64
1-1/4"	201	150	95
1-1/2"	249	182	110
2"	332	242	140
2-1/2"	527	401	205
3"	683	493	250
3-1/2"	831	573	325
4"	972	638	370

*Weight specifications are for comparative purposes only. They are not a requirement of U.L. 1242.

Figure 5.5 Dimension and weight comparison chart. (Courtesy of Allied Tubing and Conduit Corporation)

bend before you measure to see if it is going to be too long or too short when completed.

When making the bend, place the conduit on the floor, and hold the hickey handle straight with the conduit so the bend will be straight.

Intermediate Metal Conduit (IMC)—NEC Article 345

Intermediate metal conduit is a steel conduit of circular cross section. The wall thickness is less than those specified for rigid metal conduit but more than those for electrical metallic tubing (Figure 5.5). Actually, there is very little difference from rigid conduit. The wall thickness requires about 30 percent less steel. The nominal external diameter for IMC may also differ from that of either rigid conduit or electrical metallic tubing. Couplings for IMC may be either of threaded or threadless design, according to the NEC. IMC may be used in trade sizes from ½ inch to 4 inches in all atmospheric conditions and occupancies, including hazardous locations. It is subject to the same restrictions with respect to possible galvanic action or use in areas of known severe, corrosive influences as are other metal conduits; however, its use for conductors carrying voltages over 600 is not recognized.

The new intermediate metal conduit can be bent with hand benders to sizes ½ inch, ¾ inch, and 1 inch, providing properly engineered tools are used.

Keep in mind that bending IMC is like fabricating a double thick thin-wall conduit. IMC diameters are large, and size for size you must move twice the metal to make any bend or offset. Size for size thin-wall benders are not the answer; they are not strong enough.

To safely bend IMC new tools have been engineered. The new tools have stronger hooks, heavier wall sections, and in some cases, a larger diameter

IMC Bending Equipment Guide

Available Equipment	*For bending IMC Trade Sizes:*									
	½	¾	1	1¼	1½	2	2½	3	3½	4
HAND BENDERS*										
Appleton—Benfield	IM-50R	IM-75R	IM-100R							
Ideal—Benfield	74-016	74-017	74-018							
Gardner	940	941								
Greenlee	841	842	843							
Enerpac	B-7	B-10								
Ensley	E714	E715								
MECHANICAL BENDERS										
Lidseen of N.C. (Chgo. Type) 5100	●	●	●							
Lidseen of N.C. (Chgo. Type) 5200				●	●					
Ensley (K.C.) E777	●	●	●	●						
Greenlee 1800	●	●	●							
Greenlee 1818	●	●	●	●	●					
Chicago Misc.	●	●	●							
Enerpac (Milwaukee) BR-10	●	●	●							
Enerpac (Milwaukee) BR-15	●	●	●	●	●					
Enerpac (Milwaukee) BR-12				●	●					
Enerpac (Milwaukee) BR-27	●	●	●	●	●					
Enerpac (Milwaukee) BR-30	●	●	●	●	●					
POWER BENDERS Electric & Hydraulic										
Greenlee 550	●	●	●							
Greenlee 555 For 1½" thru 2" add Attachment Set 555-IMC	●	●	●	●	○	○				
Ensley 666	●	●	●	●						
Greenlee 782				●	●	●				
Greenlee 881 For 1¼" thru 2" add Attachment Set 881-B				○	○	○	●	●	●	●
Greenlee 882 plus attachment (Use Attachment Set 882-IMC-G1)				●	●	●				
Greenlee 884 plus attachment (Use Attachment Set 884-G22)				●	●	●	●	●	●	●
Greenlee 885 plus attachment (Use Attachment Set 885-G24)				●	●	●	●	●	●	●
Greenlee 885T plus attachment (Use Attachment Set 885T-IMC-G4)				●	●	●	●	●	●	●
Greenlee 892				●	●	●				
Enerpac-Mini Eegor B-200 Series			●	●	●	●				
Enerpac-Eegor B-448 For 2" add B2-20 2" Shoe, B2-22 2" Follow Bar, B2-24 2" U-Stop						○	●	●	●	●

For EMT benders, the next size larger model number was used.

This guide compiled from catalog material furnished by the manufacturers listed. For exact specifications and capabilities of each manufacturer's products, reference should be made to their published material. No warranty or guarantee is made by Allied Tube & Conduit Corporation.

Figure 5.6 IMC bending equipment guide. (Courtesy of Allied Tubing and Conduit Corporation)

handle has been used for extra strength. Most new model benders are for size ½ inch and ¾ inch IMC, not for size 1 inch (Figure 5.6). The 1-inch size is quite a different story because it has a wall heavy enough to hold its round during the bending operation. A standard 1¼ inch EMT bender (the type with a booster step) will do an excellent job. Hydraulic or mechanical powered benders should be used for IMC sizes 1¼ inches and above.

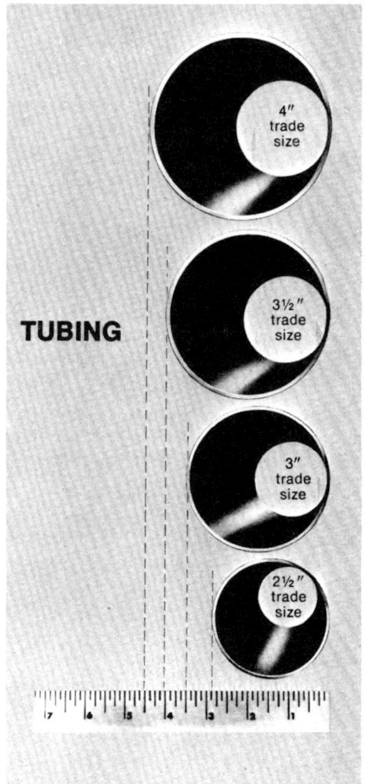

Figure 5.7 Electrical metallic tubing, trade sizes. (Courtesy of Republic Steel Corporation)

Bending IMC right the first time is important. Electricians apply their thin-wall precision bending technique to the new IMC benders, and because IMC benders also bend rigid conduit, the entire hand bending art is now the same for all three: IMC, EMT, and rigid metal conduit.

Electrical Metallic Tubing (EMT)— NEC Article 348

Electrical metallic tubing is similar to rigid metal conduit and IMC; however, it is much lighter, weighing approximately 40 percent as much as rigid conduit of the same nominal size. EMT, like the other metal conduits, is made to conform to Underwriters Laboratories Standard. It is produced in trade sizes from ½ inch to 4 inches (Figure 5.7). The rules that apply to rigid metal conduit and IMC, such as the number of wires permitted, bending, and supports, apply also to EMT. Compression, indention, setscrew, or push-on type connectors and couplings are used, because the ends of EMT are not threaded (Figure 5.8). EMT, like rigid metal conduit and IMC, must be supported within 3 feet of every box, fitting, and cabinet, and at intervals not exceeding 10 feet regardless of size.

EMT may be used in most locations where rigid metal conduit or IMC is permitted, but because of its lighter weight and threadless connections it is restricted to areas where it will not be subjected to severe mechanical damage or systems exceeding 600 volts. In general it is restricted to nonhazardous locations except those identified in Class II, Division 2 (NEC Section 502-4b).

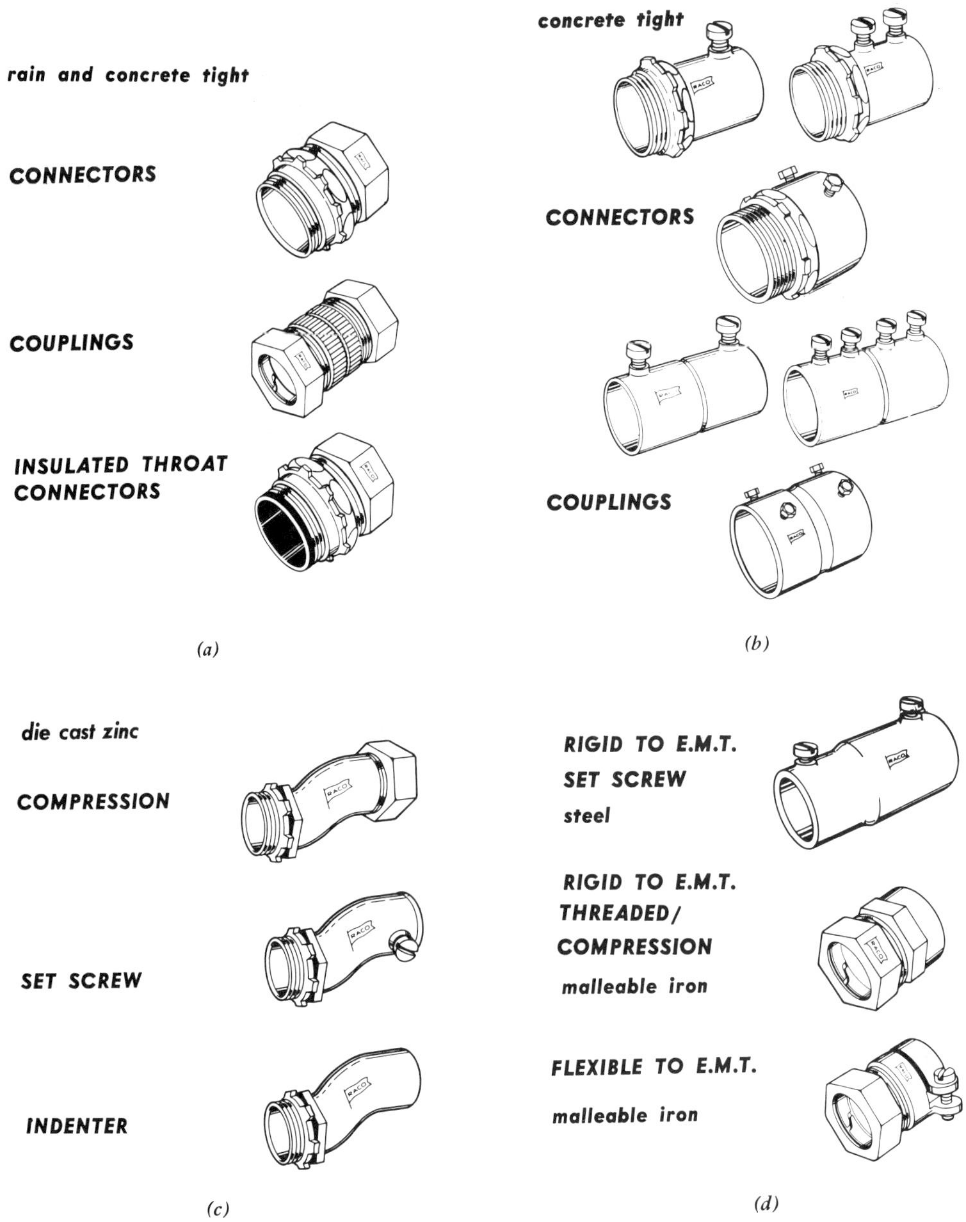

Figure 5.8 Various types of EMT fittings. (*a*) Compression connectors. (*b*) Steel set screw connectors and couplings. (*c*) Offset connectors. (*d*) Combination couplings. (Courtesy of Raco Inc.)

Bending the Smaller Sizes of EMT

EMT is easier to bend than rigid conduit because of the thin wall. The bender is pulled through the complete bend without moving it after it is set. If the bender is moved from its original position, it will crimp the tubing. Do not make the mistake of improvising a bender on EMT. Invariably it will crimp the tubing.

To make a 90° bend for a stub-up, measure the distance of the stub, subtract the length of the "bender shoe" (usually it is 5 inches for ½-inch tubing, 6 inches with ¾-inch tubing, and 8 inches with 1-inch tubing) from this distance, and place the toe of the bender B at this point. Place one foot on the tubing to hold it secure, place the other foot on the heel of the bender, and bend. Pressure is applied to the heel of the bender with the foot at the

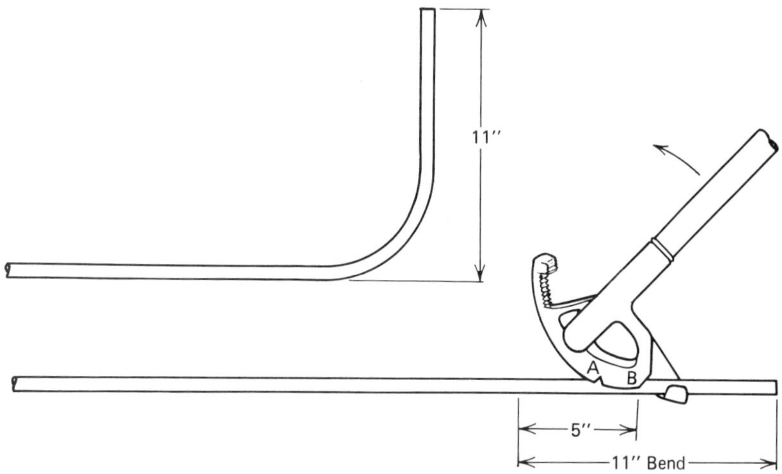

Figure 5.9 Making a 90° bend.

same time the bender handle is being pulled. Make a full swing to make any bend (see Figure 5.9).

In bending ½-inch tubing, the bender has a standard take-up of 5 inches. If a 14-inch stub is required, figure

14 inches − 5 inches = 9 inches

Set the bender with point B 9 inches from the end of the tube and the complete stub will be exactly 14 inches high. To produce stubs of ¾ inch allow 6-inch take-up; there is 8-inch take-up on 1-inch size.

To make a 45° bend, take measurements as for making 90° bends, subtract the length of the bender shoe from the distance (5 inches for ½ inch, 6 inches for ¾ inch and 8 inches for 1 inch), place the toe of the bender on the B mark, and swing the bender handle to a vertical position. See Figure 5.10.

To make an offset, measure back from the end you desire to make the offset in. Make a 45° bend, then determine the length of the offset and make another 45° bend in the opposite direction. See Figure 5.11. It is always desirable to keep the angles of an offset as small as possible by letting the length of the offset be as long as conditions permit.

To make a back-to-back bend: start by making a 90° bend; measure off the length from X to Y; mark and reverse bender on the tube; place A on bender at point Y; using outside measurements, line up conduit with first 90° bend; and swing the bender handle, keeping foot pressure on the step of

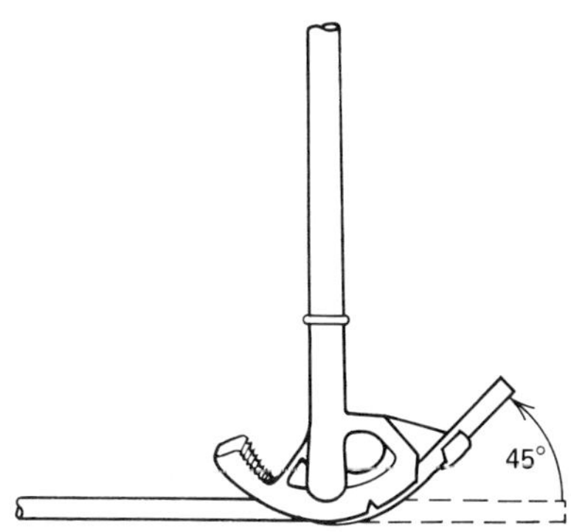

Figure 5.10 Making a 45° bend (a 45° bend is reached when bender handle is at a right angle to the tube).

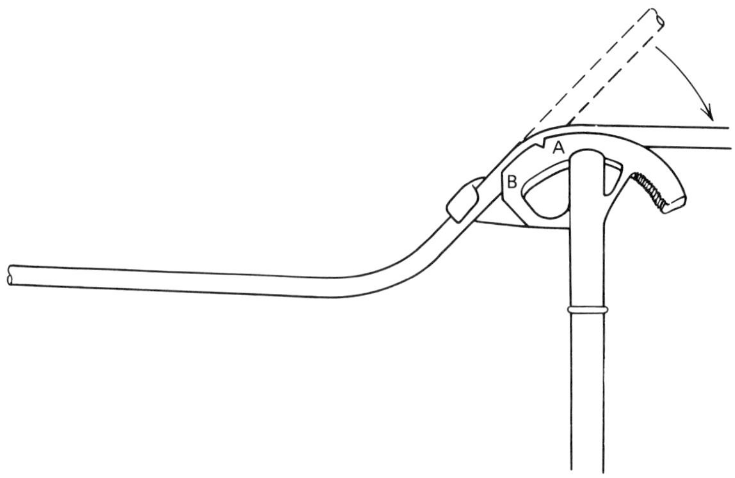

Figure 5.11 A true offset.

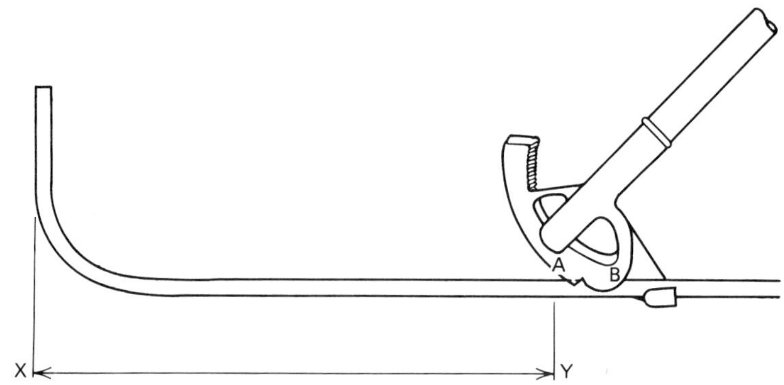

Figure 5.12 A back-to-back bend.

the bender and the second bend will be completed. See Figure 5.12. Should you over-bend, place the handle of the bender over the stub, or use a piece of pipe that will fit inside and push to the desired position.

Flexible Metal Conduit—NEC Article 350

Flexible metal conduit consists of a single strip of galvanized steel spirally wound on itself and interlocked in such a manner as to provide a round cross section of high mechanical strength and great flexibility (Figure 5.13).

Figure 5.13 Flexible metal conduit. Shows the proper way for flexible conduit to be fastened to the box. Notice hole is drilled at an angle through wood stud, and a 16 penny nail is driven into stud, bent around flex to hold in place. (Courtesy of Albert Lea Voc Tech)

Its flexibility together with its continuous length (25 to 250 feet, depending on the size) makes its use practical whenever other types of metal conduit and tubing are permitted except in wet locations, hoistways, storage-battery rooms, hazardous locations, and where rubber covered conductors are exposed to oil, gasoline, and so forth.

SCHEDULE A THIN WALL, PV-DUIT

Carlon PV-Duit, Schedule A, is 90°C UL listed for underground encased use. Conduit conforms to NEMA TC-2 and UL-651 Standards.

All joints shall be solvent cemented in accordance with the recommendations of the manufacturer.

NOM. SIZE	CATALOG NO.	O.D.	I.D.	WALL	WT. PER 100′ PLAIN END	FEET PER BUNDLE
½″	49105	.840	.720	.060	11	100
¾″	49107	1.050	.930	.060	13	100
1″	49108	1.315	1.195	.060	17	100
1¼″	49109	1.660	1.520	.070	25	50
1½″	49110	1.900	1.740	.080	32	50
2″	49111	2.375	2.175	.100	51	10
2½″	49112	2.875	2.655	.110	67	10
3″	49113	3.500	3.250	.125	89	10
4″	49115	4.500	4.200	.150	141	10

SCHEDULE 40 HEAVY WALL, PV-DUIT PLUS™

Carlon PV-Duit Plus, Schedule 40, is 90°C UL listed and UL listed for aboveground and underground uses. Conduit conforms to NEMA TC-2 and UL-651 Standards. It is UL listed in conformity with Article 347 of the National Electrical Code.

PV-DUIT PLUS is manufactured from Carlex 300, an improved PVC compound that substantially reduces smoke and HCL emissions in a fire situation and offers superior weatherability.

All joints shall be solvent cemented in accordance with the recommendations of the manufacturer.

NOM. SIZE	CATALOG NO.	O.D.	I.D.	WALL	WT. PER 100′ PLAIN END	FEET PER BUNDLE
½″	49005	.840	.622	.109	17	100
¾″	49007	1.050	.824	.113	22	100
1″	49008	1.315	1.049	.133	33	100
1¼″	49009	1.660	1.380	.140	44	50
1½″	49010	1.900	1.610	.145	53	50
2″	49011	2.375	2.067	.154	71	50
2½″	49012	2.875	2.469	.203	111	10
3″	49013	3.500	3.068	.216	145	10
3½″	49014	4.000	3.548	.226	173	10
4″	49015	4.500	4.026	.237	205	10
5″	49016	5.563	5.047	.258	277	10
6″	49017	6.625	6.065	.280	359	10

SCHEDULE 80 EXTRA HEAVY WALL, PV-DUIT

Carlon PV-Duit, Schedule 80, is designed for aboveground and underground applications where an extra heavy wall PVC is needed. Typical applications would include those subject to physical abuse such as pole risers. Schedule 80 is manufactured from Carlon 2000 Compound, is UL listed and conforms to UL and NEMA Standards.

NOM. SIZE	CATALOG NO.	O.D.	I.D.	WALL	WT. PER 100′ PLAIN END	FEET PER BUNDLE
½″*	49405	.840	.546	.147	21	100
¾″*	49407	1.050	.742	.154	28	100
1″*	49408	1.315	.957	.179	41	100
1¼″	49409	1.660	1.278	.191	57	50
1½″	49410	1.900	1.500	.200	69	50
2″	49411	2.375	1.939	.218	96	10
2½″	49412	2.875	2.323	.276	145	10
3″	49413	3.500	2.900	.300	194	10
4″	49415	4.500	3.826	.337	283	10
5″	49416	5.563	4.813	.375	393	10

*Meets NEMA Standards but not UL Listed.

SNAP STRAP™ CONDUIT WALL HANGER

New, high-strength Clamp solves bowing problems resulting from the expansion and contraction of conduit caused by varying temperature changes. mount with standard wall hanger.

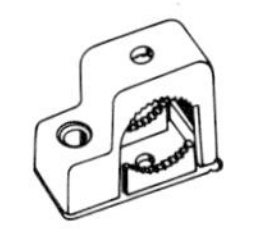

PART NO.	Size	Pkg. Qty.
E978D	1/2	100
E978E	3/4	100
E978F	1	100

NOTE: These products conform to NEMA TC-2, Federal Specification WC-1094A and UL-651 Standards.

(a)

Figure 5.14 PVC rigid plastic conduit. (*a*) Shows schedule A thin-wall PVC-Duit, schedule 40 heavy wall, and schedule 80 extra heavy wall PVC-Duit. (Courtesy of Carlon Company)

Its installation is easier and quicker than the installation of rigid metal conduit, IMC, and EMT. The conduit is easily fished and requires no elbow fittings; however, care should be exercised in properly fastening the conduit.

Flexible conduit must be supported within 12 inches of every box, cabinet, or fitting, and at regular intervals not more than 4½ feet throughout the run.

Rigid Polyvinyl Chloride Conduit (PVC)—NEC Article 347

Until recently, all plastic PVC rigid conduit had going for itself was its light weight, extreme strength, ease of handling, and impact and corrosion resistance (Figure 5.14). Now, with the growing acceptance of this material and its recognition by the National Electrical Code, both conduit manufacturers and electrical contractors have devised new techniques which slash installation time.

This reduced installation time coupled with the material's other inherent benefits puts PVC rigid conduit on the list of materials to consider when planning a rigid conduit installation. Recent field experience has indicated that PVC can replace conventional rigid conduit in all installations except: hazardous locations; for the support of fixtures or other equipment; where subjected to physical damage (unless identified for the purpose); and where subjected to ambient temperatures exceeding those for which the conduit is approved (NEC 347-3).

Figure 5.14 PVC rigid plastic conduit. (*b*) Shows PVC schedule 40, heavy wall for underground in accordance with Article 347 of NEC. (Courtesy of Carlon Company)

Figure 5.14 PVC rigic plastic conduit.
(*c*) Schedule 40 installed in exposed installation.
(Courtesy of Carlon Company)

PV-DUIT FITTINGS

COUPLINGS. Socket type.
For joining PV-DUIT.

UL LISTED

PART NO.	SIZE	PKG. QTY.
E940D	½	100
E940E	¾	80
E940F	1	50
E940G	1¼	30
E940H	1½	20
E940J	2	25
E940K	2½	15
E940L	3	20
E940M	3½	15
E940N	4	10
E940P	5	5
E940R	6	5

FEMALE ADAPTERS For adapting PV-DUIT to threaded fittings, metallic systems. Female threads on one end, socket end on other.

UL LISTED

PART NO.	SIZE	PKG. QTY.
E942D	½	100
E942E	¾	75
E942F	1	40
E942G	1¼	25
E942H	1½	20
E942J	2	25
E942K	2½	25
E942L	3	20
E942M	3½	15
E942N	4	15
E942P	5	5
E942R	6	5

TERMINAL ADAPTERS For adapting PV-DUIT to boxes, threaded fittings, metallic systems. Male threads on one end, socket end on other.

UL LISTED

PART NO.	SIZE	PKG. QTY.
E943D	½	100
E943E	¾	100
E943F	1	50
E943G	1¼	30
E943H	1½	20
E943J	2	30
E943K	2½	30
E943L	3	20
E943M	3½	15
E943N	4	15
E943P	5	5
E943R	6	5

BELL ENDS For termination of PV-DUIT into a manhole.

PART NO.	SIZE	PKG. QTY.
E997F	1	25
E997G	1¼	25
E997H	1½	20
E997J	2	25
E997K	2½	10
E997L	3	30
E997M	3½	25
E997N	4	20
E997P	5	10
E997R	6	10

BOX ADAPTERS Adapts plastic conduit to all electrical boxes by inserting adapter through knockout and cementing into PV-DUIT couplings.

UL LISTED

PART NO.	SIZE	PKG. QTY.
E996D	½	100
E996E	¾	100
E996F	1	100
E996G	1¼	25 •
E996H	1½	25
E996J	2	20
E996K	2½	25
E996L	3	25
E996M	3½	25
E996N	4	25

CAPS AND PLUGS
Caps available ½" to 1½".
Plugs available 2" to 6".

PART NO.	SIZE	PKG. QTY.
E958D	½	100
E958E	¾	100
E958F	1	75
E958G	1¼	50
E958H	1½	50
*P258J	2	75
*P258L	3	25
*P258M	3½	25
*P258N	4	25
*P258P	5	15
*P258R	6	15

EXPANSION COUPLINGS
Provides 6" of expansion and contraction. In warm weather, install closed. In cold weather, install open.

UL LISTED

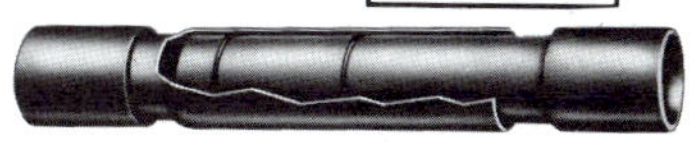

PART NO.	SIZE	PKG. QTY.
E945D	½	10
E945E	¾	10
E945F	1	10
E945G	1¼	10
E945H	1½	10
E945J	2	10
E945K	2½	5
E945L	3	5
E945M	3½	5
E945N	4	5
E945P	5	5
E945R		5

SHORT EXPANSION COUPLINGS
Provide a maximum of 2" expansion and contraction.

UL LISTED

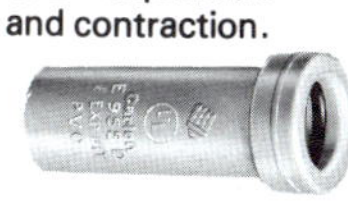

PART NO.	SIZE	PKG. QTY.
E955D	½"	40
E955E	¾"	40
E955F	1"	25
E955G	1¼"	15
E955H	1½"	10
E955J	2"	6

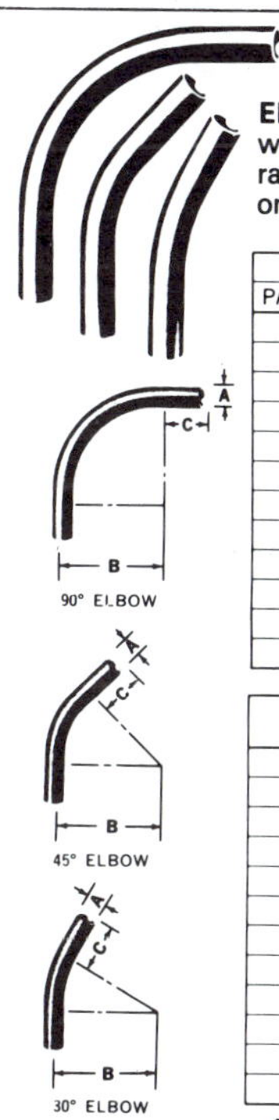

ELBOWS For use with PV-DUIT solvent weld fittings. Both ends plain. Special radius elbows available on request.

UL LISTED

90° PART NO.	45° PART NO.	30° PART NO.	SIZE	PKG. QTY.
E780D	E760D	E740D	½	20
E780E	E760E	E740E	¾	20
E780F	E760F	E740F	1	20
E780G	E760G	E740G	1¼	20
E780H	E760H	E740H	1½	20
E780J	E760J	E740J	2	Bndle 5
E780K	E760K	E740K	2½	Bndle 5
E780L	E760L	E740L	3	Bndle 5
E780M	E760M	E740M	3½	Bndle 5
E780N	E760N	E740N	4	Bndle 5
E780P	E760P	E740P	5	1
E780R	E760R	E740R	6	1

SIZE	A	B MIN.	C MIN.
½"	.840	4"	1½"
¾"	1.050	4½"	1½"
1"	1.315	5¾"	1⅞"
1¼"	1.660	7¼"	2"
1½"	1.900	8¼"	2"
2"	2.375	9½"	2"
2½"	2.875	10½"	3"
3"	3.500	13"	3⅛"
3½"	4.000	15"	3¼"
4"	4.500	16"	3⅜"
5"	5.563	24"	3⅝"
6"	6.625	30"	3¾"

Dimensions Applicable to all sweeps 90°-45°-30°

(*d*)

Figure 5.14 PVC rigid plastic conduit. (*d*) PVC-Duit fittings. (Courtesy of Carlon Company)

BOXES, COVERS AND ACCESSORIES

JUNCTION BOXES WITH SCREW DOWN COVER.
4 x 4 x 4 PVC, gasket included. Outlets may be cut-in with circular hole-saw.

PART NO.
4x 4x4 E987N*
6x 6x4 E987R
8x 8x4 E989N
12x12x6 E989R

*U.L. Listed

ROUND JUNCTION BOXES
PVC box. Four knock-out type socket openings, 90° spacing. Available with ½" or ¾" socket outlets.

PART NO.
½ E970D
¾ E970E

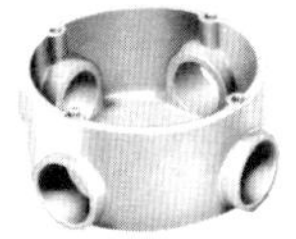

FSS SERIES (PVC)
All sizes take standard covers, switches or receptacles.

PART NO.
½ E982D
¾ E982E

FSC SERIES (PVC)
All sizes take standard covers, switches or receptacles.

PART NO.
½ E981D
¾ E981E

FSE SERIES (PVC)
All sizes take standard covers, switches or receptacles.

PART NO.
½ E980D
¾ E980E

BLANK COVERS
PVC covers for FS series boxes. Gasket included.

PART NO.
E980C

PLASTIC FLOOR BOXES
Features exclusive Rocker Ring for fast leveling. Handles any standard single receptacle. Cut off mounting ears and insert receptacle into slot in outer ring of black plastic top. Center ring of the box has standard 1¼" and 1" conduit thread to accommodate any floor stand or service fitting. Outlet sizes ½", ¾" and 1".

UL LISTED

PART NO.
Base . . E971Q
Top Assembly . . E971QC

COVERS, ROUND JUNCTION BOXES. PVC.
Gasket included.

PART NO.
E970C

VAPORPROOF FIXTURES
Phenolic. Fits round junction box only. Gasket included. Threaded 7" clear globe. Stainless steel mounting screws.

PART NO.
H982T

UL LISTED

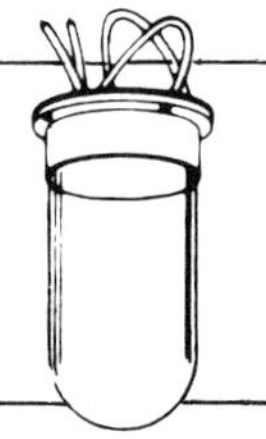

PVC CONCRETE BOXES (MUD BOXES)
Nail to ceiling form for embeding in slab. Alternate rows of knockouts for ¾" or 1" conduit. Accepts standard 4" diameter ceiling fixture.

PART NO.
(HEIGHT)
2½" E972K
3" E972L
4" E972N
6" E972R

WEATHERPROOF RECEPTACLE COVERS
Phenolic. Fits FS Series boxes, gasket included. Designed for single or duplex receptacle. Stainless steel mounting screws.

PART NO.
H985W
H985P

UL LISTED

ACCESS FITTINGS UL LISTED

T LL LR LB C E

SIZE	TYPE T PART NO.	TYPE LL PART NO.	TYPE LR PART NO.	TYPE LB PART NO.	TYPE C PART NO.	TYPE E PART NO.
½"	E983D	E984D	E985D	E986D	E987D	E988D
¾"	E983E	E984E	E985E	E986E	E987E	E988E
1"	E983F	E984F	E985F	E986F	E987F	E988F
1¼"	E983G	E984G	E985G	E986G	E987G	E988G
1½"	E983H	E984H	E985H	E986H	E987H	E988H
2"	E983J	E984J	E985J	E986J	E987J	E988J

(e)

Figure 5.14 PVC rigid plastic conduit. (*e*) PVC boxes, covers, and accessories. (Courtesy of Carlon Company)

According to revised sections of the National Electrical Code and Federal Specifications L-C-00704b, PVC rigid conduit can be installed by direct burial, or in concrete walls, floors, and ceilings, in wet locations, in severe corrosive applications of cinder fill, and in exposed work where it will not be subjected to physical damage.

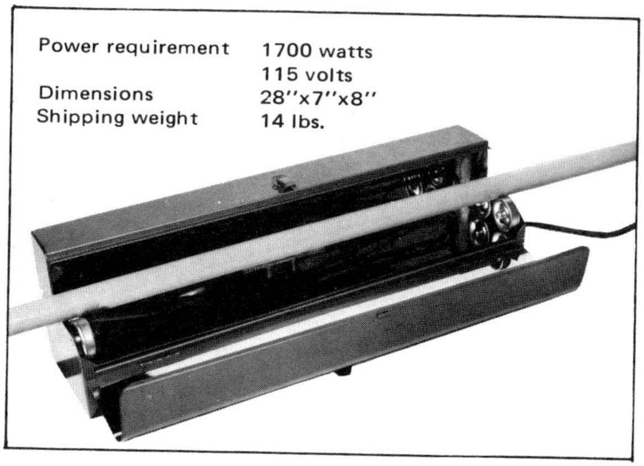

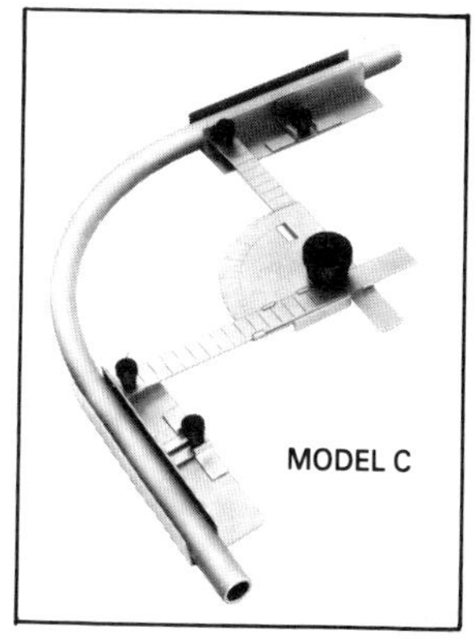

(a) (b)

Figure 5.15 Bending equipment. (a) Hotbox bender, a fast method of heating to produce field bends of plastic conduit of ½ through 1½ inch diameters. Heater provides sufficient heating length to form up to 90° elbows, offsets, kicks, stubs, and saddles. (b) Bending guide, is a variable jig for field bending plastic stubs (or line of conduit) and locking them in position. Heat conduit and place in preset bend guide. (Courtesy of Carlon Company)

Field Bending PVC Conduit

Little practice is required to master the three steps in bending plastic conduits and ducts: (1) heating, (2) forming the bend, and (3) cooling.

Heating The conduit section to be bent must be heated evenly over the entire length of the curve. There are electric heaters designed specifically for the purpose in sizes to accommodate all conduit diameters. These devices employ infrared heat energy which is most quickly absorbed by the conduit. Small sizes are ready to bend after a few seconds in the "hot box" (Figure 5.15). Larger diameters require two or three minutes, or more, depending on conditions (Figure 5.16). The use of torches or other flame type devices is not recommended. PVC conduit exposed to excessively high temperatures may take on a brownish color. Sections showing evidence of such scorching should be discarded.

Forming the Bend (½- Through 1½-inch Diameter) When properly heated, the conduit is very flexible and can be shaped to almost any configuration. For the production of most bends a variable jig is available which adjusts for any radius from 5 through 12 inches for any segment of bends from 0° to 90°. It can also be adjusted for offsets (Figure 5.17).

After adjusting the jig for the desired bend, place the heated conduit in the clamp sections to provide a smooth curve. The conduit is then cooled by sponging with water, and the bend is ready to install.

Two-Inch and Larger Diameters Larger sizes of conduits and ducts require internal support to prevent "crimping" or deforming during the bending process. Bending plugs are available which are inserted in each end of the conduit section before heating. The plugs are expanded to provide an airtight seal.

As the conduit is heated, the retained air expands, and the increased internal pressure allows the conduit to be bent without deforming. The conduit must be cooled by sponging with cold water before the plugs are removed (Figure 5.18).

Methods of Joining After cutting to size with a hack saw, solvent cement is either painted or sprayed onto the conduit end. Then the slip coupling,

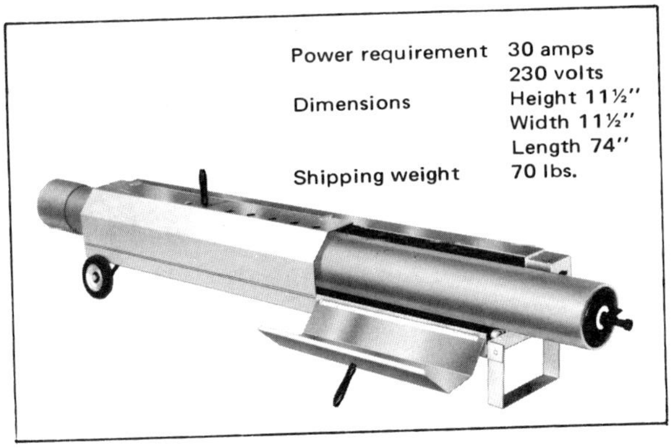

(a)

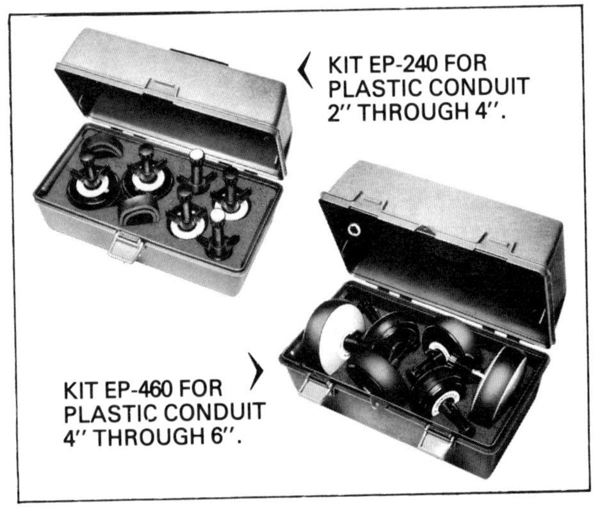

(b)

(c)

Figure 5.16 Bending large diameter PVC, 2 through 6 inches. (*a*) A hotbox bender offers advanced feature for field bending of all sizes of plastic conduit and ducts. It is extra fast, larger, and easily portable. Heats conduit lengths quickly and evenly to sufficiently form 48 inch radius, degrees, elbows, stubs, kicks, offsets, and saddles. (*b*) Hotbox expansion plugs. (*c*) Bending guide, variable jig simplifies field bending of 2 through 6 inch plastic conduit. Used to form elbows, kicks, and offsets. (Courtesy of Carlon Company)

BENDING SMALL DIAMETER PV-DUIT (½″ THROUGH 1½″)

1. HEATING. Rotate the conduit to provide even heating. Time varies with conduit size.

2. FORMING THE BEND. Use bend guide for small diameters, or a jig, or a line drawn on the work surface.

3. COOLING. The bend will "set" when cooled. Use a wet rag or sponge to speed up the cooling process. When "set", remove plugs. The bend is now ready to install.

Figure 5.17 Forming the bend, ½ through 1½ inch diameter. (Courtesy of Carlon Company)

BENDING LARGE DIAMETER PV-DUIT (2" THROUGH 6")

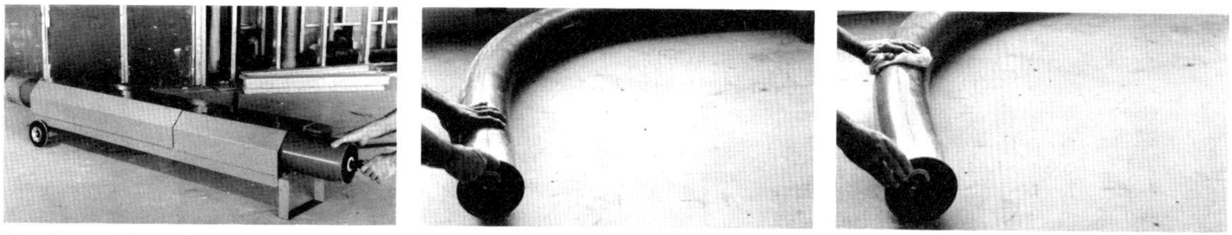

1. HEATING. Insert bending plugs as shown before heating conduit.

2. FORMING THE BEND. Large sizes can be formed using a corner for an "extra set of hands". Plugs must remain tight during handling.

3. COOLING. The bend will "set" when cooled. Use a wet rag or sponge to speed up the cooling process. When "set", remove plugs. The bend is now ready to install.

Figure 5.18 Forming the bend for PVC 2 inch and larger diameters. (Courtesy of Carlon Company)

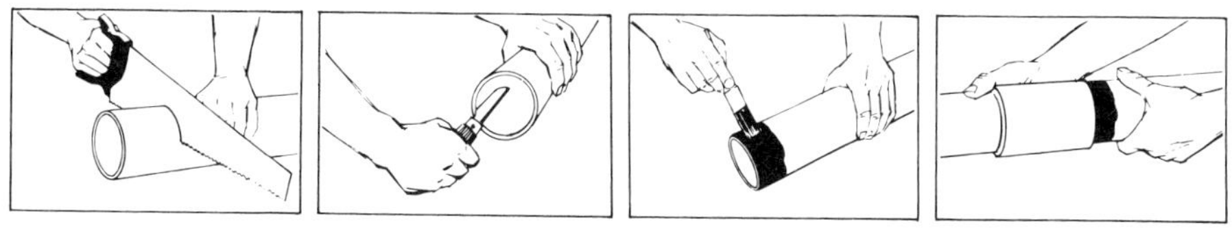

(1) (2) (3) (4)

Figure 5.19 Methods of joining PVC. (1) Cut square (using a PVC snipper or fine tooth handsaw). (2) Deburr with knife. (3) Wipe pipe clean and dry, apply coat of solvent cement to the end of conduit, the length of socket to be attached. (4) Push conduit together, rotating slightly (about one-half turn) to spread cement evenly. Allow joint to set approximately 10 minutes. (Courtesy of Carlon Company)

adapter, or electrical box is attached. The joint is set up in a matter of minutes and a short time later becomes as strong as the conduit itself (Figure 5.19).

Advantages One of the major advantages of rigid conduit is that it is not necessary to thread either the conduit or the coupling. PVC may be adapted to fit any area or box by cutting it with a saw and solvent-welding adapters to the PVC. Since the joints are solvent-welded they are leak proof.

Supports PVC rigid conduit is a thermoplastic material. As such it should be supported at certain specified intervals when installed in exposed applications.

The following table gives recommended support spacing for a wide variety of conduit sizes.

Diameter of Conduit, Inches	Support Spacing Intervals, Feet
½–1	3
1–2	5
2½–3	6
3½–5	7
6	8

In addition to the requirements for support along the span of rigid conduit, the material should be supported within 3 feet of each termination, as specified in the National Electrical Code (NEC 347-8).

There can be considerable expansion and contraction of PVC rigid conduit when it is in an exposed installation and subject to temperature variation. A good rule of thumb is to allow up to 4 inches per 100 feet of conduit for expansion or contraction.

These conditions can usually be handled by inserting elbows in the system. For best results, however, it is recommended that expansion couplings be used.

To allow this expansion and contraction to take place freely, it is advisable not to tighten the supports or clamps that hold the conduit in place.

The Surface Metal Raceway

Surface metal raceways are defined by the U.S.A. Standards Institute as "an assembly of backing and capping made of metal suitably coated to be corrosion resistant and having an interior finish designed to avoid abrasion of the electrical conductors which it contains."

This definition separates wiremold raceways and similar systems from rigid metal conduit, IMC, EMT, nonmetallic tubing, and similar products which are grouped together in Article 100 of the NEC under definitions of a raceway.

Any electrical raceway regardless of type is used to hold wiring and to protect it from mechanical injury. Metal raceways offer maximum protection. The basic reasons for using surface metal raceways versus all other raceways can be summed up as follows: appearance and accessibility.

Appearance. Only surface metal raceways are expressly designed to blend as inconspicuously as possible with surfaces on which they are installed.

Accessibility. Although all surface wiring systems offer a degree of accessibility, wiremold surface systems are specifically engineered for easy entrance at any point whenever wiring requirements change.

Surface metal raceways will fill the need for wiring systems that make it easy to bring conductors from any point to the point of use. Whether for power, light, or communications wiring, they offer a practical way to wire all types of buildings. See Figure 5.20.

According to Article 352 of the NEC, surface metal raceways may be installed in dry locations, and wherever they will not be subjected to severe physical damage or corrosive vapors.

Underfloor Steel Raceways—NEC Article 354

Underfloor steel raceways are used extensively in office buildings because they allow great flexibility in office layouts. Use of movable office partitions make doubly important the availability of ample underfloor raceways for lighting, power, and communication systems, which can be used to supply outlets with few limitations as to their location.

Underfloor raceway systems (Figure 5.21) have three major parts: the raceway proper with capped outlet openings on 24-inch centers; inserts on floor outlets for receptacles; and junction boxes. The system may consist of single ducts for lighting and power systems only, or multiple ducts in parallel to accommodate the required electrical and communication systems as well as to provide spare ducts.

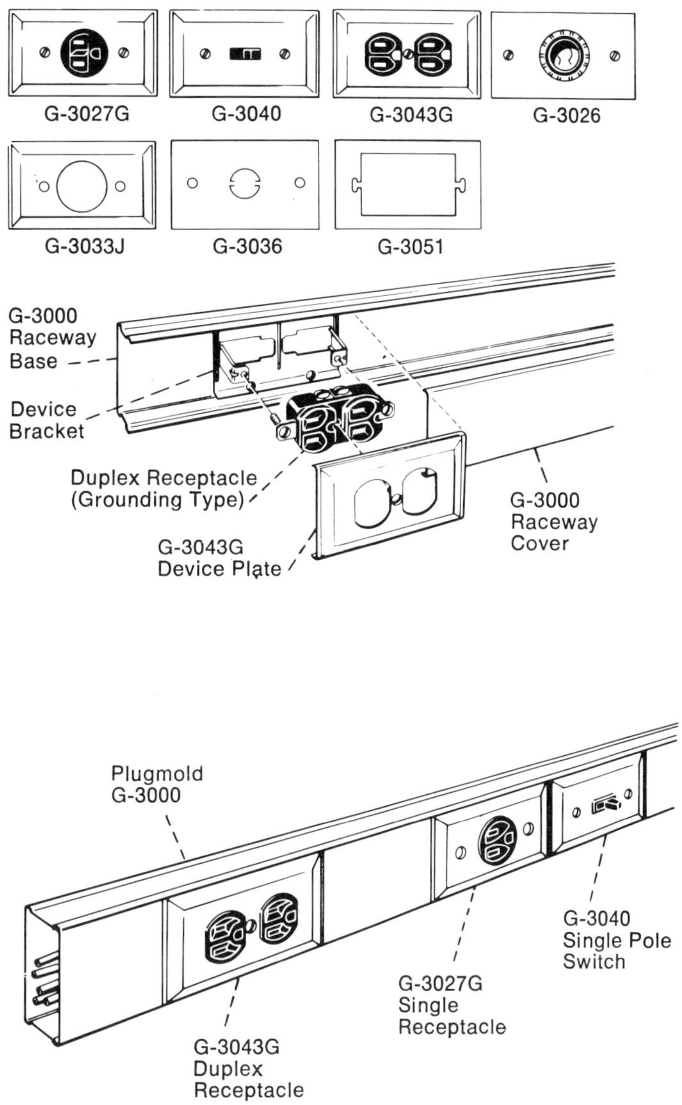

Figure 5.20 Plugmold has a broad range of fittings for any multioutlet system. Devices most commonly installed are switches, single and duplex receptacles, and sign receptacles. (Courtesy of The Wiremold Company)

Flat Cable Assemblies—NEC Article 363

A flat cable assembly, type FC, consists of a specially constructed four-conductor, three-phase flat cable assembly in parallel alignment within an insulated web (Figure 5.22). Flat cable is designed for field installation in a standard 1⅝ × 1⅝ inch U-shaped framing strut, UL-listed as surface metal raceway. See Figure 5.23. The cable consists of four PVC insulated No. 10 AWG stranded copper conductors, three-color coded phases plus a neutral. The system is rated at 20 amperes, three-phase for a total of 60 amperes on a run by splitting the load between phases, or 20 amperes per phase in single phase. Consequently, three lighting circuits may be accommodated by using various taps to balance the load between the three phases. Also, each phase may be connected to its own breaker for control of a group of lights.

Splices are made in junction boxes having color-coded terminals: white

TYPICAL "SMALL JOB" AND BILL OF MATERIAL

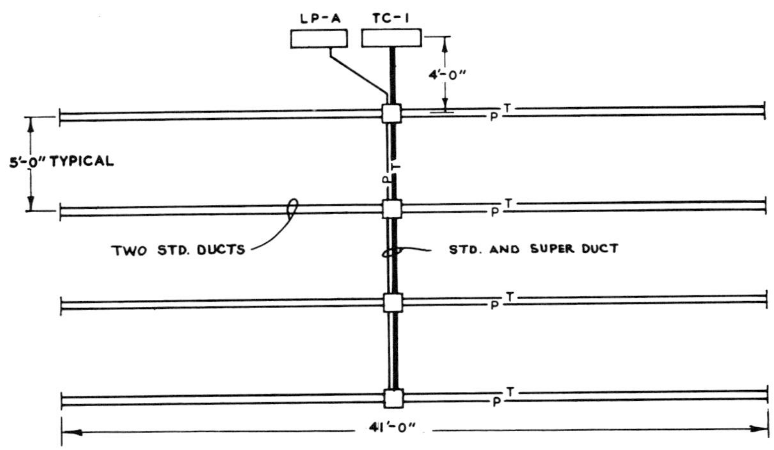

BILL OF MATERIAL

Quantity	Catalog No.	Description
3	GD100	Blank Duct,10′
3	GD200A	Blank Super Duct,10′
32	GD107A	Std Insert Duct,10′
4	GS33-07-P1-2	Junction Boxes
32	G1538-11	Support-Coupler
2	G1538-2	Support-Coupler
3	G1538-1	Support-Coupler
1	G1401	Box Opening Plug
1	G1402	Box Opening Plug
8	G1401-1	Box Plug and Separator
16	G1411	Duct End Plug
1	G1603	Std. Vertical Elbow
1	G1604	Super Vertical Elbow
1	G1621	Std. Cabinet Connector
1	G1622	Super Cabinet Connector
4	G1615	Horizontal Elbow
2	G1463	Sleeve Coupler
32	G155	Marker Screw
2	G1470	Sealing Compound

NOTES: Be sure to include material called for in specifications. Double check the scale on the drawing. Check slab depth and construction for proper levelling leg height on boxes, support-couplers.

CARPET PANS FOR JUNCTION BOXES

Catalog Number	Height	Junction Box	Weight Each
G312	¼″ to ⅝″	GS11, GS12, GT11, GT12	1.0
G322	¼″ to ⅝″	GS22, GT22, GH22	2.0
G333	¼″ to ⅝″	GS33	4.0

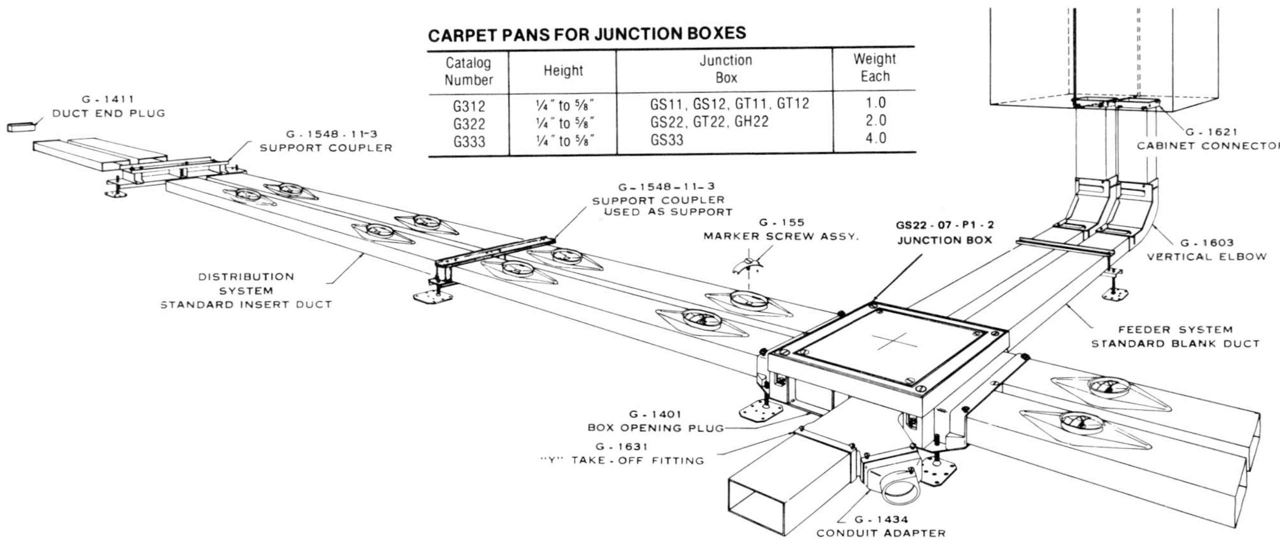

Figure 5.21 Underfloor duct installation. (Courtesy of Square D Company)

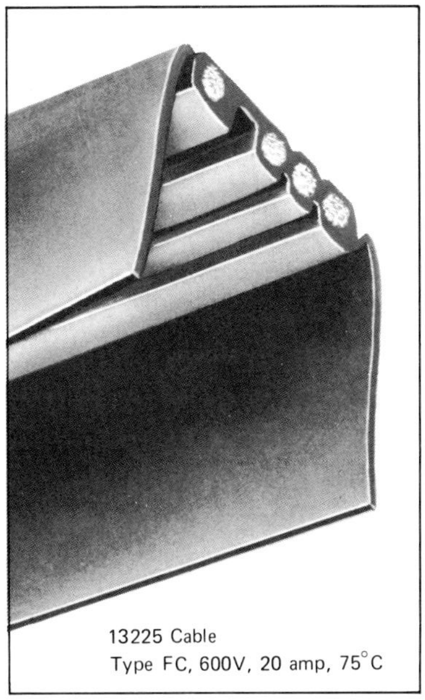

Figure 5.22 Flat cable assemblies, type FC. (Courtesy of The Wiremold Company)

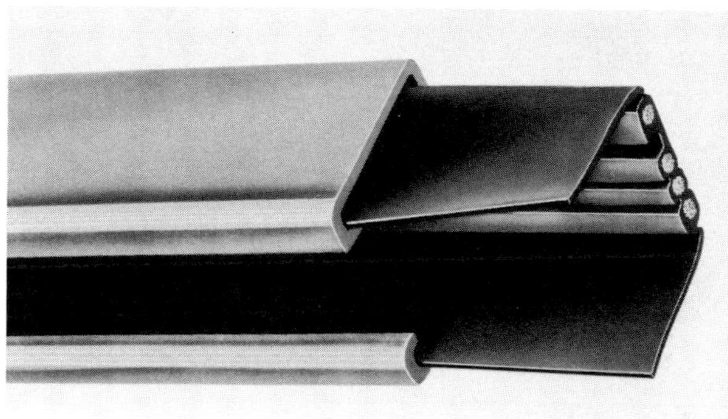

Figure 5.23 Flat cable assemblies installed in a standard U-shaped framing strut. (Courtesy of The Wiremold Company)

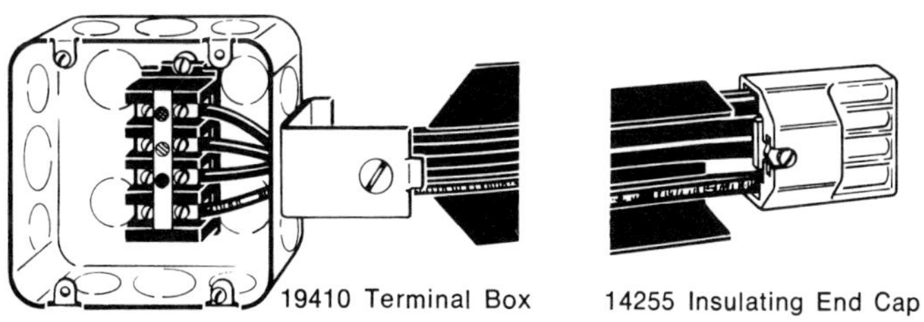

Figure 5.24 Junction box, showing color-coded terminal block and insulating end cap. (Courtesy of The Wiremold Company)

(neutral), black, red, and blue, to distinguish conductors and phases and to correspond with tap coding. An insulated end cap is required at the dead end of each run (Figure 5.24).

Tap fittings for connecting lighting fixtures, appliances, and small power loads consist of devices with manganese bronze pins which puncture the cable insulation to connect on phase and neutral. The self-healing insulation does not leave exposed wires after removal of a tapped load. Tap bodies are color-coded (as are their hot wires) in the sequence of black for Phase A, red for Phase B, and blue for Phase C (neutral wires are white), corresponding to terminal block coding (Figure 5.25). To assume precise contact of tap pins

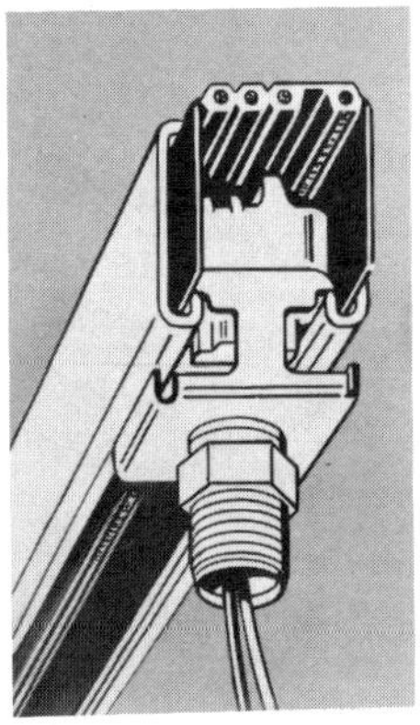

Figure 5.25 Color-coded tap body. (Courtesy of The Wiremold Company)

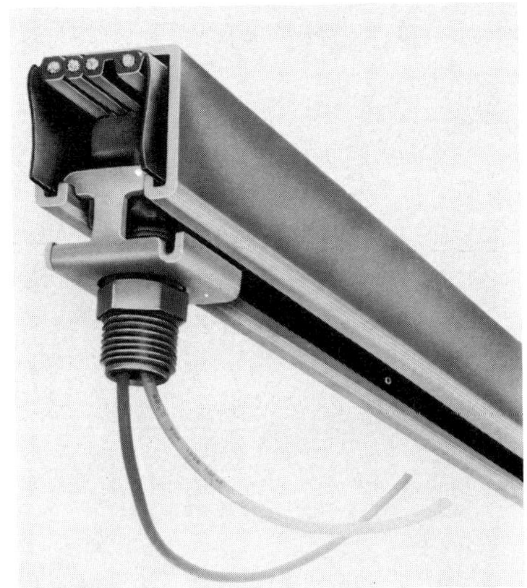

Figure 5.26 Tap-body, showing how keyed for proper alignment in strut. (Courtesy of The Wiremold Company)

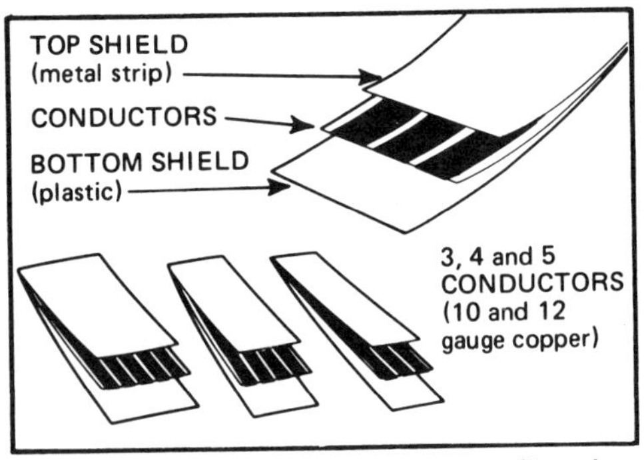

Cable types will be in 3, 4 and 5 wire configurations, allowing multiple circuits or phased voltage.

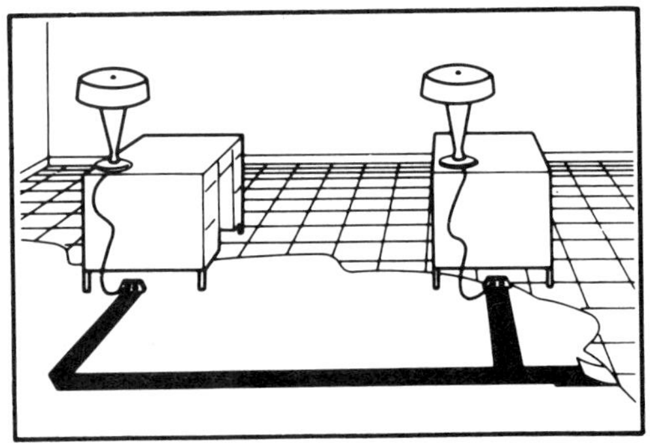

Branching from a common line, connectors maintain pattern of current flow and grounding.

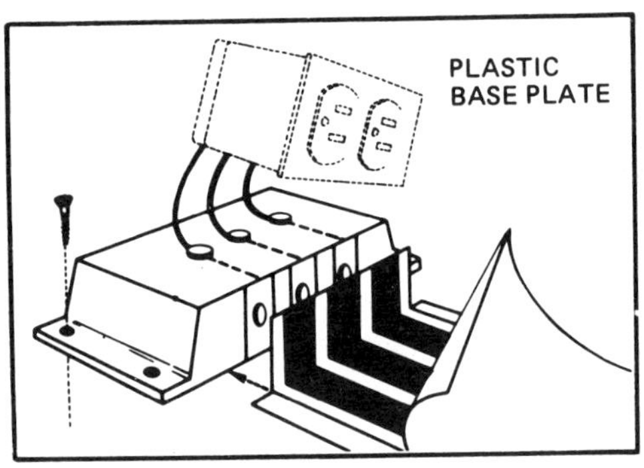

Power is available through duplex receptacles in floor tombstones.

Carpet is cut around the tombstone to allow its placement.

Figure 5.27 Flat conductor cable. Showing installation procedures. (Courtesy of Thomas & Betts)

with the flat cable conductors, taps are keyed for proper alignment in the strut (Figure 5.26).

Flat cable assemblies are approved for exposed locations only where they are not subjected to severe physical damage. They are not permitted in hazardous areas or in hoistways. They may be installed in corrosive or wet locations only if approved for the purpose.

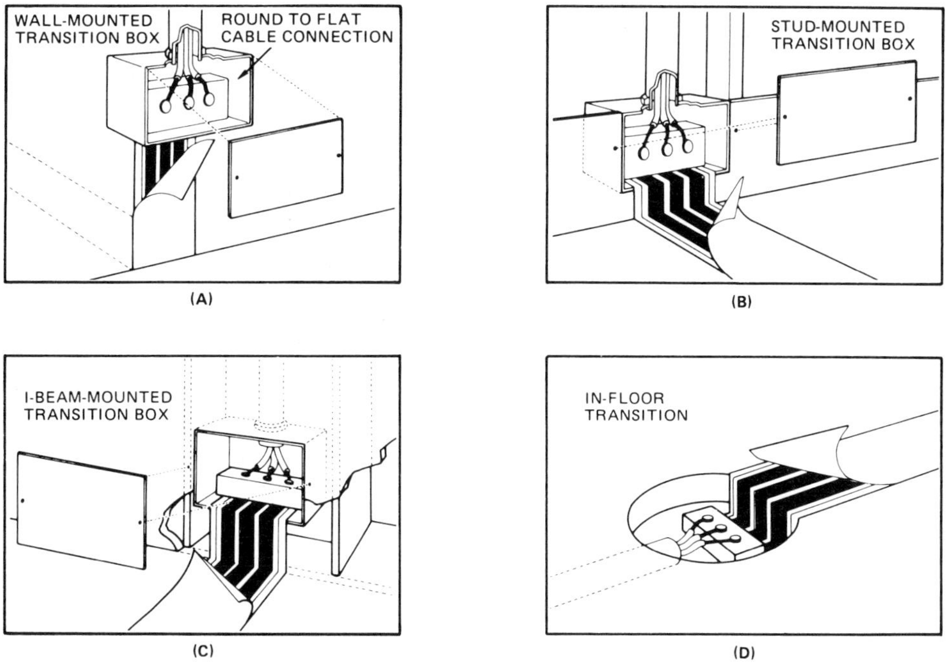

Figure 5.28 Flat conductor cable. Showing transition box. (Courtesy of Thomas & Betts)

Flat Conductor Cable (Type FCC)— NEC Article 328

This is a new system of branch wiring that supplies floor outlets in office areas and other commercial buildings. The system is an outcome of a joint effort by the National Aeronautics and Space Administration (NASA) and ten private manufacturers.

The wiring system consists of a flat conductor cable enclosed in a raceway and is provided with special designed boxes, fitting, and surface metal raceways to permit connection with conventional, round wire, wiring systems (Figure 5.27).

The method may be used for new buildings or for modernized or expansion in existing interiors. The flexibility and ease of installation of this surface-mounted flat-cable wiring system meets the need that arises from the fact that the average floor power outlet in an office area is relocated every two years. See Figure 5.28.

The flat conductor cable under-carpet wiring system is rated 15 or 20 amperes, single-phase, two wires with ground, maximum 150-volts AC between conductors and ground.

Type FCC wiring may supply floor outlets for power and communication at any location on the floor.

5-2 METHODS OF INSTALLATION

The materials and assemblies used in the construction of commercial buildings have an important impact on the ease of the installation.

In this section, we will consider the types of construction and then the methods of installation.

Poured Concrete Construction

If conduit, boxes, sleeves or cases, and straps for electrical work are placed by the electrician in a form to be filled with concrete, coordination is required with the general contractor since some of the above materials should be installed before the reinforcing rods are positioned.

Sleeves are often installed before the reinforcing rods, therefore they must be securely fastened to the forms so they will not be disturbed by the steelworkers or by the concrete when it is poured. Sleeves are the holes left in the concrete so that the conduit may pass through the wall or floor.

Embedding conduit in concrete presents no problem except that care must be taken in pouring the concrete so that the conduit is not moved or damaged.

Metal underfloor raceways may be installed in a poured concrete floor to allow the passage of electrical and communication wires and additional conduit. The raceway may be either level with or below the surface of the floor.

Precast Concrete

When precast concrete construction is used, conduits are either placed on top of the precast concrete and embedded in a concrete topping, or they are placed below the concrete structural member, usually below when a soffit or suspended ceiling is used.

Conduit and boxes installed below the precast concrete are usually attached by a clip and an insert. See Figure 5.29. When entire walls are precast, conduit and boxes are often cast right into the wall.

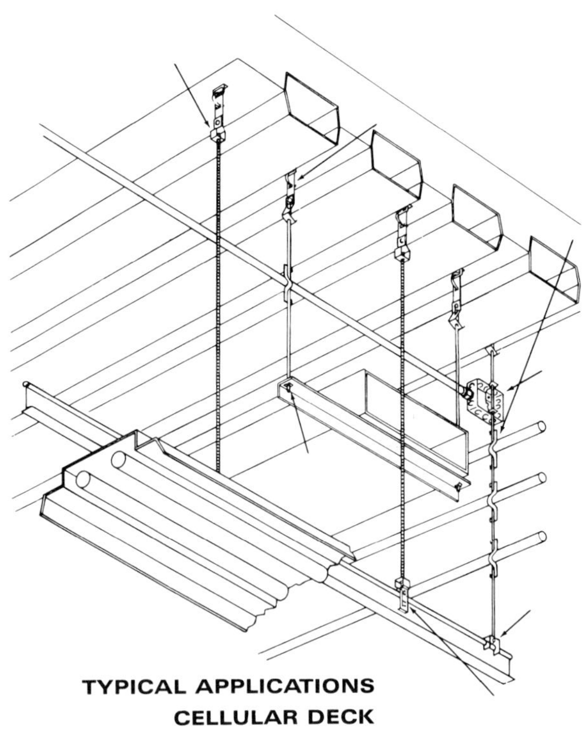

TYPICAL APPLICATIONS
CELLULAR DECK

Figure 5.29 Shows use of beam clamps to support boxes, fixtures, and conduits from cellular deck. (Courtesy of Caddy, Fastener Div.)

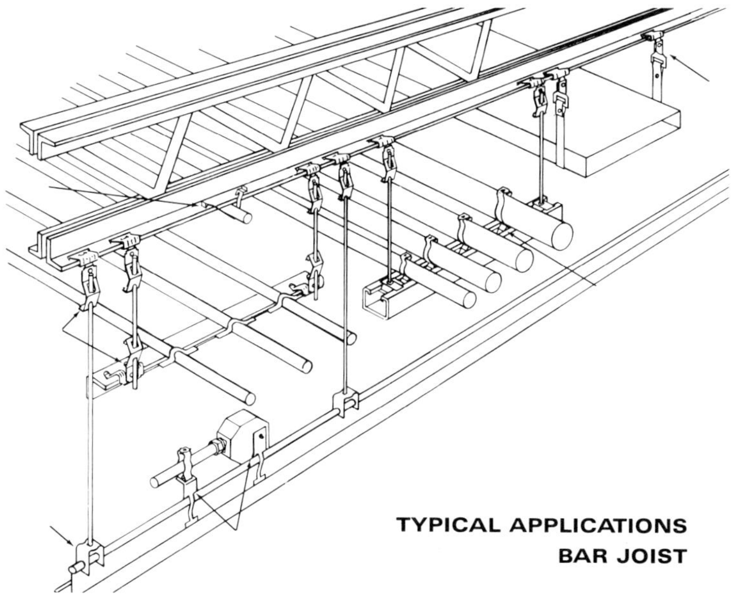

Figure 5.30 Typical installation of conduits from bar joist. Shows freedom and flexibility of conduit run. (Courtesy of Caddy, Fastener Div.)

Steel Joists

Steel joist construction offers some of the most economical solutions for installations. The space available in joist design offers flexibility in installing conduit. See Figure 5.30.

A suspended ceiling allows freedom in the layout of conduit in terms of directions, because the shortest distances and diagonal runs can be used.

Wood Construction

Some modern buildings are constructed with laminated wood beams and arches with a wooden deck between the pieces of laminated wood. Beams are easily cut to allow for conduit runs. When runs for lighting fixtures are required on the ceiling, the decking or beams may be routed (grooved) parallel to the edge to conceal the conduit.

Partitions

Some developers of commercial buildings require inside partition to be of a metal stud drywall construction. Some solid metal studs are available with prepunched holes or with knockouts 24 in O.C. Should the stud be of a solid material, a bar hanger punch is available to electricians through a supplier. See Figures 5.31 and 5.32 for metal stud installation.

Masonary wall construction requires cooperation. The electrician must work closely with the mason laying the blocks. EMT is used with watertight compression box connectors fastened to masonary boxes. Full-length conduits are not used during the laying of blocks. The mason would have to slip the blocks over 10 foot lengths and this is impractical. Conduit lengths are coupled together as the job progresses.

When conduits are extended beyond the top of the finished walls, they are bent into the suspended ceiling area for coupling together and completing the runs to ceiling outlets or pull boxes.

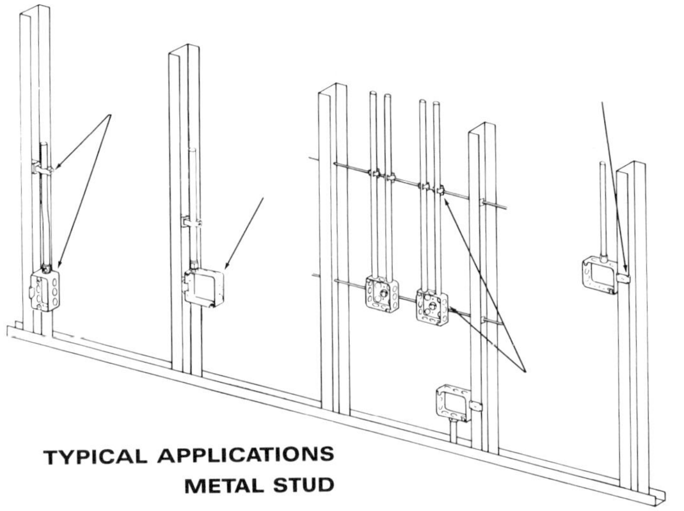

Figure 5.31 Typical metal stud installation. (Courtesy of Caddy, Fastener Div.)

CAT. NO. PSH

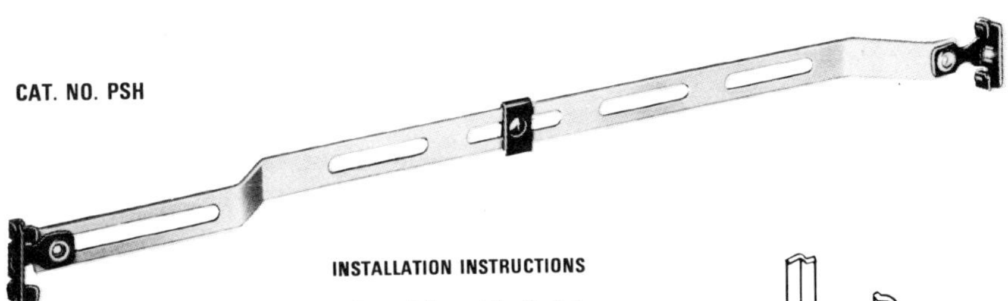

INSTALLATION INSTRUCTIONS

1. Place end clips on studs with pliers.
2. Mount box with clip and screw provided.

CAT. NO. PSH-1
(screw provided)

FEATURES

1. Provides a bar for 1½" deep electrical boxes.
2. The bar adjusts from 14" to 18" to easily handle studding usually on 16" centers.
3. Simple installation.

SIZING CHART

CATALOG NUMBER	DESCRIPTION	WEIGHT PER BOX	PACKAGED PER BOX
PSH	Plaster stud box hanger	18 lbs.	50
PSH-1	Box mount clip with screw	1½ lbs.	100

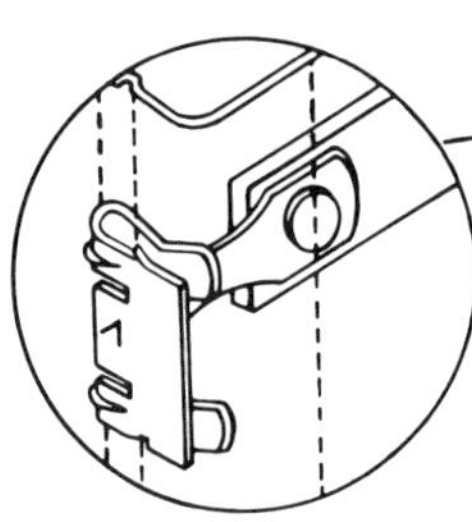

APPLICATIONS

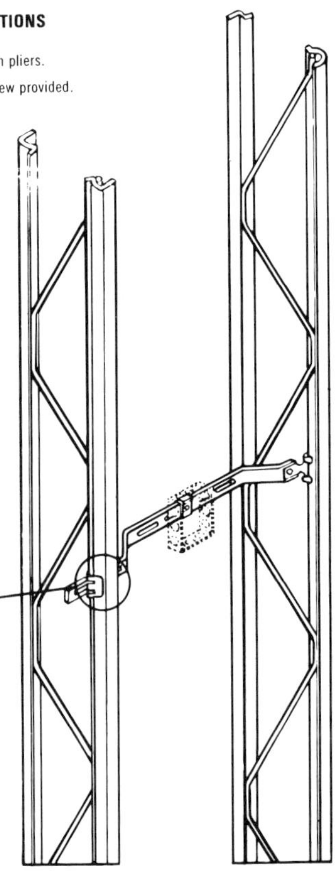

Figure 5.32 Typical plastered wall, showing stud box hanger and mounting clip. (Courtesy of Caddy, Fastener Div.)

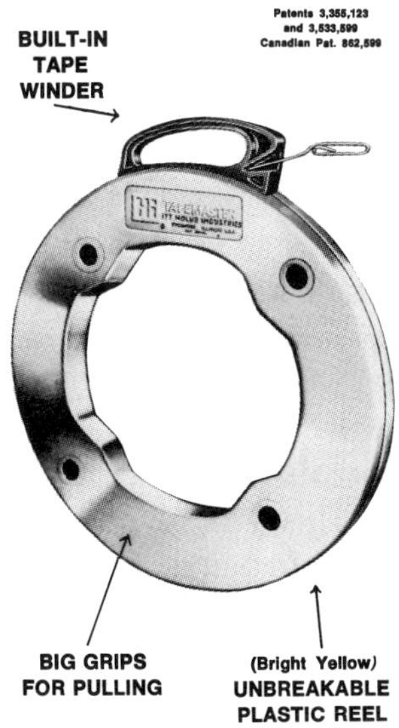

Figure 5.33 Tapemaster fish tape reel. (Courtesy of ITT Holub Industries)

5-3 FISHING CONDUCTORS IN A RACEWAY

In a straight run raceway, if the number of conductors is not too great, conductors may be pushed from one outlet to another. If it is not possible to push the conductors through a run, the conductors are "fished" through, usually by using a flat flexible steel tape called a fish tape (Figure 5.33). One end of the fish tape is bent into a hook to keep the fish tape from hanging on couplings and bends as well as to provide a means for attaching the conductors. If the conduit has several bends, the fish tape may buckle. Too many bends makes it hard to get the tape and conductors through and will damage the insulation on the conductors when pulling them. For this reason, the NEC specifies that a run between outlet and outlet or fitting and fitting shall not have more than the equivalent of four quarter bends or a total of 360°, including the offset at the fitting or outlet.

When the tape is hard to push through the conduit, the conduit may be jarred with a pair of pliers to get the tape around bends or couplings. The best method, however, is to use a spiral fish tape leader attached to the hook end of the fish tape. Leaders are designed to withstand normal pulling.

After the fish tape has been pushed through the conduit, prepare the conductors for attaching as illustrated in Figure 5.34. The conductors are

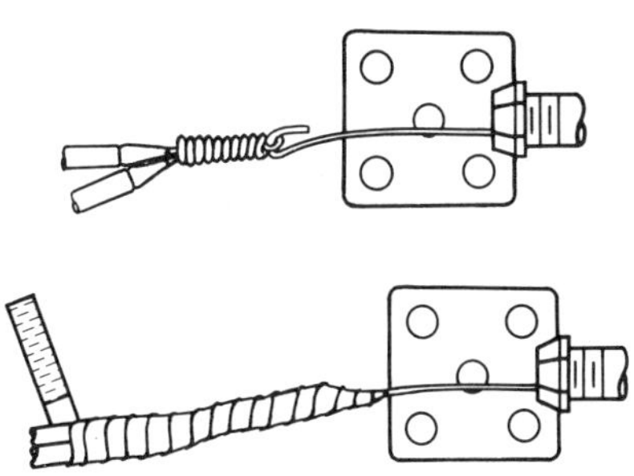
Figure 5.34 Illustrates method of wrapping and taping wires before pulling.

hooked to the bent hook end of the tape, then twisted around that part of the conductor with the insulation stripped off. After the conductors are secured to the fish tape, they are wrapped with plastic tape to prevent the hook ends of the conductor from hanging-up on connectors, couplings, and bends.

One person should never attempt to fish wires through a conduit alone. There should be a person at the pulling end and one at the outlet where the conductors are entering the raceway. The person feeding the conductors into the raceway should keep the conductors straight in respect to the position they enter the raceway, keeping them free from kinks and bends. The person at the pulling outlet should pull with steady pulls 8 to 12 inches at a time at regular intervals, the intervals giving the person at the feeding outlet time to do the job. The two must work together in order for the person doing the pulling to get full benefit of the pushing or feeding at the other end.

Should it become difficult to pull the conductors through the raceway, a wire pulling lubricant may be applied. Some manufacturers recommend a wax-base formula for use on rubber, vinyl, polyethelene, or other plastic or lead covered wire or cable. The lubricant spreads out, giving a superslick coating that reduces the strain on insulation and preventing abrasions throughout the pulling process.

SUMMARY

1. Rigid metal conduit may be used under all conditions and occupancies.
2. The use of dissimilar metals in contact anywhere in a system should be avoided to eliminate the possibility of galvanic action.
3. No conduit smaller than ½-inch electrical trade size shall be used.
4. Steel conduit shall have an interior coating of a character and appearance so as to readily distinguish it from ordinary pipe.
5. Electrical metallic tubing shall not be used where it will be subject to severe physical damage.
6. Couplings of a watertight type are required wherever electrical metallic tubing is used on the exterior of buildings.
7. A conduit run between outlet and outlet, or between outlet and fitting, shall not contain more than the equivalent of four quarter bends including those located immediately at the outlet or fitting.
8. Surface metal raceways may be extended through dry walls, dry partitions, and dry floors if unbroken lengths of raceway are used where passing through.
9. Materials and assemblies used in the construction of commercial buildings have an important impact on the ease of conduit installation.
10. Pull wires shall not be installed until the raceway system is in place.

PROBLEMS

5-1 For each of the following raceways, how many supports must be installed? Both ends of each raceway are terminated in outlet boxes.

a. 16 feet of ½-inch EMT ______

b. 60 feet of ½-inch EMT ______

c. 48 feet of ¾-inch IMC ______

d. 18 feet of 1-inch rigid metal conduit ______

e. 25 feet of 1-inch PVC ______

5-2 A length of 1-inch electrical trade size rigid steel conduit is to be bent 90° and stubbed up 18 inches from the base. Make a sketch of a length of conduit, and mark off for bending the place you would start the bend.

5-3 Intermediate metal conduit has a wall thickness ______ than those specified for rigid conduit, ______ than those for EMT.

5-4 The rules that apply to rigid conduit, IMC, such as the number of wires permitted, bends, supports, *apply* or *do not apply* (circle one) to EMT.

5-5 Name at least three types of connectors used with EMT. ______, ______, ______.

5-6 In bending ½-inch EMT the bender has a take-up of ______ inches.

5-7 Flexible conduit is supported with a strap or wire nail within ____ inches of every box and at regular intervals not more than ____ feet.

5-8 PVC rigid conduit can be installed by: ______ ______, ______ ______, ______, and ______, ______ ______, ______, ______.

5-9 How do you determine the spacing of supports for PVC? ______
______.

5-10 Flat cable assemblies are approved for ______ locations only.

chapter 6

CALCULATING NUMBER OF CONDUCTORS IN CONDUIT

Instruction Objectives

1. To calculate the allowable fill for specific conductors in conduit.
2. To identify the conduit size based on the allowable fill value.
3. To explain and define the need for calculating the number of conductors allowed in a conduit.
4. To use the tables in the NEC to calculate conductor size.
5. To become familiar with types of conductor insulation.

Self-Evaluation Questions

Test your prior knowledge of the information in this chapter by answering the following questions. Watch for the answers as you read the chapter. Your final evaluation of whether you understand the material is measured by your ability to answer the questions. When you have completed the chapter, return to this section and answer the questions again.

1. Where is capacity information for conduit fill located?
2. How many conductors can be installed in a raceway?
3. How do you calculate the size of the conduit for a combination of wire sizes?
4. Is it necessary to know what type of insulation covers the conductor when calculating conduit fill?
5. Is conduit larger or smaller than the numerical size given?
6. Why is it impractical for the electrician to measure all of the conductors and conduit to find its area?
7. What size conduit will take care of eight conductors?

The number of permissible conductors in a conduit are given for any combination of conductors of the same size in the tables given in the NEC. In order to calculate the size of conduit for a combination of different sizes, the apprentice must be capable of interpreting the meaning of the tables and using mathematics.

Table 3A in Chapter 9 of the NEC gives the maximum number of conductors in trade sizes of conduit or tubing. See Table 6-1. Assume you are preparing to install a run of conduit in a commercial building. First, the electrician must determine what type and number of conductors are to be pulled in the conduit. For example, you may need nine No. 12-TW conductors. The problem is what size conduit must be used. First, find the listing of type TW conductors from Table 3A. This table has conductors with other types of insulation as well as type TW. The first column on the left gives the insulation type. The second column gives the wire size from No. 14 AWG to No. 4/0 AWG, numerically and from 4/0 to 750 MCM. Follow the column down until you come to No. 12 AWG. Check to see if you are in the TW section. Now read across from left to right. The conduit trade size in inches shows 7 is the maximum number of conductors in a ½-inch conduit and 12 is the maximum number in a ¾-inch conduit. Therefore, a ¾-inch electrical trade size conduit must be used to accomodate nine No. 12 type TW conductors.

Often the problem arises of having a combination of conductors of different sizes in one conduit. When this problem arises, the overall area of the conductor, including the insulation, must be calculated against the inside area of the conduit. See Table 6.2.

The sum of the cross-section areas of the conductors should not exceed the percentage of the interior cross-sectional area of the conduit or tubing shown in Table 1 of the NEC. See Table 6.2.

Electrical trade size conduit and tubing is slightly larger than the numerical size given. See Table 4 of the NEC (Table 6.3). For this reason it is termed "trade size." For example, ½-inch electrical trade size conduit has an internal diameter of 0.622 inches.

Table 6.3 illustrates Table 4 of the NEC, giving the nominal size of conductors and conduit recommended for use in computing the size of conduit or tubing for combinations of conductors.

The interpretation of Table 4 is as follows: The column marked "Trade Size" lists the conduit sizes as ½ inch, ¾ inch, 1 inch, up to and including 6 inches; the column marked "Total 100 Percent" gives the total internal area of the conduit in square inches. An example is given to simplify the explanation.

One-half inch electrical trade size conduit.

Internal diameter is 0.622 inches.

Total area of the conduit is 0.30 sq in.

If one conductor is to be pulled in a conduit, it should not exceed 53 percent of the area of the conduit.

$$0.30 \times 0.53 = 0.159 \text{ or } 0.16 \text{ sq in.}$$

TABLE 6.1 Maximum Number of Conductors in Trade Sizes of Conduit or Tubing (NEC Table 3A) (Based on NEC Table 1, Chapter 9)

Conduit Trade Size (Inches)		½	¾	1	1¼	1½	2	2½	3	3½	4	4½	5	6
Type Letters	Conductor Size AWG, MCM													
TW, T, RUH,	14	9	15	25	44	60	99	142						
RUW	12	7	12	19	35	47	78	111	171					
XHHW (14 through 8)	10	5	9	15	26	36	60	85	131	176				
	8	2	4	7	12	17	28	40	62	84	108			
RHW and RHH	14	6	10	16	29	40	65	93	143	192				
(without outer	12	4	8	13	24	32	53	76	117	157				
covering),	10	4	6	11	19	26	43	61	95	127	163			
THW	8	1	3	5	10	13	22	32	49	66	85	106	133	
TW,	6	1	2	4	7	10	16	23	36	48	62	78	97	141
T,	4	1	1	3	5	7	12	17	27	36	47	58	73	106
THW,	3	1	1	2	4	6	10	15	23	31	40	50	63	91
RUH (6 through 2),	2	1	1	2	4	5	9	13	20	27	34	43	54	78
RUW (6 through 2),	1		1	1	3	4	6	9	14	19	25	31	39	57
FEPB (6 through 2),	0		1	1	2	3	5	8	12	16	21	27	33	49
RHW and	00		1	1	1	3	5	7	10	14	18	23	29	41
RHH (without	000		1	1	1	2	4	6	9	12	15	19	24	35
outer covering)	0000			1	1	1	3	5	9	10	13	16	20	29
	250			1	1	1	2	4	6	8	10	13	16	23
	300			1	1	1	2	3	5	7	9	11	14	20
	350				1	1	1	3	4	6	8	10	12	18
	400				1	1	1	2	4	5	7	9	11	16
	500				1	1	1	1	3	4	6	7	9	14
	600					1	1	1	3	4	5	6	7	11
	700					1	1	1	2	3	4	5	7	10
	750					1	1	1	2	3	4	5	6	9

TABLE 6.2 Percentage of Cross-Sectional Area of Conduit and Tubing for Conductors (NEC Table 1)

Number of Conductors	1	2	3	4	Over 4
All conductor types except Lead-covered (new or rewiring)	53	31	40	40	40
Lead-covered conductors	55	30	40	38	35

Thus far we have established all information necessary for calculating a combination of conductors in conduit or tubing except the diameters of various conductors.

It would be impractical for the electrician to measure all conductors, find their area, measure the conduit, and find its area; however, it is relatively simple to use the tables located in Chapter 9 of the NEC.

6-2 CALCULATIONS

Determine the size conduit needed for four No. 14 type TW conductors, two No. 10 type TW conductors, and two No. 12 type TW conductors to be pulled in one conduit. By using the tables illustrated in this chapter we will solve this problem.

We realize that any one group of these conductors could be determined by Table 3A if they were of relative size, that is, all No. 14's, No. 12's, or No. 10's. The problem above specifies a combination of the three sizes; therefore, Table 3A (Table 6.1) cannot be used.

First, we must determine the total area of the conductors in square inches. By consulting NEC, Table 5, we find it lists the areas of type TW conductors.

1—No. 14 has an area of 0.0135 sq in. see Column 5.
4—No. 14's have an area of 4 × 0.0135 = 0.0540 sq in.
1—No. 10 has an area of 0.0224 sq in.
2—No. 10's have an area of 2 × 0.0224 = 0.0448 sq in.
1—No. 12 has an area of 0.0172 sq in.
2—No. 12's have an area of 2 × 0.0172 = 0.0344
Total area of the eight conductors 0.1332 sq in.

This should not be confused with the area of the wire itself; it is the area of the total of the insulation and the conductor.

It is an established fact that the conductors in question have a total area of 0.1332 sq in. By consulting Table 1 (Table 6.2) we see that for all conductor types, except those that are lead-covered (new work) and have over four conductors, the conductors should not exceed 40 percent of the area of the conduit. What size conduit will adequately take care of eight conductors without exceeding 40 percent of the conduit area? It could be calculated mathematically; however, it is already calculated in Table 4 (Table 6.3). Look under "Conductors not lead-covered" and find the column with four conductors and over 40 percent; follow down the column until you find the area, or the next size larger area. The total area of the eight conductors we are using is 0.1332 sq in. or 0.14 sq in. which is not listed; however, 0.21 sq in. is

TABLE 6.3 **Dimensions and Percent Area of Conduit and of Tubing (NEC Table 4). (Areas of Conduit or Tubing for the Combinations of Wires Permitted in NEC Table 1, Chapter 9.)**

			Area, Square Inches							
			Not Lead Covered			Lead Covered				
Trade Size	Internal Diameter Inches	Total 100 percent	2 Conductor, 31 percent	Over 2 Conductor, 40 percent	1 Conductor, 53 percent	1 Conductor, 55 percent	2 Conductor, 30 percent	3 Conductor, 40 percent	4 Conductor, 38 percent	Over 4 Conductor, 35 percent
½	0.622	0.30	0.09	0.12	0.16	0.17	0.09	0.12	0.11	0.11
¾	0.824	0.53	0.16	0.21	0.28	0.29	0.16	0.21	0.20	0.19
1	1.049	0.86	0.27	0.34	0.46	0.47	0.26	0.34	0.33	0.30
1¼	1.380	1.50	0.47	0.60	0.80	0.83	0.45	0.60	0.57	0.53
1½	1.610	2.04	0.63	0.82	1.08	1.12	0.61	0.82	0.78	0.71
2	2.067	3.36	1.04	1.34	1.78	1.85	1.01	1.34	1.28	1.18
2½	2.469	4.79	1.48	1.92	2.54	2.63	1.44	1.92	1.82	1.68
3	3.068	7.38	2'29	2.95	3.91	4.06	2.21	2.95	2.80	2.58
3½	3.548	9.90	3.07	3.96	5.25	5.44	2.97	3.96	3.76	3.47
4	4.026	12.72	3.94	5.09	6.74	7.00	3.82	5.09	4.83	4.45
4½	4.506	15.94	4.94	6.38	8.45	8.77	4.78	6.38	6.06	5.56
5	5.047	20.00	6.20	8.00	10.60	11.00	6.00	8.00	7.60	7.00
6	6.065	28.89	8.96	11.56	15.31	15.89	8.67	11.56	10.98	10.11

listed. Following a straight edge under 0.21 sq in. to the left, we see that a ¾-inch conduit would take care of the combination of the eight conductors.

This application may be reversed. If the conduit is already installed, the maximum number of conductors for a combination may be determined.

Conduit sizes for lead covered cables may be determined in the same manner by using the columns under "lead covered."

See Table 5, Chapter 9 of the NEC for rubber-covered and thermoplastic-covered conductors, and Table 7 for asbestos-varnished-cambric insulated conductors.

Table 8 is referred to for additional information and so that the electrician will not confuse the area of the metal itself with the overall area of the conductor and insulation. The area of the conductor is given in circular mils, while the area of the overall dimensions is given in square inches. One reason for this is to prevent confusion. Since both the conductor and the conduit are circular, it would be just as easy to calculate for combination on the circular mils basis.

SUMMARY

1. The sum of the cross-sectional areas of the conductors should not exceed the percentage of the interior cross-sectional area of the conduit.
2. Electrical trade size of conduit is slightly larger than the numerical size.
3. It would be impractical for the electrician to measure all conductors, find the area, measure the conduit, and find its area when it is relatively simple to use the tables located in the NEC.
4. In order to calculate the size of conduit for a combination of different sizes, the apprentice must be capable of interpreting the meaning of the tables and using some mathematics.

PROBLEMS

6-1 What size conduit would be used to accommodate eight No. 12 RHW conductors?

6-2 What size conduit must be used to accommodate a combination of two No. 6 and three No. 4 type TW conductors?

6-3 If seven No. 6 type RH conductors are run in the same conduit, what size conduit shall be used?

6-4 If the conductors in Problem 6-2 were type THHN, what size conduit would you use?

6-5 How many No. 10 THW conductors can be pulled in a 2-inch conduit?

6-6 What is the minimum size of conduit required for six No. 10 THHN wires?

6-7 What size conduit is the minimum for use with four No. 6 RHH wires?

6-8 What is the minimum size conduit required for four 500 MCM THHN wires?

6-9 It would be impractical for the electrician to measure all conductors, find their area, measure the conduit, and find its area; however, it is relatively simple to use the tables located in ______________ .

6-10 When there is a combination of conductors of different sizes in one conduit, the overall area of conductor, including the ______________ must be calculated against the inside of the ______________ .

chapter 7

OUTLETS AND WIRING DEVICES

Instructional Objectives

1. To identify the NEC sections relating to outlets.
2. To list and describe the different types of boxes used in commercial construction.
3. To describe the methods of mounting different type boxes.
4. To recommend the type box used for masonary construction.
5. To determine the maximum number of wires allowed in an outlet or junction box.
6. To describe wiring devices correctly according to their electrical and mechanical characteristics.

Self-Evaluation Questions

Test your prior knowledge of the information in this chapter by answering the following questions. Watch for the answers as you read the chapter. Your final evaluation of whether you understand the material is measured by your ability to answer the questions. When you have completed the chapter, return to this section and answer the questions again.

1. What is an outlet?
2. Is a wall switch an outlet?
3. How many types of outlet boxes are there?
4. Why do some boxes have brackets, and some do not?
5. How many outlets require outlet boxes?
6. How many wires are you allowed to have in an outlet box?
7. Does the NEC define a receptacle?
8. What do wiring devices include?
9. What is the grade of a receptacle?
10. How is the flow of an electric circuit controlled?
11. How do you control the lighting in a large corridor?

12. What is a safety-ground receptacle?
13. How high do you mount wall convenience receptacles at a work bench?
14. What is the difference between a three-way and a four-way switch?

7-1 OUTLETS

Refer to the NEC for a definition of an outlet.

> Article 100—Definitions—an "Outlet." A point on the wiring system at which current is taken to supply utilization equipment.

In construction terms, any opening is considered an outlet; however, all openings are not outlets by the NEC definition. In simple terms, any point that supplies an electric load is called an outlet. A duplex wall receptacle, a ceiling light box, an electric heater junction box and the like are all outlets. A wall switch is not an outlet, although in the field you will hear electricians and other crafts frequently refer to it as a "switch outlet." When counting the number of outlets on a circuit, switches are not included. They are only included when counting outlets (openings) for the purpose of making a cost estimate or ordering materials. Generally, people in the industry consider an outlet to consist of a small metal box (outlet box) into which a conduit or raceway extends (Figure 7.1).

Outlet Boxes

Outlet boxes vary in size and shape and are specifically suited for different kinds of construction (Figure 7.2).

Types of Boxes

Handy Box Intended for single gang wall device.

Octagon Box Intended for ceiling outlets.

Four-Inch Square Box General purpose box.

Masonary Box Used in cinder or concrete block construction.

Boxes intended for fastening to wall studs are furnished with brackets, spurs, and slotted holes; holes are for toe-nailing brackets to studs. For ceiling outlets, boxes with brackets may be used for areas where boxes need not be centered between joists. Others are mounted on steel straps that in turn fasten to ceiling joists, allowing the box to be centered in position.

Boxes are equipped with knockouts to receive conduit or conduit connectors. See Figure 7.3.

The term "gang" when referring to a box indicates the number of wiring device positions supplied. For example, if three switches are to be mounted side by side, a three gang box would be used. If three receptacles are mounted side by side, a three gang box would be used. If two receptacles are to be mounted side by side, a two gang box would be used, and so on.

The important fact to remember is that every outlet requires a box; however, there is only one situation that requires a box that is not an outlet. This is the 4-inch square box, used as a general purpose junction box. When conduit is intalled and wire is run in a building, it frequently becomes

Figure 7.1 Cutaway box showing application of EMT connecters. (Courtesy of Crouse-Hinds Company, Arrow-Hart Div.)

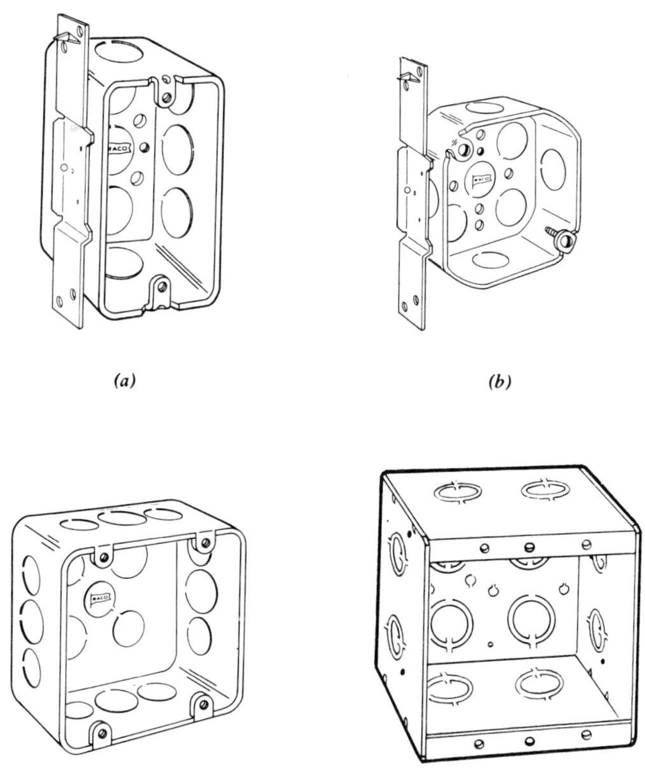

Figure 7.2 Outlet boxes used in commercial construction. (*a*) Handy box. (*b*) Octagon box. (*c*) 4-inch square box. (*d*) Masonary box. (Courtesy of Raco, Inc.)

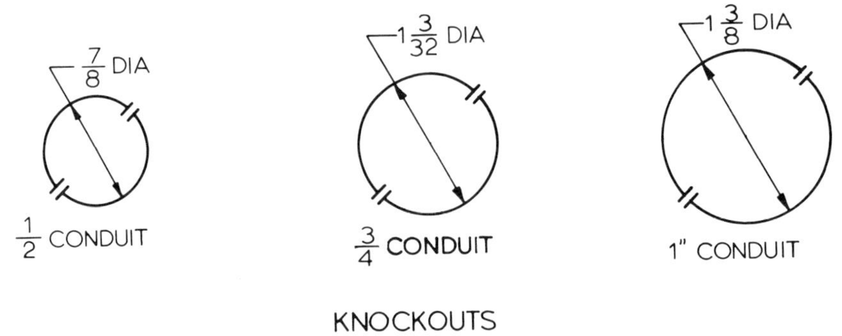

(a)

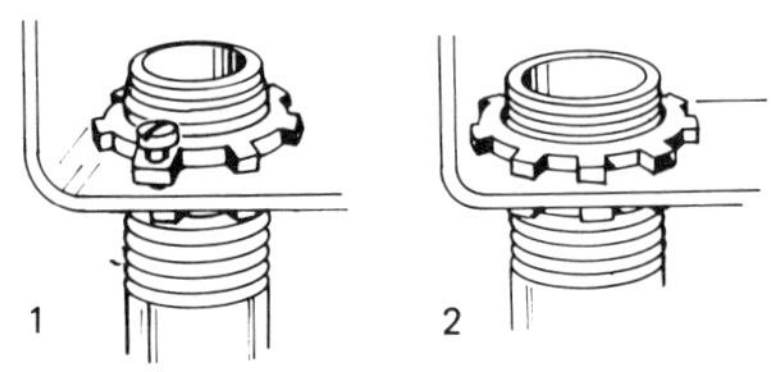

(b)

Figure 7.3 Knockouts. (*a*) Shows diameter of knockouts for ½, ¾, and 1-inch conduit. (*b*) Shows application of conduit in metal box knockout. (1) Grounding locknut. (2) Double locknut on IMC or rigid conduit.

necessary to make a splice in the wiring to take power off to another point, or if more than 360° of bends are made in the conduit run, it will be necessary to install a general purpose box, which becomes a pull box. There are times in construction when junction boxes are unavoidable; however, such a box is not an outlet, since by the definition, it does not supply current to a utilization device.

A few details of outlet boxes and their mountings are given in Figure 7.4. You will notice in each case we have basically the same material-conduit, a metal box, plaster-ring, and cover, and the mounting arrangement.

In the last few years, PVC (polyvinyl chloride) conduit has gained popularity with many electricians in the construction industry. Review Chapter 5, Section 5-1. Nonmetallic boxes and fittings are being manufactured to be installed with PVC. It should be noted that the application of this raceway is somewhat specialized. A check in Article 347-3 of NEC shows that

> PVC shall not be permitted in hazardous locations, for the support of fixtures, where subject to physical damage, where subject to ambient temperatures and for conductors whose insulation temperature limitations would exceed those for which the conduit is approved.

7-2 CUBIC CAPACITY AND NUMBER OF WIRES IN A BOX

The number of wires and devices permitted in a box is limited by the cubic inch capacity of the box. Unless superseded by local electrical codes. Article 370-6 of the NEC governs the number of wires allowed in a box.

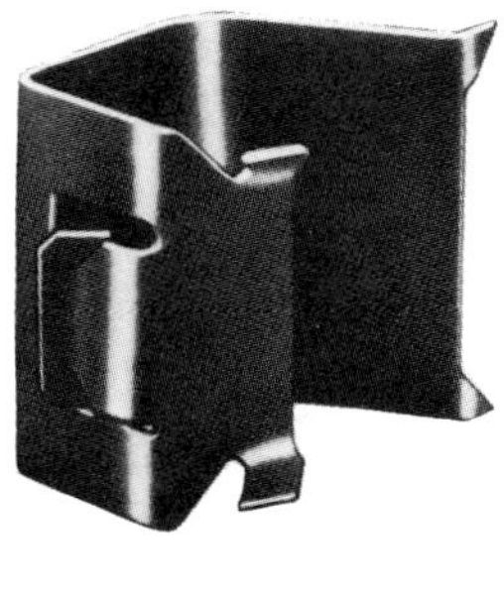

CAT. NO. MSC

INSTALLATION INSTRUCTIONS

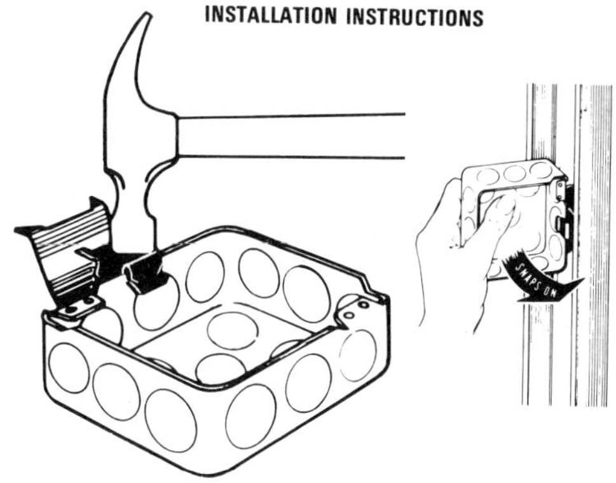

FEATURES

1. Fits all electrical boxes requiring plaster rings.
2. Fits all ribbed metal studding.
3. Requires only a hammer to install.
4. One piece installation.
5. No screws required.

SIZING CHART

CATALOG NUMBER	DESCRIPTION	PACKAGED PER BOX	WEIGHT PER 100 (LBS.)
MSC	Metal Stud Box Support	100	6

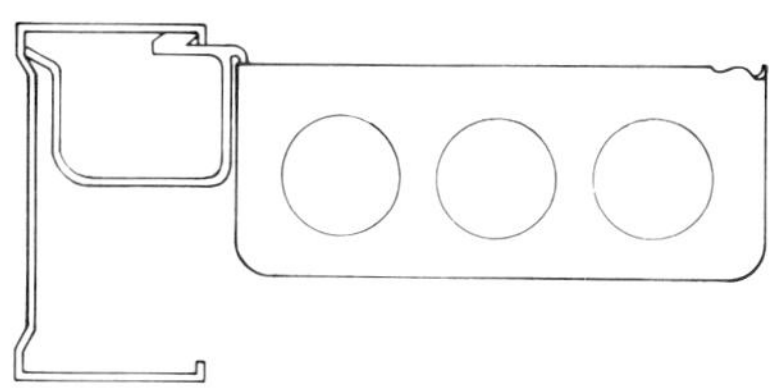

APPLICATIONS

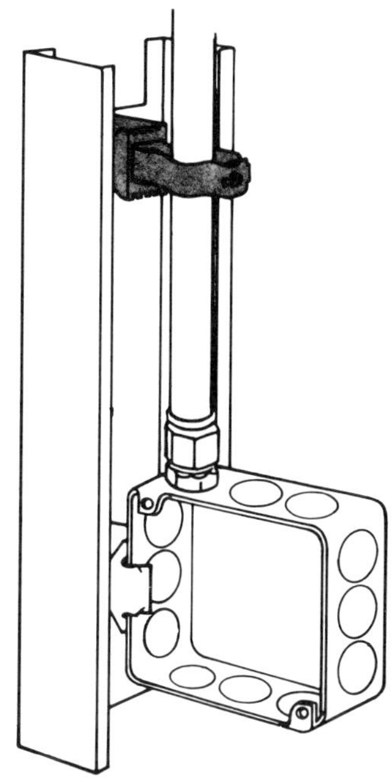

(a)

Figure 7.4 Methods of installing metal boxes. (*a*) Shows installing metal box in ribbed metal studding. (*b*) Shows installing metal box to metal stud. Requires only a hammer to install. (Courtesy of ERICO Products, Inc.)

Article 370-6. Number of conductors in switch, outlet, receptacle, device, and junction boxes. Boxes shall be of sufficient size to provide free space for all conductors enclosed in the box.

(a) *Standard Boxes.* The maximum number of conductors, not counting fixture wires permitted in standard boxes, shall be as is listed in Table 370-6(a) [reproduced as Table 7.1].

(1) Table 370-6(a) shall apply where no fittings or devices, such as fixture studs, cable clamps, hickeys, switches, or receptacles, are contained in the box and where no grounding conductors are part of the wiring within the box. Where one or more fixture studs, cable clamps, or hickeys are contained in the box, the number of conductors shall be one less than shown in the tables; an additional deduction of one conductor shall be made for each strap containing one or more devices; and a further deduction of one conductor shall be made for one or more grounding conductors entering the box. A conductor running through the box shall be counted as one conductor, and each conductor originating outside the box and terminating inside the box is counted as one conductor. Conductors, no part of

INSTALLATION INSTRUCTIONS

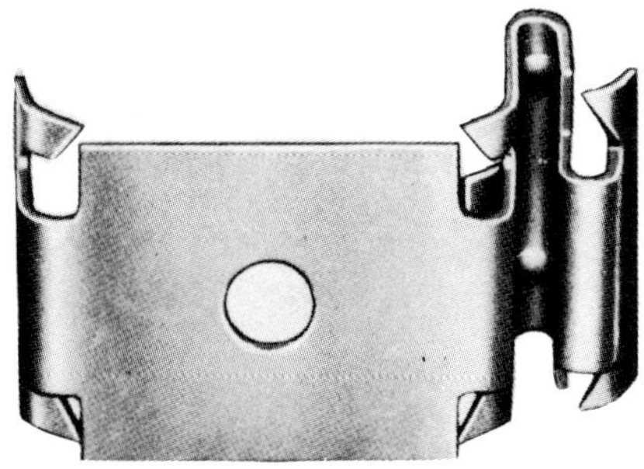

CAT. NO. MSF

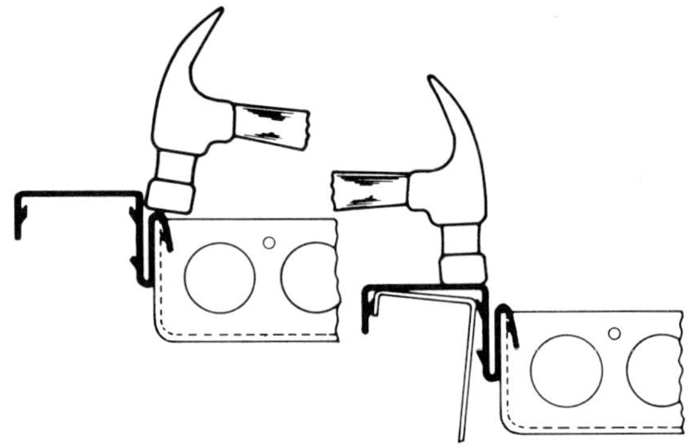

FEATURES

1. Fits all electrical boxes requiring plaster rings.
2. Fits all metal studding—does not require reinforced rib in metal stud.
3. Will not move on stud.
4. Requires only a hammer to install.
5. No screws required.

SIZING CHART

CATALOG NUMBER	DESCRIPTION	WEIGHT PER 100 (LBS.)	PACKAGED PER BOX
MSF	Metal Stud Box Support	6	100

APPLICATIONS

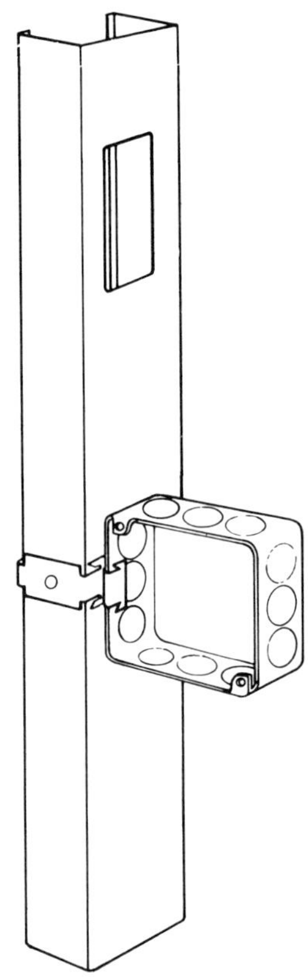

(b)

which leaves the box, shall not be counted. The volume of a wiring enclosure shall be the total volume of the assembled sections, and where used, the space provided by plaster rings, domed covers, extension rings, etc., that are marked with their volume in cubic inches shall be added to the total volume of the box.

(2) For combination of conductors sizes shown in Table 370-6(a), the volume per conductor listed in Table 370-6(b) [reproduced as Table 7.2] shall apply. The maximum number and size of conductors listed in the Table 370-6(a) [Table 7.1] shall not be exceeded.

(b) *Other Boxes.* Boxes 100 cubic inches or less, other than those described in Table 370-6(a), and conduit bodies having provision for more than two conduit

TABLE 7.1 Metal Boxes [NEC Table 370-6(a)]

Box Dimension, Inches, Trade Size or Type	Minimum Cubic Inch Capacity	Maximum Number of Conductors				
		No. 14	No. 12	No. 10	No. 8	No. 6
4 × 1¼ round or octagonal	12.5	6	5	5	4	0
4 × 1½ round or octagonal	15.5	7	6	6	5	0
4 × 2⅛ round or octagonal	21.5	10	9	8	7	0
4 × 1¼ square	18.0	9	8	7	6	0
4 × 1½ square	21.0	10	9	8	7	0
4 × 2⅛ square	30.3	15	13	12	10	6*
4¹¹⁄₁₆ × 1¼ square	25.5	12	11	10	8	0
4¹¹⁄₁₆ × 1½ square	29.5	14	13	11	9	0
4¹¹⁄₁₆ × 2⅛ square	42.0	21	18	16	14	6
3 × 2 × 1½ device	7.5	3	3	3	2	0
3 × 2 × 2 device	10.0	5	4	4	3	0
3 × 2 × 2¼ device	10.5	5	4	4	3	0
3 × 2 × 2½ device	12.5	6	5	5	4	0
3 × 2 × 2¾ device	14.0	7	6	5	4	0
3 × 2 × 3½ device	18.0	9	8	7	6	0
4 × 2⅛ × 1½ device	10.3	5	4	4	3	0
4 × 2⅛ × 1⅞ device	13.0	6	5	5	4	0
4 × 2⅛ × 2⅛ device	14.5	7	6	5	4	0
3¼ × 2 × 2½ masonary box/gang	14.0	7	6	5	4	0
3¼ × 2 × 3½ masonary box/gang	21.0	10	9	8	7	0

*Not to be used as a pull box. For termination only.

entries shall be durably and legibly marked by the manufacturer with their cubic inch capacity and the maximum number of conductors permitted shall be computed using the volume per conductor listed in Table 370-6(b) [Table 7.2]. Boxes described in Table 370-6(a) [Table 7.1] that have a larger cubic inch capacity than is designated in the table shall be permitted to have their cubic inch capacity marked as required by this section and the maximum number of conductors permitted shall be computed using the volume per conductor listed in Table 370-6(b).

(c) *Conduit Bodies.* Conduit bodies enclosing No. 6 conductors or smaller shall have a cross-sectional area not less than twice the cross-sectional area of the largest conduit to which it is attached. The maximum number of conductors permitted shall be the maximum number permitted by Table 1, Chapter 9, for the conduit to which it is attached. [See Table 6.2, Chapter 6.]

Conduit bodies having provisions for no more than two conduit entries shall not contain splices, taps, or devices.

7-3 WIRING DEVICES

The general classification of wiring devices includes all receptacles, wall switches, small dimmer switches, and outlet mounted pilot lights. Essentially, wiring devices are those found in common use in lighting and receptacle circuits, which normally do not exceed 30 amperes in rating.

Receptacle

The NEC Article 100—Definitions, defines a receptacle.

> NEC Article 100—A receptacle is a contact device installed in an outlet, for the connection of a single attachment plug.

TABLE 7.2 Volume Required per Conductor [NEC Table 370-6(b)]

Size of Conductor	Free space within box for each conductor
No. 14	2 cubic inches
No. 12	2.25 cubic inches
No. 10	2.5 cubic inches
No. 8	3 cubic inches
No. 6	5 cubic inches

Since the normal wall receptacle will take two attachment plugs, it is referred to as a duplex receptacle. See Figure 7.5.

The common wall outlet is properly called a convenience outlet. The term "wall plug" which is heard so often by newcomers as well as seasoned electricians, is really incorrect. A plug is another name for the attachment cap on the wiring coming from a device such as a lamp or an appliance. It is plugged into the wall; hence the name "plug." However, since the term wall plug is used so often, it has been accepted in the trade circles.

Receptacles are described and identified by poles and wires. The number of poles equals the number of hot contacts, thus excluding the grounding pole. The number of wires includes all connections to the receptacle, including the green ground wire.

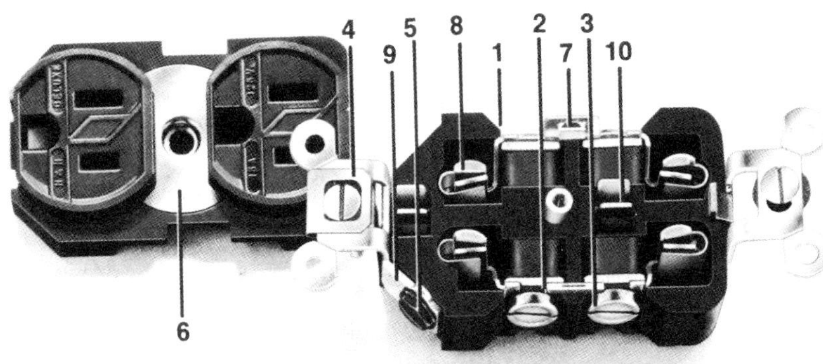

Figure 7.5 Specification grade duplex receptacle designed for heavy commercial, industrial, and institutional applications where long electrical and mechanical life and foolproof grounding are essential. (Courtesy of Crouse-Hinds Company, Arrow-Hart Div.)

1. Back wiring, with eight holes for feed-through wiring and handy wire strip gauges.
2. Spring-loaded, back wire clamps hold wires while terminal screws are being tightened. Speeds up installation by eliminating wire fishing and holding problems.
3. Side wiring, by means of 10–24 captive type terminal screws with extra-large slotted heads.
4. New safety-ground self-grounding clip is riveted to the steel ground strap for extra strength, and eliminates the need for a bonding jumper. The grounding path is automatically established, and is safe, sure, and foolproof.
5. A grounding terminal is also provided, in addition to the safety-ground clip, for use with nonmetallic boxes. The green hex head slotted screw with wire retention hole allows fast, secure wiring.
6. Steel central support plate is designed to allow maximum adjustment of wall plate, and provides support for the cover.
7. Break-off fins with open access slot permit optimal two-circuit installations.
8. Double-wipe 90-10 bronze power contacts and spring steel backup clips assure long-life contact pressure, and easily meet NEMA WD1-1974 plug retention requirements.
9. Wraparound mounting strap is locked to bakelite base for superior assembly strength.
10. Bronze "U" ground contacts, 0.30 inches thick, with double-wipe design ensure optimum current continuity.

Receptacles are available in ratings of 10 to 400 amperes, two to four poles, and 125 to 600 volts. The typical duplex wall convenience receptacle is two-pole, three-wire, 125-volt, and 15- or 20-ampere. The quality or grade of the receptacle is usually specified in the job specifications and can be economy, standard, or specification.

Other receptacles, new to the industry and used in commercial wiring, are the safety-ground receptacle, and the isolated ground duplex receptacle.

1. *Safety-Ground Receptacle.* Safety-ground, self-grounding receptacles comply with NEC Article 250-74 Exception 2. They have a special grounding clip to establish the ground circuit between the mounting strap and metallic flush-type boxes without the use of a bonding jumper as permitted by Exception 2. They are available in 15 and 20 amperes, 125- and 250-volt ratings and are manufactured in deluxe, heavy duty, and standard grades. See Figure 7.6.

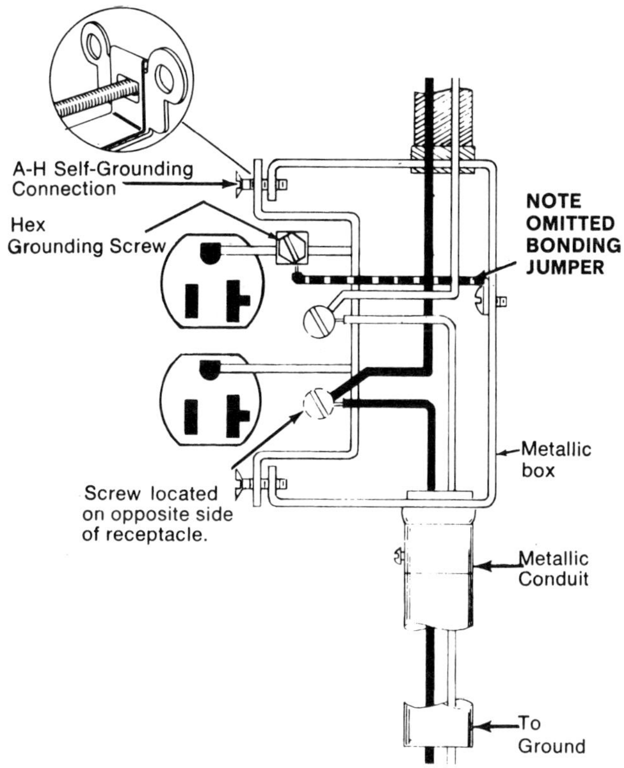

(a)

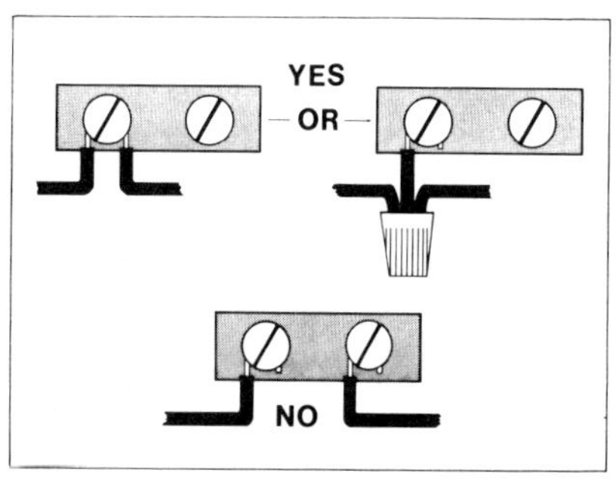

(b)

Figure 7.6 (*a*) Safety-ground receptacle, built with a special grounding clip to establish the grounding circuit between the mounting strap and metallic flush type box without the use of a bonding jumper. (Courtesy of Crouse-Hinds Company, Arrow Hart Div.) (*b*) Shows loop or pigtail at receptacle. (Courtesy of Pass and Seymour)

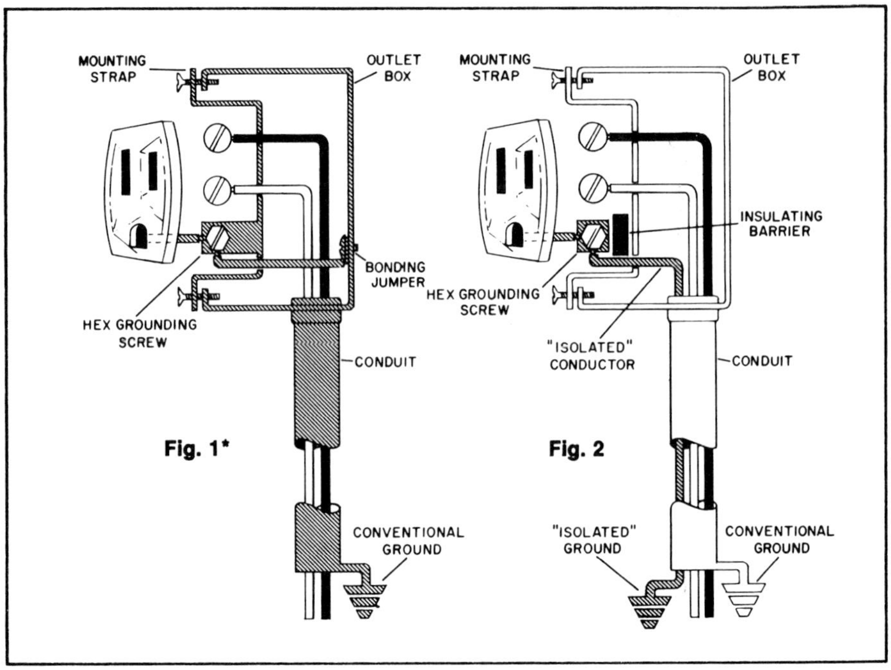

Figure 7.7 Isolated grounding receptacle. Fig. 2 shows how the grounding path within the grounding receptacle is separated from the mounting strap and the rest of the grounding system in the building. (Courtesy of Pass and Seymour)

2. Figure 7.7 shows how the grounding path within the isolated grounding receptacle is separated from the mounting strap and the rest of the grounding system in the building.

Wall convenience duplex receptacles are normally installed vertically, between 12 and 18 inches above the finished floor. Some developers prefer horizontal mounting and if this is desired it must be so specified, preferably on the drawings, along with the mounting height. In work areas, shops and the like, where the receptacle outlets must be accessible above tables and work benches, mounting height is 42 inches and horizontal installation is preferable so that cords do not hang on top of each other. This, too, must clearly be specified on the drawings.

Wall Switches

The flow of electric current in the various lighting circuits of a commercial building must be controlled. This is done by using a variety of switches capable of opening and closing the circuits.

The basic function of the switch is to open and close a circuit. To do this the switch has a combination of fixed and movable contacts which are usually made of brass. Figure 7.8 shows the design and methods for moving the contacts.

Switches designed to carry a high current have larger and stronger contacts than switches designed to operate only one or two light fixtures. Nearly all, however, have some form of spring inside to open and close the contacts quickly. If the contacts are slow to open and close there is danger of current arcing—a spark jumping across the contacts as they part thus burning the contacts.

1 Both movable and stationary contact buttons made of a silver alloy for high electrical conductivity and long life.

2 Movable contact arm of special copper alloy for maximum conductivity and long life.

3 Trunnions and spring assembly lubricated for the life of the switch.

4 Urea arc barrier for extra safety — an integral part of every double-pole switch housing.

5 Arcing contacts are located far away from the trigger opening for dielectric strength.

6 Long-life neoprene bumpers stay resilient for the life of the switch.

LISTED BY U.L. — MEET NEMA PERFORMANCE STANDARDS — MEET FEDERAL SPECIFICATIONS

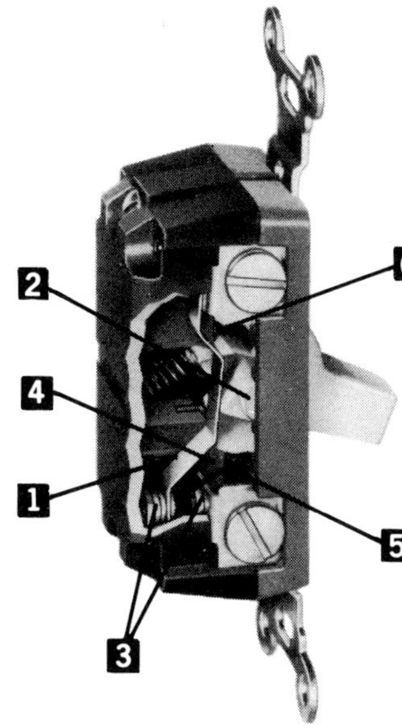

Figure 7.8 Quiet AC wall switch. (Courtesy of General Electric Company)

The main difference between a good quality switch and a competitive quality unit is the size and strength of the contacts and springs.

A switch is usually a one gang size wiring device, available in ratings from 15 to 30 amperes, with generally one or two poles and single or double throw. From outward appearance it is frequently difficult to distinguish between different units, since they are generally made to fit, as mentioned, in a single gang box.

Offices with two entrances often have two switches controlling the fixture, referred to as three-way switches. When the fixture is to be controlled from three locations (as a corridor), it will require the use of two 3-way switches and one 4-way switch (Figure 7.9).

A dimmer control is used when it is desirable to vary the intensity of the light.

Low-voltage switching will be discussed in Chapter 11.

SUMMARY

1. Outlet boxes vary in size and shape and are specifically suited for different kinds of construction.
2. Switches are not included when counting the number of outlets on a circuit.
3. It is important to remember that every outlet requires a box.
4. The number of wires and devices permitted in a box is limited by the cubic inch capacity of the box.
5. Conduit bodies having provisions for no more than two conduit entries shall not contain splices, taps, or devices.
6. The common wall outlet is properly called a convenience outlet.
7. Receptacles are described and identified by poles and wires.
8. Wall convenience duplex receptacles are normally installed vertically, between 12 and 18 inches above the finished floor.

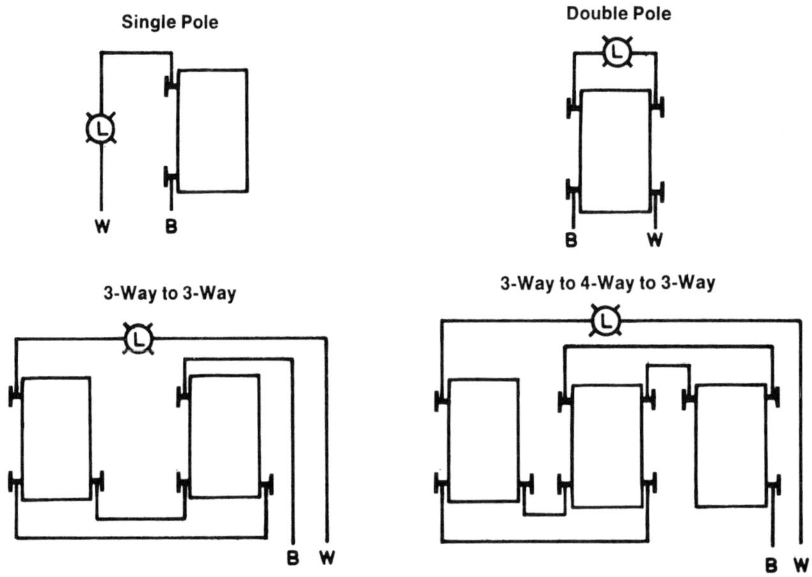

Figure 7.9 Wiring diagrams.˴(Courtesy of Crouse-Hinds Company, Arrow Hart Div.)

9. The flow of electric current in various lighting circuits of commercial buildings is usually controlled by using a variety of switches capable of opening and closing the circuits.
10. The main difference between a good quality switch and a competitive quality unit is the size and strength of the contacts and springs.

PROBLEMS

7-1 List three types of boxes used in commercial construction.

1. ____________
2. ____________
3. ____________

7-2 What limits the number of wires in a box? Explain.

____

____

____

____

7-3 How many conductors are allowed in the following boxes:

a. No. 14 conductor in 4 inch square × $2\frac{1}{8}$ inches ________
b. No. 14 conductor in 4 inch square × $1\frac{1}{2}$ octagon ________
c. No. 8 conductor in 4 inches × $1\frac{1}{4}$-inch square ________
d. No. 10 conductor in $4\frac{11}{16}$ inches × $1\frac{1}{2}$-inch square ________
e. No. 12 conductor in 3-inch × 2-inch × $3\frac{1}{2}$-inch device ________
f. No. 6 conductor in 4 inches × $2\frac{1}{8}$-inch round ________

7-4 What volume is allowed by the NEC for each conductor within a box?

a. No. 12 conductor ____________

b. No. 6 conductor ____________

c. No. 14 conductor ____________

d. No. 10 conductor ____________

e. No. 8 conductor ____________

7-5 Draw a wiring diagram showing a three-way to four-way to three-way quiet AC switch. Feed switch with hot wire.

7-6 An outlet is defined as any point that ____________________ .

7-7 List two examples for the above question. ____________________ ,

____ .

7-8 A conductor running through the box shall be counted as ________ conductor(s).

7-9 The general classification of wiring devices includes ____________

____

____

7-10 The common wall outlet is properly called ____________________ .

chapter 8

BRANCH CIRCUIT WIRING METHODS

Instructional Objectives

1. To learn how to calculate the occupied floor area of a single-family dwelling.
2. To become familiar with the NEC to determine the basic requirements for branch circuits.
3. To become aware of the minimum number of lighting circuits required in a dwelling.
4. To learn how to install and support the cable for the outlets.
5. To understand the need for proper grounding connections when using nonmetallic cable.
6. To develop the appliance branch circuits.
7. To learn how to use the National Electric Code to determine the requirements for installing outlets and receptacles.
8. To understand the operation of a dimmer switch.
9. To learn how to install and make connections for three- and four-way switches.
10. To develop an ability to calculate the floor space of a residence for calculating branch lighting circuits.
11. To understand the need for the installation of a ground-fault circuit interrupter.

Self-Evaluation Questions

Test your prior knowledge of the information in this chapter by answering the following questions. Watch for the answers as you read the chapter. Your final evaluation of whether you understand the material is measured by your ability to answer these questions. When you have completed the chapter, return to this section and answer the questions again.

1. What dimensions are used when measuring the area of a dwelling?
2. How is the total lighting load in amperes determined?

3. What is unit load per square foot for a single-family dwelling?
4. According to the NEC, on what basis is the minimum number of receptacle outlets determined for most occupied rooms in a home?
5. What is the appliance circuit?
6. Is an unoccupied basement area included in the calculations?
7. Would you install a lighting outlet in an attic?
8. Where are receptacles to be located in the kitchen?
9. Is it necessary to install lighting outlets in every habitable room?
10. What type of circuits must be provided for receptacle outlets in the kitchen, pantry, breakfast room, and family room?

In the proceeding chapters we discussed the working drawing, specifications, symbols, and the need for safety standards. We are now ready to calculate branch circuits. It is usually standard practice for the electrician to plan for and lay out the circuits. However, they must conform to the standards established by the NEC and local code requirements.

8-1 CALCULATING OCCUPIED FLOOR AREA

General Lighting Load

The NEC in Article 220 and Section 2(b) states: Unit lighting load for dwelling units shall not be less than 3 watts per sq ft [Table 220-2(b)]. In determining this load on a watts per sq ft basis, use the outside dimensions of the dwelling (Figure 8.1). The computed floor area does not include open porches, garages, unused or unfinished spaces unless adaptable for future use.

According to Section 3(d) Article 220 NEC: "Watts per sq ft" load shall be apportioned evenly among branch circuits, according to their capacities, by using the examples in Chapter 9 of NEC for general ilumination. One 15-

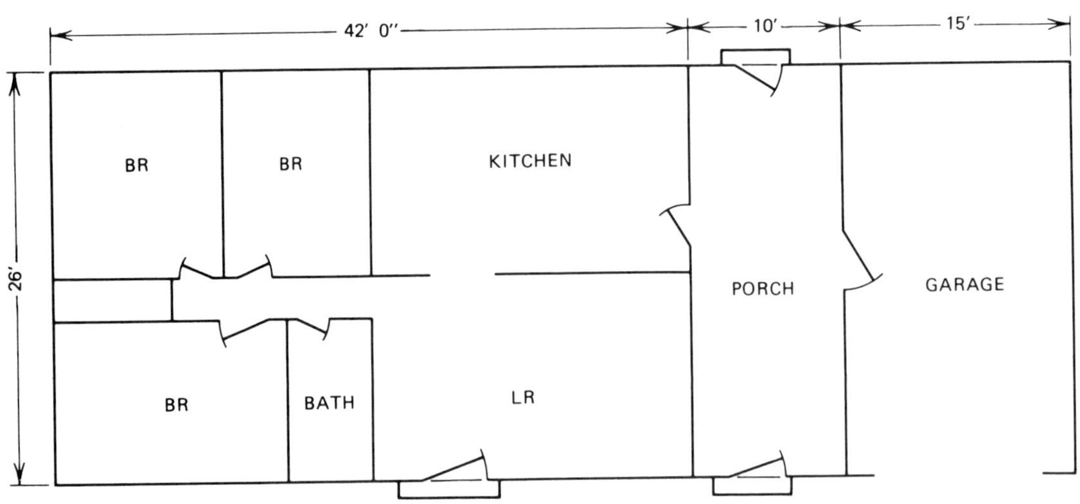

Figure 8.1 Floor plan.

ampere, 115-volt branch circuit would be required for every 575 sq ft of occupied floor area.

EXAMPLE 1

3 watts per sq × 575 sq ft = 1725 watts ÷ 115 volts = 15 amps. A calculated load for a total occupied area of 30 ft × 55 ft = 1650 sq ft × 3 watts per sq ft = 4950 watts.

8-2 CALCULATING MINIMUM NUMBER OF LIGHTING CIRCUITS

The total required amperes is equal to amperes $= \dfrac{\text{watts}}{\text{volts}}$

$$\text{Amperes} = \frac{4950}{115} = 43.04 \text{ amperes}$$

As a result the minimum number of circuits required for general lighting load is obtained by dividing the total required amperes by the maximum amperage rating of each circuit.

$$\frac{43.04}{15} = 2.86 \text{ or 3 circuits minimum required.}$$

Convenience Outlets

In addition to general lighting load, the NEC Table 220-2(b) states "receptacles other than those of the two small appliance circuits are considered part of general illumination and require no allowance for additional load."

8-3 DETERMINING THE NUMBER OF SMALL APPLIANCE BRANCH CIRCUITS

Kitchen

The NEC rules here, Article 220-3(b), for small appliance load, including refrigeration equipment in kitchen, pantry, breakfast room, dining room, and family room, two or more 20-ampere appliance circuits shall be provided. Such circuits shall have no other outlets. The outlets in the kitchen must be supplied by at least two 20-ampere circuits. An electric clock outlet may be installed on the general lighting circuit.

NEC 220-16(a): the feeder load for the two small appliance circuits of above Article 220-3(b) is to be taken at 3000 watts or 1500 watts each two-wire circuit.

Laundry

At least one additional 20-ampere branch circuit is needed to supply the laundry receptacle outlet. With no other outlets on this circuit, NEC 220-3(c), this could be calculated at 1500 watts NEC 220-16(b).

Outlets for Branch Circuits

As stated in Section 6-1, convenience outlets are connected to the general lighting branch circuits with the exception of the receptacles in the kitchen (see Section 6-2). These outlets are connected to the 20-ampere small appliance circuit. Usually the location of convenience receptacles are up to the owner/builder. However, the code does give general requirements for the location of these outlets (Section 210-52(b) NEC).

Receptacles

Section 210-52(b) requires that, in dwellings, receptacles must be installed:

1. in all rooms of general occupancy, no space along a wall is to be more than 6 feet from a receptacle outlet measured horizontally.
2. in any wall space 2 feet or more in width.
3. in kitchen where each counter space is wider than 12 inches.
4. insofar as practicable, spaced equal distance apart.
5. with at least one wall outlet in the bathroom near basin.
6. with at least one outlet outdoors.
7. with at least one outlet in basement in addition to laundry outlet.
8. with at least one outlet in each attached garage.
9. within 6 feet of an intended location of an appliance.
10. in addition to any receptacles located or 5½ feet above the floor.

Lighting Outlets

Section 210-70(a) Dwelling Units, Requirements

1. At least one wall switch controlled lighting outlet must be installed in every habitable room, bathrooms, hallways, stairways, attached garages, and outdoor entrances.
2. At least one lighting outlet is to be installed in an attic, underfloor space, utility room, and basement area used for storage or space containing equipment requiring service.

8-4 CALCULATIONS BASED UPON CODE SECTIONS

Area of floor space—30 ft × 55 ft = 1650 sq ft		NEC 220-2(b)
Minimum wattage, general lighting = 1650 × 3 =	4950 watts	220-2(b)
Two appliance circuits	3000 watts	220-16(a)
Laundry circuit	1500 watts	220-16(b)
Total Light-Appliance-Laundry	9450 watts	220-16(a) (b)

Table 220-11 demand factors	
First 3000 watts at 100 percent	3000 watts
Remaining 6450 watts at 35 percent	2257 watts
Load requirements for above	5257 watts

Two—20-ampere appliance circuits	2
One—20-ampere laundry circuit	1
Three general lighting 3 15-ampere	3
	6 circuits

The above calculations are only minimum requirements; other circuits will be discussed in Chapters 4 and 5.

8-5 BRANCH CIRCUIT WIRING METHODS

One of the most important parts of any residential electrical system is the conductor that makes up the wiring circuit.

Conductor Materials

All residential circuit conductors consist of an insulated length of wire, usually copper. However, aluminum and copper-clad aluminum have made considerable inroads into the electrical industry in the last few years. Aluminum oxidizes if exposed to the air; this oxidation produces a layer of oxide on the surface of the conductor. If a conductor is allowed to oxidize at the terminal point or splice, the current-carrying ability of the conductor will be seriously reduced, because the oxide layer does not conduct electricity.

Copper

Because copper is an excellent conductor, it is easy to make and handle and does not oxidize as much as aluminum. It is the conductor most often used in electrical installations.

Aluminum

Aluminum is lighter than copper but not as good a conductor. To obtain the same current-carrying capacity, an aluminum conductor must be slightly larger than one made of copper. As stated above, aluminum oxidizes rapidly. The aluminum oxide formed acts as an insulator and reduces the flow of electrical current.

Conductor Sizes

In the United States (prior to any change to metric sizing) the copper or aluminum conductor used in electrical installations is graded for size according to the American Wire Gauge Standard, simply called AWG (Figure 8.2). This gauge is used only to measure solid conductors. The outer edge of the gauge has slots that are numbered. The smallest slot into which the wire will fit is the gauge number of the wire. The wire diameter is expressed as a whole number rather than as a fractional or decimal dimension, such as Nos. 14, 12, 10, 8, and so on (Figure 8.2).

Ampacity

The purpose of a conductor is to carry current from one place in a circuit to another. The size of wire used for any circuit depends on the maximum amount of current (amperes) to be carried. Ampacity is the term used to

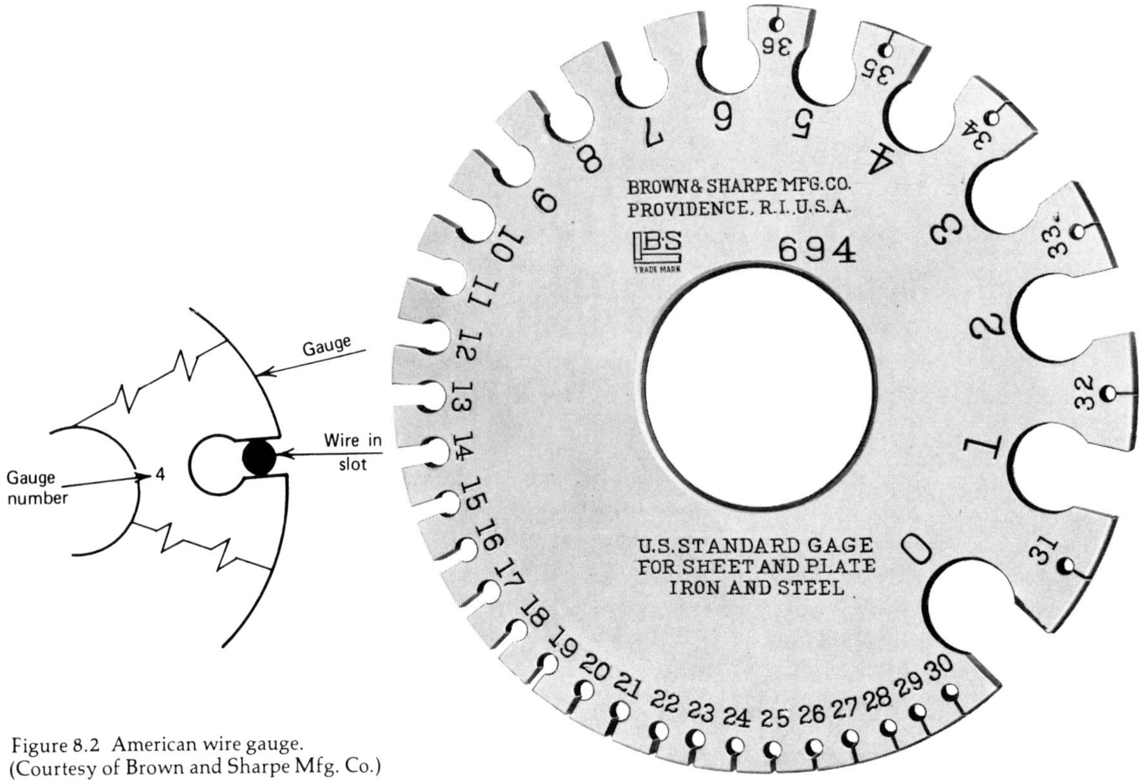

Figure 8.2 American wire gauge.
(Courtesy of Brown and Sharpe Mfg. Co.)

express the current-carrying capacity of a wire in amperes and depends on the material, size, and insulation of the wire. See Table 8.1.

Material

The material of which the conductor is made determines how easily it will carry current. Copper is a better conductor than aluminum and will, therefore, carry more current.

Size

The larger the conductor, the more current it will carry without heating.

Insulation

A conductor with insulation capable of withstanding heat will have a higher ampacity rating than a conductor of the same size with a lower insulator temperature rating. All conductors are insulated to prevent contacting each other and short-circuiting. The insulation used on building wires and cables is usually a rubber compound with an outer braid or a thermoplastic material. Table 310-13 of the NEC lists the various types of insulation available.

The NEC specifies in Section 210-19(c) "that the minimum conductor size permitted in house wiring is No. 14 AWG."

Let us examine a branch circuit, by referring to Figure 8.12. In the drawing, the branch circuit consists of lines connecting the wall receptacle outlets in the bedrooms. In actual practice, the wiring method for this branch circuit can consist of:

TABLE 8.1 **Allowable Ampacities of Insulated Conductors Rated 0–2000 Volts, 60° to 90°C. Not More Than Three Conductors in Raceway or Cable or Earth (Directly Buried), Based on Ambient Temperature of 30°C (86°F) (NEC Table 310-16).**

Size	Temperature Rating of Conductor, See Table 310-13								Size
	60°C (140°F)	75°C (167°F)	85°C (185°F)	90°C (194°F)	60°C (140°F)	75°C (167°F)	85°C (185°F)	90°C (194°F)	
AWG MCM	Types RUW, T, TW, UF	Types FEPW, RH, RHW, RUH, THW, THWN, XHHW, USE, ZW	Types V, MI	Types TA, TBS, SA, AVD, SIS, †FEP, †FEPB, †RHH, †THHN, †XHHW*	Types RUW, T, TW, UF	Types RH, RHW, RUH, THW, THWN, XHHW, USE	Types V, MI	Types TA, TBS, SA, AVB SIS, †RHH, †THHN, †XHHW*	AWG MCM
	Copper				Aluminum or Copper-Clad Aluminum				
18	—	—	—	21	—	—	—	—	—
16	—	—	22	22	—	—	—	—	—
14	15	15	25	25	—	—	—	—	—
12	20	20	30	30	15	15	25	25	12
10	30	35	40	40	25	30	30	35	10
8	40	45	50	50	30	40	40	40	8
6	55	65	70	70	40	50	55	55	6
4	70	85	90	90	55	65	70	70	4
3	80	100	105	105	65	75	80	80	3
2	95	115	120	120	75	90	95	95	2
1	110	130	140	140	85	100	110	110	1
0	125	150	155	155	100	120	125	125	0
00	145	175	185	185	115	135	145	145	00
000	165	200	210	210	130	155	165	165	000
0000	195	230	235	235	155	180	185	185	0000
250	215	255	270	270	170	205	215	215	250
300	240	285	300	300	190	230	240	240	300
350	260	310	325	325	210	250	260	260	350
400	280	335	360	360	225	270	290	290	400
500	320	380	405	405	260	310	330	330	500
600	355	420	455	455	285	340	370	370	600
700	385	460	490	490	310	375	395	395	700
750	400	475	500	500	320	385	405	405	750
800	410	490	515	515	330	395	415	415	800
900	435	520	555	555	355	425	455	455	900
1000	455	545	585	585	375	445	480	480	1000
1250	495	590	645	645	405	485	530	530	1250
1500	520	625	700	700	435	520	580	580	1500
1750	545	650	735	735	455	545	615	615	1750
2000	560	665	775	775	470	560	650	650	2000

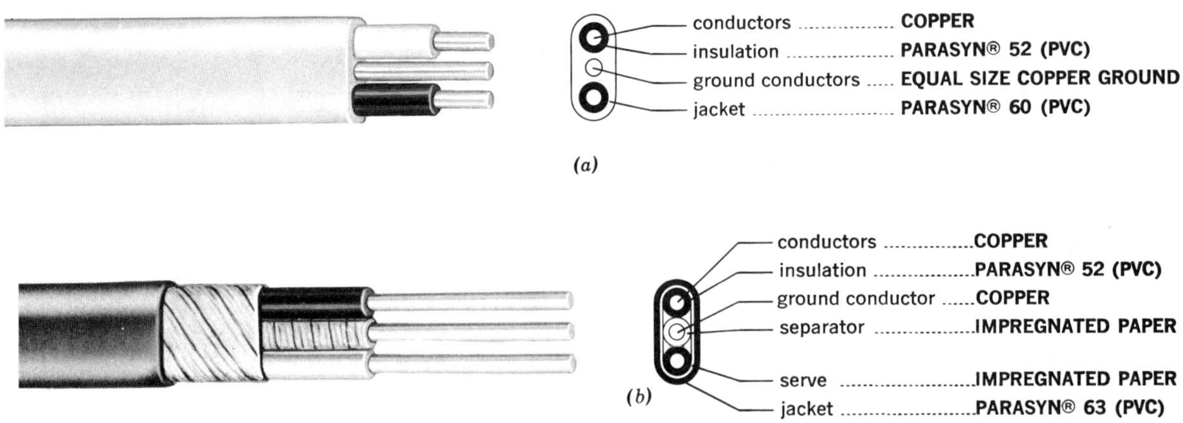

Figure 8.3 (*a*) Construction of typical NEC type NM nonmetallic sheathed cable. (*b*) UF underground feeder cable. (Courtesy of Wire and Cable Division, ESSEX INTERNATIONAL)

Type NM nonmetallic-sheathed cable, NEC Section 336. Nonmetallic-sheathed cable (NM) is used more often in residential wiring installations than any other wiring method. NM was first produced by the Rome Wire and Cable Company, which named its new product Romex. This name is still used in the trade when referring to nonmetallic cable. NM cable consists of a rubber- or plastic-insulated wire in a cloth or plastic jacket. This cable is available with two or three current-carrying conductors in sizes ranging from No. 14 through No. 2, with copper conductors, and in sizes No. 12 through No. 2 with aluminum- or copper-clad aluminum conductors. Nonmetallic-sheathed cable is available with an uninsulated conductor, called the grounding wire, which is used for grounding purposes only. It is not intended to be used as a current-carrying circuit wire (Figure 8.3).

Nonmetallic-sheathed cable is the least expensive of the various wiring methods. It is relatively lightweight and is easy to handle and install. Therefore it is widely used for residential installations.

Cable Supports

NEC Section 336-5 states "type NM cable must be fastened by staples or straps so designed and installed as not to injure the cable" (Figure 8.4). The NEC requires that "a staple or strap be placed every 12 inches from every box or fitting and 4½ feet apart on the runs between the boxes."

Fastening a Cable to the Box

NEC Section 370-7(c) states "where nonmetallic boxes are used, nonmetallic-sheathed cable shall extend into the box no less than ¼ inch through a nonmetallic-sheathed cable knockout opening. When used with a single gang box and where the cable is fastened within 8 inches of the box measured along the sheath, and extends into the box no less than ¼ inch, securing the box shall not be required" (Figure 8.5).

Cable Protection

When drilling through studs or joists, the code requires NM cable to be at least 1¼ inch from the outer edge of the 2 × 4 or wooden member. The normal procedure is to drill a ½ or ⅝ inch hole in the center of the stud or joist that the cable must pass through. Sometimes it is necessary to drill the hole

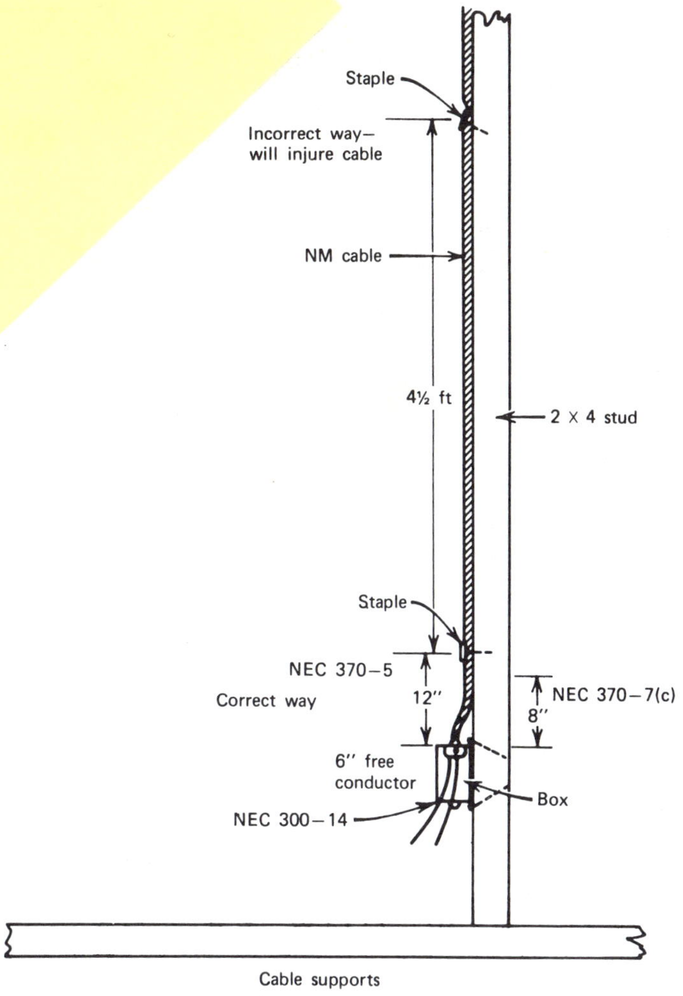

Figure 8.4 Fastening cable with staples.

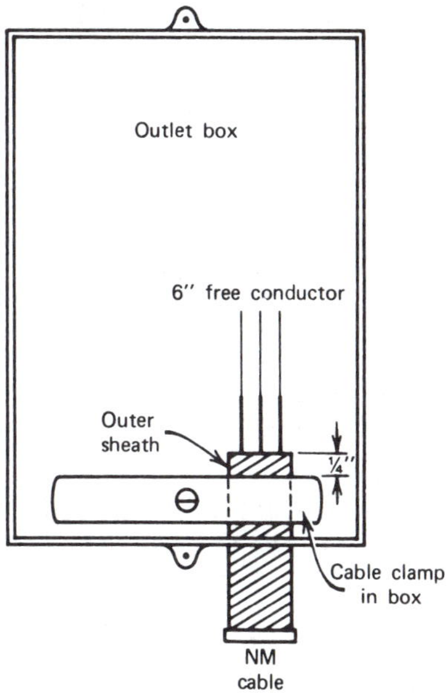

Figure 8.5 Method of fastening cable in metal box.

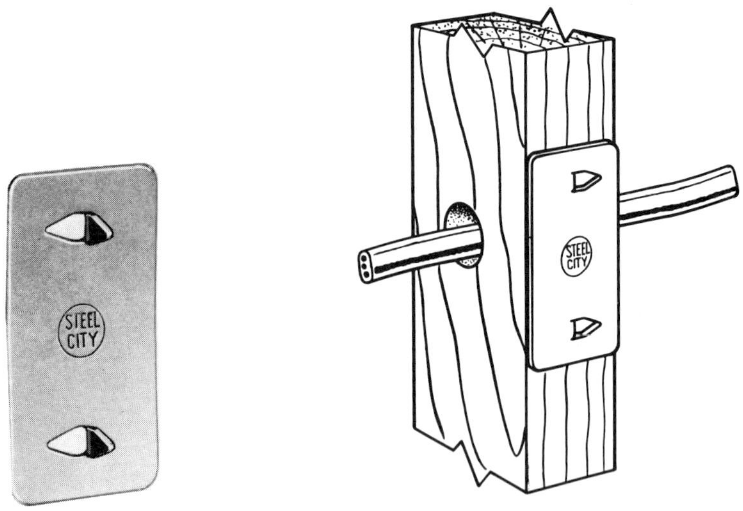

(a) (b)

Figure 8.6 (*a*) Steel plate protects cable running through wood. A steel plate protector is designed to protect electrical cable or wiring raceway running through notched or bored studs, joists, rafters, or other wood structures. (*b*) Two integral spurs supply attachment to the stud by hammer blow; no nails are required. (Courtesy of Midland-Ross Corporation)

Figure 8.7 Cable connector. Remove the knockout snap in the connector, insert cable. Half-turn with screwdriver secures cable from pullout. (Courtesy of Raco, Inc.)

closer than the 1¼-inch limit. In this case, a 1/16-inch steel plate is fastened to the wooden member in front of the cable to protect it from being pierced with a nail (Figure 8.6).

Not all electrical boxes are equipped with built-in clamps. Service panels and outlet boxes with ½-inch knockouts require a cable connector (Figure 8.7).

NEC Section 300-14 states, "at least 6 inches of free conductor must be left at each outlet and switch point." Figure 8.8 shows a cable ripper that saves time and prevents damage to the cable during the stripping process.

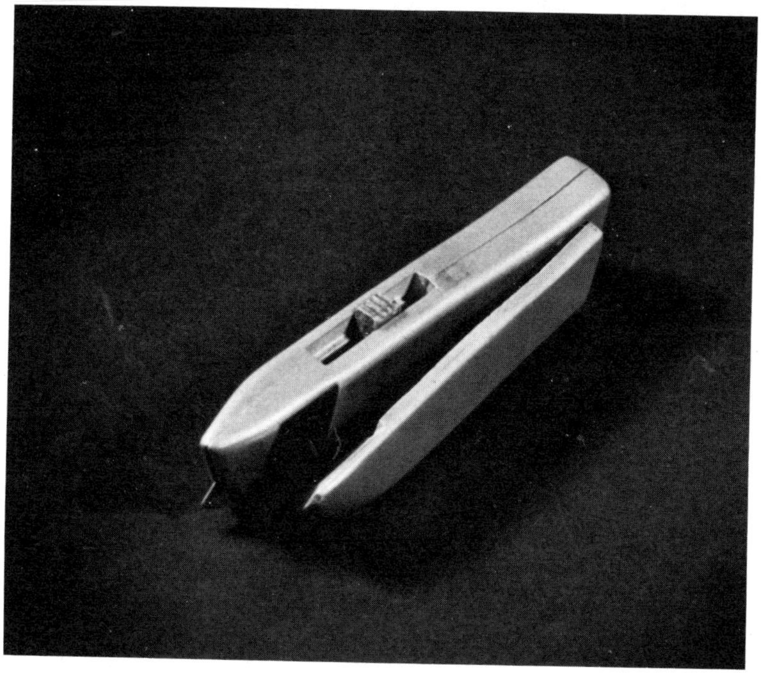

(a)

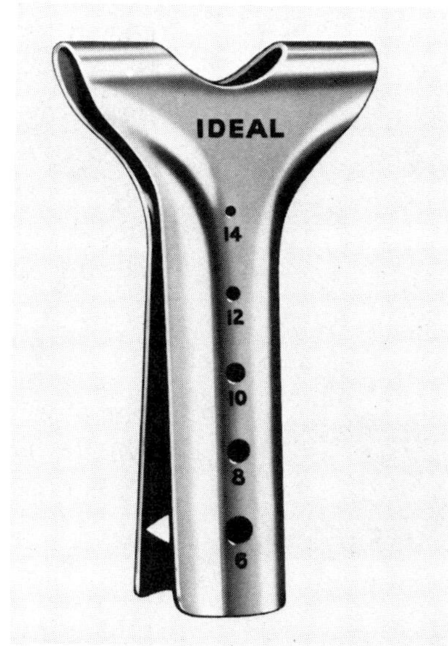

(b)

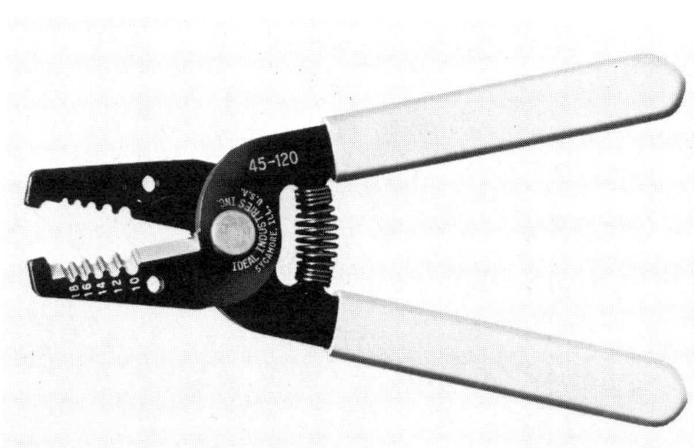

(c)

Figure 8.8 Cable strippers. (*a*) Grip stripper utility knife. (Courtesy of Blackhawk Industries) (*b*) Cable ripper with wire gauge. Cut nonmetallic sheathed duplex or lead covered cable. Squeeze onto cable and pull at least 6 inches for cut. (Courtesy of Ideal Industries, Inc.) (*c*) Handy flat wire strippers; strips sizes No. 18 to No. 10 wire easily. (Courtesy of Ideal Industries, Inc.)

8.6 LIGHTING CIRCUIT SWITCH CONTROL

The flow of electrical current in the various lighting circuits of a dwelling must be controlled. This is done by a switch capable of opening and closing the circuit.

The electrician installs and connects the various types of switches used for the lighting circuits. In addition to the installation, the electrician must understand the meanings of the current and voltage ratings marked on lighting switches as well as know the NEC requirements for the installations of these switches.

Wall Switches

Switches listed by Underwriters' Laboratories, Inc., are either AC for alternating current loads or AC/DC suitable for use on either alternating or direct current circuits. There is a common misconception that the AC/DC switch, being more versatile than the AC, is the superior switch. In fact, on AC loads, the AC-only switch performs much better than the dual-rated switch.

AC switches are rated for loads operating at 120 volts AC or 120/227 volts AC. Usually 120-volt AC-rated switches are intended for light-duty use and the 120/227-volt AC switches are for general or heavy-duty applications. The current rating of a switch should be adequate for the load involved.

Toggle Switch

The most frequently used switch in residential lighting circuits is the toggle switch. Toggle switches are available in three types.

Single Pole

A single-pole toggle switch is identified by its two terminals and the words ON and OFF on the handle. It is usually installed when a light or group of lights must be controlled from one switching point. This type of switch is connected in series with the ungrounded (hot) wire feeding the load (Figure 8.9).

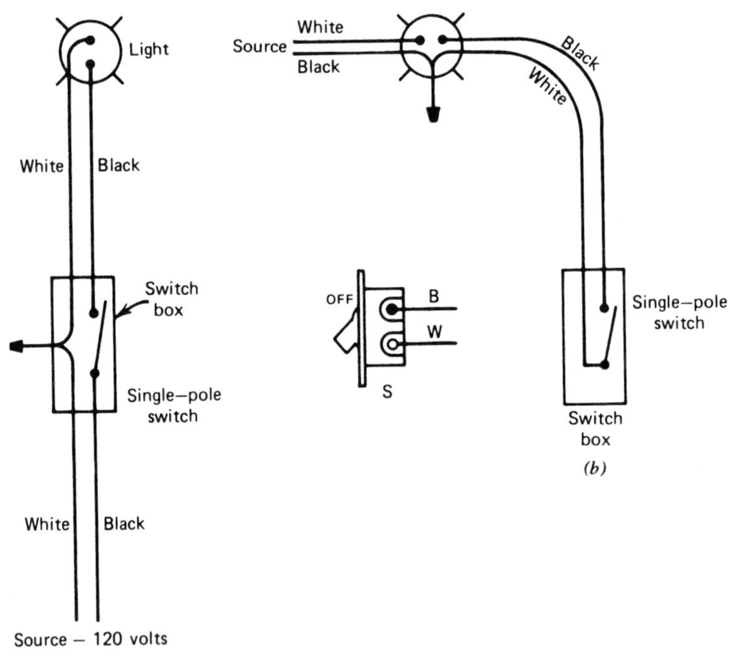

Figure 8.9 (*a*) Typical application of a single-pole switch to control a light from one switching phase. One hundred twenty volts feeds through switch, black wire is broken at the switch, white wire is fastened together with wirenut. (02b) With feed at light. The 120 volts feeds directly to the light outlet; a two-wire cable with black and white wires is used as a switch loop between the switch and the load. The white wire is connected to the black wire at the source [NEC 200-7].

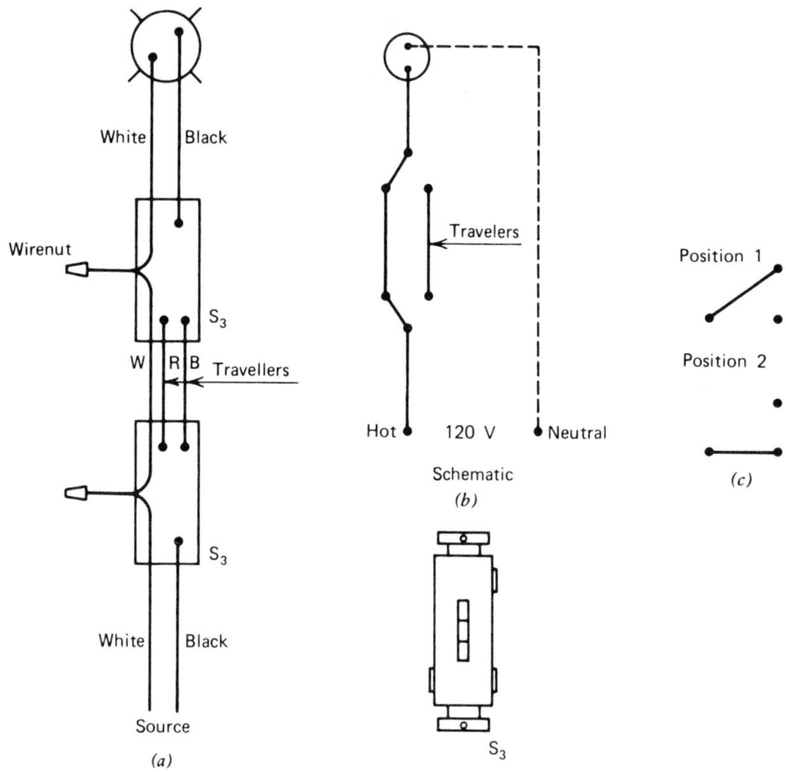

Figure 8.10 (*a*) Two three-way switch control, feed at the first switch control point. (*b*) Schematic wiring diagram. (*c*) The internal design of the three-way switch allows current to flow through the switch in either of its two positions.

Three-Way Switch

A three-way switch can be distinguished from a single-pole switch by noting that a three-way switch has three terminals. One of these terminals, the common terminal, is a darker color than the other two traveler wire terminals, which are natural brass color.

Three-way switches are used when a load must be controlled from two different switching points. Two three-way switches are used, one at each switching control point (Figure 8.10).

Four-Way Switches

A four-way switch can be identified readily by its four terminals and the fact that the toggle does not have ON or OFF on the toggle. The four-way switch has two positions; neither of these positions is ON or OFF. The internal mechanism of the four-way switch is constructed so that the switching contacts can alternate their positions as shown in Figure 8.11. Care must be used in connecting the traveler wires to the proper terminals. Notice the color of the terminal screws when making connections.

Some suggestions for using wall switches include:

1. Point-of-use control helps conserve energy and adds flexibility for the owner. All loads should be designed to be turned off when not in use.

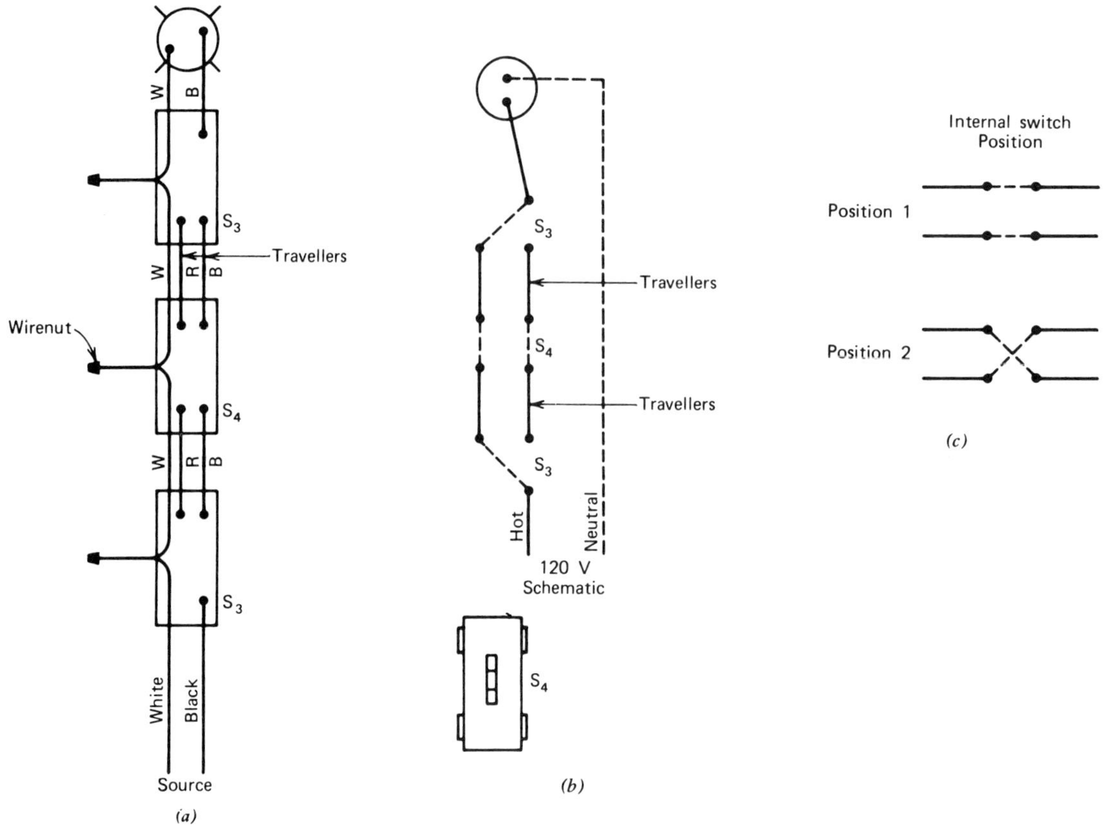

Figure 8.11 (*a*) Cable wiring diagram showing three-way and four-way switch control. (*b*) Schematic. (*c*) Internal switch position.

2. Pilot-lighted switches offer ON/OFF indication at the switching point whenever the switch is out of sight of the load.
3. Lighted switches aid persons entering dark areas, particularly if they are unfamiliar with the room, and make sense in bathrooms.

8-7 LIGHTING BRANCH CIRCUITS

Number of Outlets per Circuit

The planning of circuits in most residential installations is usually left up to the judgment of the electrician. There are many possible combinations or groupings of outlets. However, the code places no limits on the number of outlets that may be connected on one branch lighting circuit. There are few guidelines other than the National Electrical Code or your local electrical code on which to base the choice of particular outlets for a particular circuit.

Estimating Wattage for Outlets

When planning circuits, the electrician must consider the types of fixtures that may be used at the various outlets. Remember that symbols are used on circuit diagrams to indicate the type of each device and outlet and where they are located on the plans.

The code does not specify the number of outlets to be connected to a circuit. However, NEC Section 220-2(a) does state that "the continuous load supplied by a branch circuit shall not exceed 80 percent of the branch circuit rating." If you figure a 15-ampere, 115-volt branch lighting circuit, this means that the continuous load must be no more than 1380 watts.

15 amperes × 115 volts × 0.80 = 1380 watts

The 15-ampere circuit is wired with No. 14 wire and protected by a 15-ampere fuse or circuit breaker. The receptacles cannot be rated at any more than 15 amperes, which means that only the ordinary household variety of receptacle may be used. However, no appliance used on the circuit may exceed 12 amperes (1380 watts) in rating.

The electrician should keep in mind that, in rooms in which the outlets are located, it does provide indications as to their possible uses, and that all circuits should be planned accordingly.

8-8 LIGHTING BRANCH CIRCUIT FOR BEDROOMS (FIGURE 8.12)

Every bedroom should be provided with general illumination controlled by a wall toggle or dimmer switch. Some owner-builders prefer ceiling fixtures; others prefer split receptacles, half being switched or controlled by the dimmer switch for controlled lighting, while the other half is a hot receptacle.

Switches are usually installed near the door. The outlet box should be nailed to the stud so that the top of the switch box is 48 inches from the subfloor. Receptacle outlet boxes should be nailed to located studs 14 inches to top of box from subfloor.

Remember, if using nonmetallic-sheathed cable that contains an extra conductor, this conductor is used only for grounding the box. NEC Section 210-7 states: "all receptacles installed on 15-ampere and 20-ampere branch circuits must be grounding type receptacles."

8-9 LIGHTING BRANCH CIRCUIT FOR BATHROOMS

The electrician must follow very carefully the NEC requirements for the installation of lighting and receptacles installed in bathrooms.

NEC Section 210-52(c) requires at least one receptacle outlet near the

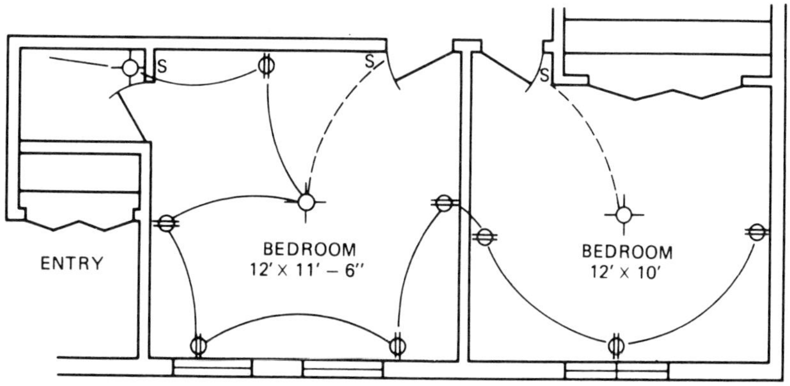

Figure 8.12 Typical circuitry of bedroom lighting and receptacles.

basin (Figure 8.13). The code further refers to Section 210-8(a), "all 120-volt, 15- and 20-ampere receptacles installed in the bathroom shall have ground-fault circuit interrupter protection for personnel."

What Is a GFCI?

A ground-fault circuit interrupter (GFCI), Figure 8.14, is an electronic device designed to protect people against ground fault (current leakage) before it can do any damage. Ground fault occurs when a person comes in contact with a hot or live line (wire). This is possible either by touching an exposed wire or by simply operating a faulty appliance or power tool having a wire defect that causes the metal housing of the product to become electrically alive.

Why a GFCI?

As a result of marked increases in the use of appliances (shavers, hair dryers, whirlpool pumps, etc.) and tools used by the average American homeowner, the NEC made ground-fault protection mandatory in many places and situations where it was not previously required. Ground fault (leakage current) is dangerous enough to cause electrocutions, electrical fires, and other serious accidents in the home.

How Does a GFCI Work?

The GFCI continuously monitors the current in the two conductors of a circuit, the hot wire and the neutral wire. These two currents should always be equal. If the GFCI senses a difference between them of more than 5 ± 1 milliamperes (mA), it assumes the difference is ground-fault current and automatically trips the circuit. Power is interrupted $\frac{1}{40}$ of a second or less, which is fast enough to prevent injury to anyone in normal health.

NEC Section 215-9 indicates that a ground-fault circuit breaker could be used to protect the feeder supplying 15- and 20-ampere receptacle branch circuits in lieu of the GFCI receptacle. However, the GFCI receptacle is usually located only a few feet from the appliance being used. People are often afraid to test a circuit breaker device located in a basement or reset it whenever it trips. Underwriters' Laboratories require that all GFCI's be tested monthly, so perhaps you'd rather install a GFCI receptacle in the bathroom.

Remember, fuses and circuit breakers are designed to protect wiring and appliances. A GFCI receptacle senses only ground fault and is designed to interrupt at a much lower current flow to protect lives (Figure 8.15).

Bathroom Lighting

To provide enough light for shaving and putting on makeup, a lighting fixture is usually installed over the mirror (controlled by a wall switch) in small bathrooms. This is enough light. However, in larger bathrooms a ceiling fixture is needed beside the light located near the mirror or medicine cabinet. For luxury and convenience, install an infrared heat lamp and exhaust fan.

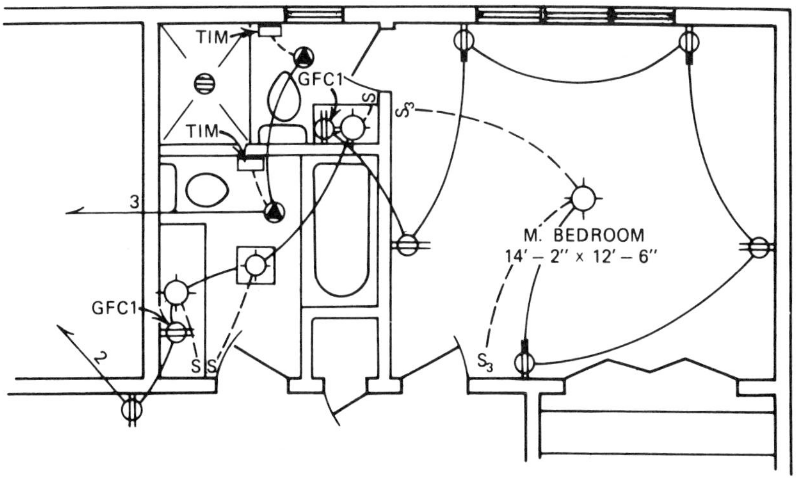

Figure 8.13 Typical circuitry of master bedroom, full bath, half bath, and hall lighting and receptacles, including ground-fault in bathrooms.

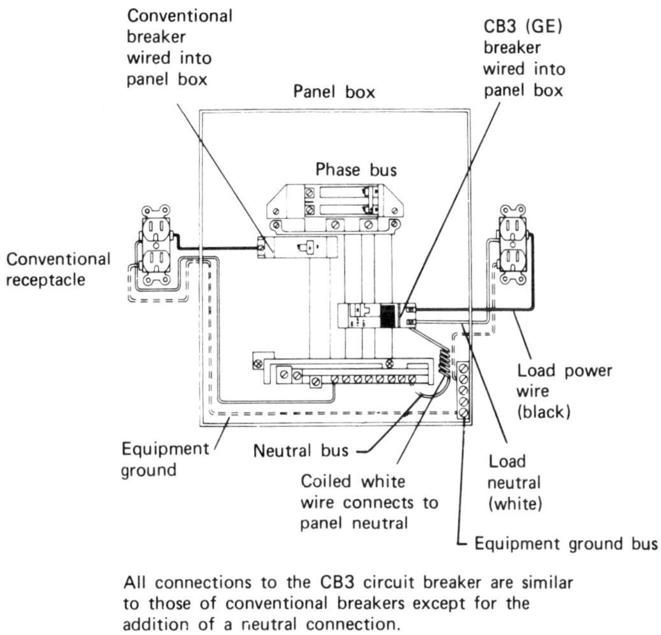

Figure 8.14 Ground-fault circuit breaker.

8-10 OUTLETS IN CLOSETS

The Code does not require a light in the clothes closet. However, should one be installed, NEC Section 410-8 has specific rules about the placement and type of lights installed.

Fixtures in closets shall be installed on the wall above the door or on the ceiling over the area unobstructed to the floor, so that clearance of stored combustibles is at least 18 inches. A flush recessed solid-lens fixture is considered outside the closet area. A pendant light (a light socket on the end of a drop cord) is not permitted. 410-8(b)states, "Pendants shall not be installed."

(a)

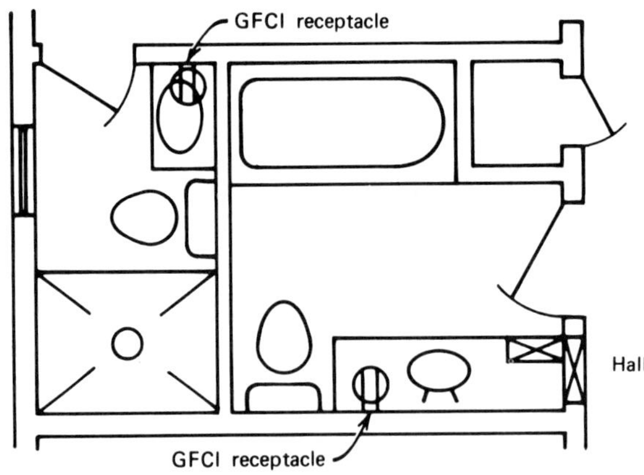

(b)

Figure 8.15 (*a*) Ground-fault receptacle located in each bathroom. Notice that the receptacle is duplex, and that it provides indication, reset, and test features. Switch could be used to control light over mirror. (*b*) GFCI wiring diagram. (Courtesy of Pass and Seymour, Inc.)

8-11 LIVING ROOM CIRCUIT (FIGURE 8.16)

Be generous with wall receptacle outlets in the living room. Install enough receptacles so that lamps can be placed anywhere without using an extension cord.

The living room is an excellent area for controlled lighting. The duplex receptacle is really two receptacles in a single device, so that two different things can be plugged in at the same time. Plan and wire the receptacles so they are split; that is, half of the receptacle is hot all of the time; the other half is switched with a dimmer switch (Figure 8.17) located at a convenient location on the wall. Three- or four-way switches and dimmer should be installed for convenience.

In a three-wire installation, extreme care must be exercised when connecting the three wires in the device. Red wire will be the switch leg to each switched receptacle, black wire to hot receptacle, and white wire to common or neutral (Figure 8.18).

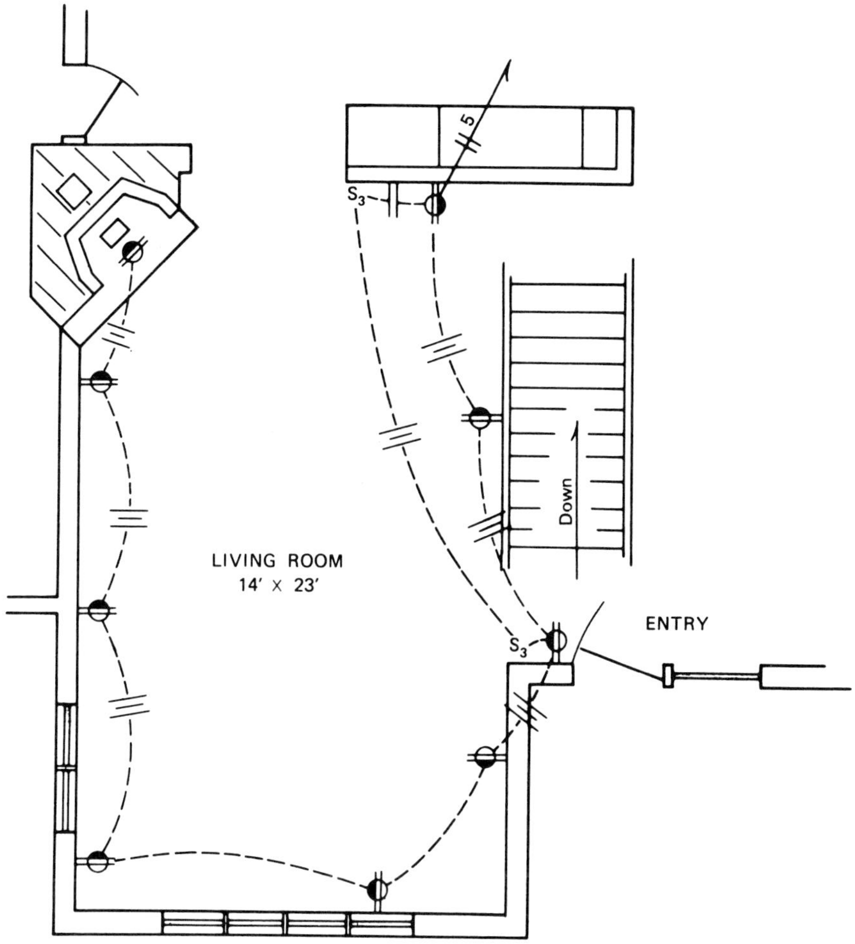

Figure 8.16 Typical living room circuitry, showing split receptacle, half hot, other half switched.

Figure 8.17 Rotary dimmer switch mounts in any standard electrical switch box and can replace any standard flush mounted switch; standard incandescent light bulbs are turned on and off by pressing the control knob, level of light is obtained by rotating the control knob. (Courtesy of Ideal Industries, Inc.)

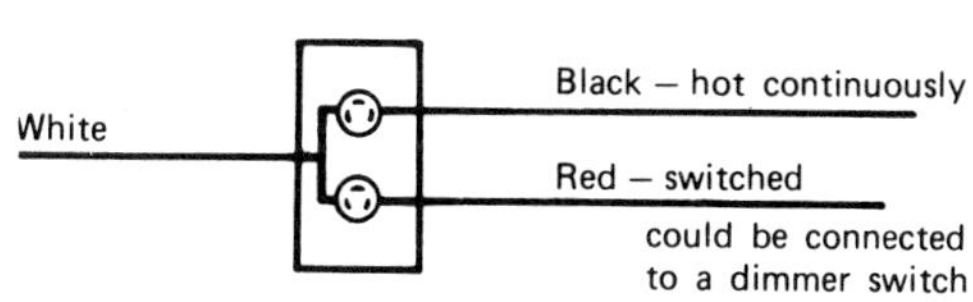

Figure 8.18 Method of connecting convenience receptacle in living room. Duplex receptacles are designed so that the bridge linking the two terminal points, one on each side, can be removed. This electrically separates the two portions of the receptacle.

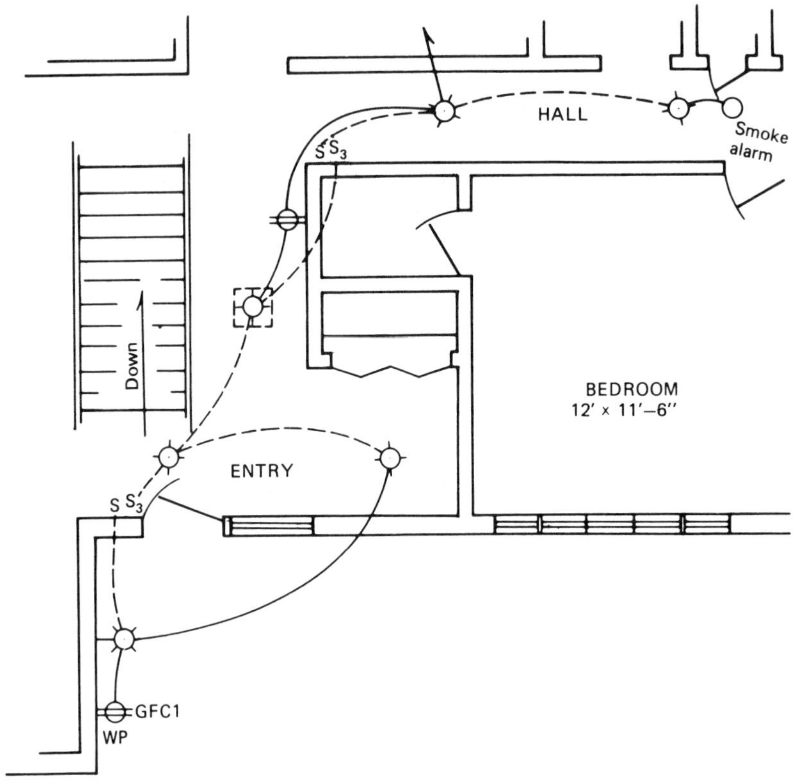

Figure 8.19 Typical circuitry for entry hall, and outside GFCI receptacle.

8-12 HALL, FRONT ENTRY, AND OUTSIDE

The hall, entry hall, and outside lights are grouped into a circuit. Ceiling lights should be installed in all halls. Should the halls be rather long, two or more ceiling lights or recessed fixtures should be installed and controlled by three-way switches at each end. Figure 8.19 shows that at the front door two 75-watt high-boy recessed fixtures are installed with a 150-watt recessed fixture near the center of the hall, controlled by three-way switches. A receptacle is installed for convenience in the hall for a vacuum cleaner outlet. The outside light could be a decorative bracket fixture or a ceiling fixture installed in the overhang. Outside is a ground-fault receptacle with a watertight cover. See Section 8-8 for ground-fault circuit interrupter protection.

8-13 KITCHEN AND DINING AREA LIGHTING (FIGURE 8.20)

For general lighting in the kitchen, there should be ample ceiling fixture outlets controlled by wall toggle switches located at each entrance to the kitchen. The prime requirement for good lighting design is to eliminate shadows. A light over the sink is essential, for without this light, a person would be standing in a shadow at this location. Control this light with a wall switch located near the sink. Perhaps you could gang two switches together here, one for the light, the other for the garbage disposal under the sink. (Garbage disposal will be discussed in Chapter 9, Section 9-7.)

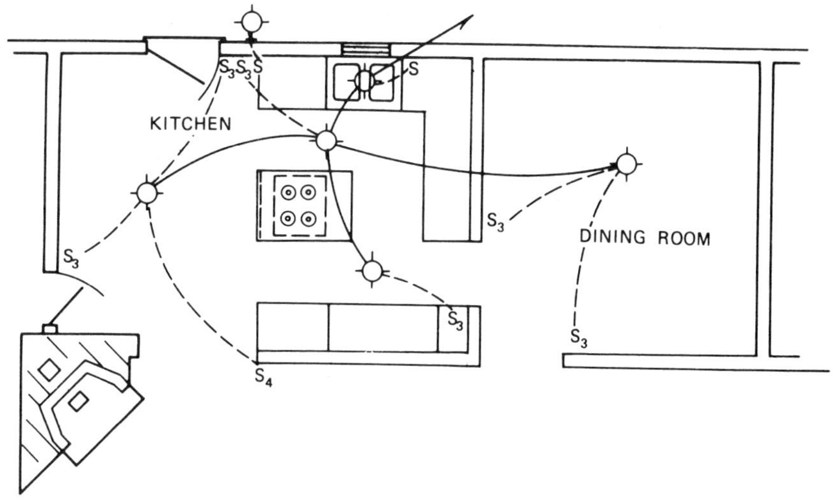

Figure 8.20 Typical lighting/switch circuitry in kitchen and dining room.

Figure 8.21 This outlet supports the clock. The receptacle is in a "well" so that the cord and plug of the clock are completely concealed. (Courtesy of Pass and Seymour, Inc.)

A clock-hanger type receptacle, Figure 8.21, could be installed on either a lighting circuit or on the 20-ampere small appliance circuit.

Locate and install a ceiling outlet in the eating area. Care should be taken in locating the outlet over the breakfast table, which is also controlled by three-way and four-way wall switches.

When locating the ceiling fixture outlet in the dining area, try to visualize a table located in the room. Center the outlet where you think the table would be and not at the center of the room. Control this light by three-way wall switches or control lighting by installing dimmer switches.

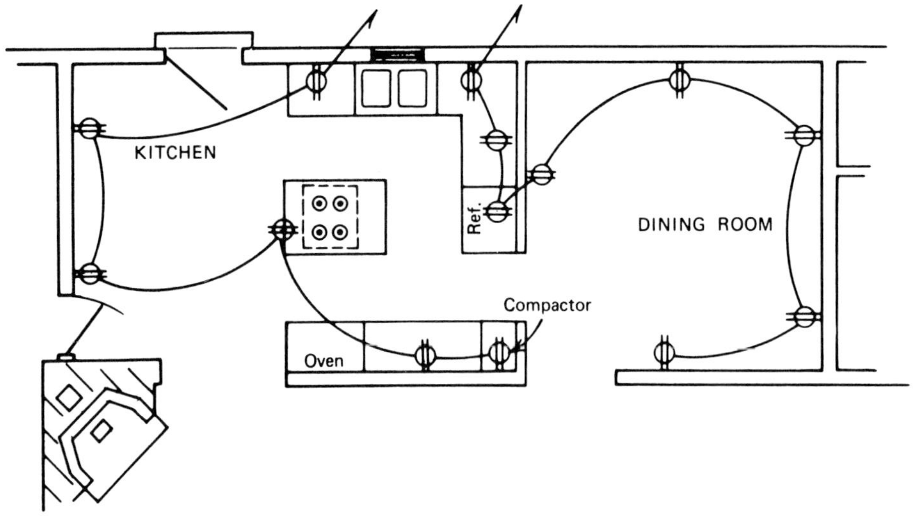

Figure 8.22 Typical small appliance circuitry, with two circuits in the kitchen.

8-14 KITCHEN SMALL APPLIANCE CIRCUITS (FIGURE 8.22)

The kitchen needs lots of receptacles. The NEC recognizes this need and requires two 20-ampere small appliance circuits to be installed in this area. See Section 8-3.

8-15 GARAGE (FIGURE 8.23)

There are some changes in the Code for receptacles in the garage. NEC 210-8 states: "all 120-volt 15–20 ampere receptacles installed in the garage shall have ground-fault circuit interrupter protection." (See exceptions.)

For convenience, garage lights are controlled by three-way and four-way wall switches located at the entrance doors.

The outside bracket lights are controlled by three-way switches, one switch located inside the breakfast room and the other near the garage door.

Installing three fixture outlets makes for good lighting and eliminates shadows between the automobiles.

The outside receptacle is on GFCI and has a weatherproof cover.

8-16 BASEMENT

The requirements for the basement vary. It depends on how elaborate the finish in the basement is to be. However, we are to have one side of the basement area as a finished family room. This will be wired like the living room upstairs and will provide for plenty of receptacles. Because the ceilings are lower in the basement, it is recommended that recessed fixtures be installed, controlled by wall switches or dimmer switches for controlled lighting.

The rest of the basement will be wired as ceiling lights are required; each storage area requires a light.

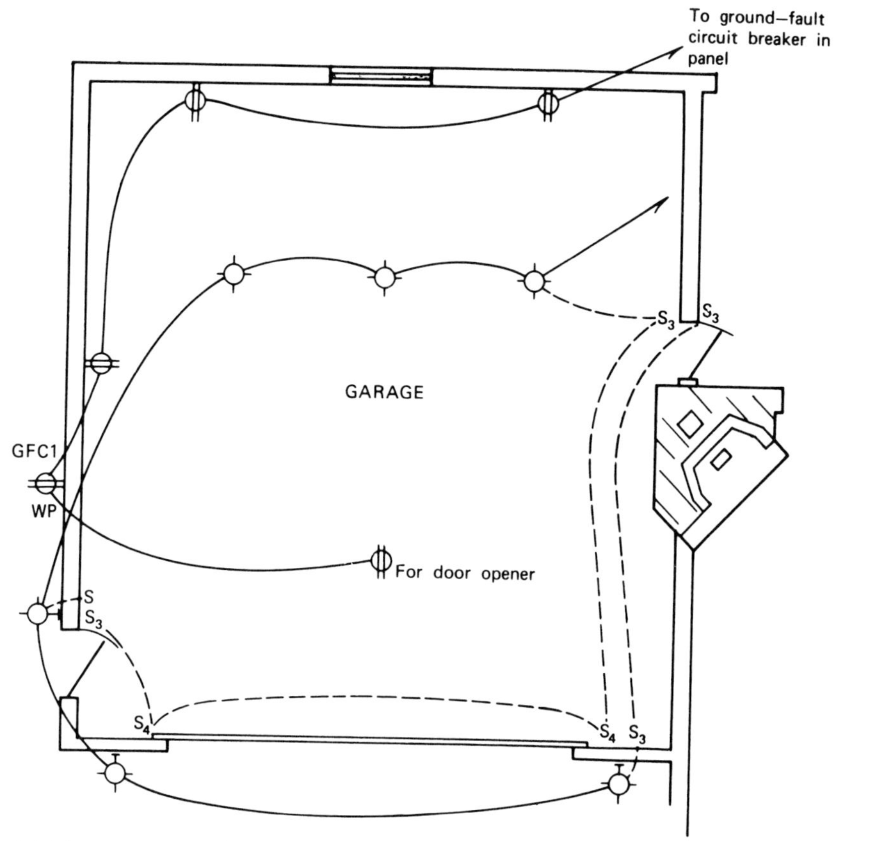

Figure 8.23 Garage circuitry showing all garage receptacles on a ground-fault breaker in panel. There is an outside GFCI with water proof cover. Light circuit is controlled by three-way and four-way switching. The 1981 NEC does not require that the garage door automatic door opener be on GFCI.

The appliances—washer, dryer, hot water heater, water conditioner, and so on—will be installed here. See Chapter 9.

The washer requires a circuit.
The clothes dryer requires its own circuit.
The water heater is on a separate circuit.
If a furnace is installed, some inspectors require a separate circuit.

A light to illuminate the stairs should be controlled by a three-way switch at the top and bottom of the stairs.

SUMMARY

1. Copper is usually the material used in residential circuit conductors.
2. Conductors used in electrical installations are graded for size according to the American Wire Gauge Standard.
3. The purpose of a conductor is to carry current.
4. Ampacity is the term used to express the current-carrying capacity of a wire in amperes.

5. Nonmetallic-sheathed cable is used more often in residential wiring than any other wiring method.
6. The NEC requires nonmetallic-sheathed cable to be supported.
7. The flow of electrical current in lighting circuits must be controlled by a switch.
8. The most frequently used switch is the toggle switch.
9. The planning of electrical circuits is usually left up to the judgment of the electrician.
10. Care must be taken when installing closet lights.
11. A ground-fault circuit interrupter is a device to protect people against shock.
12. Dimmer switches are often installed to control lighting.
13. Be generous with receptacles in the living room, kitchen, and dining area.
14. Room lights are usually controlled by wall switches near the entrance of the room.

PROBLEMS

8-1 Draw a one-line schematic of a ceiling fixture controlled by two 3-way switches; hot wire is fed into the fixture outlet.

8-2 Plan and draw the layout of outlets for the small appliance circuit for the floor plan shown in Figure 8.24.

8-3 Draw in the outlets for the dwelling shown in Figure 8.25. Show lighting and receptacles, oven, range, dishwasher, washer, dryer, and water heater. Panel is located behind the door in hall across from water heater.

8-4 Explain briefly how a ground-fault circuit interrupter works.

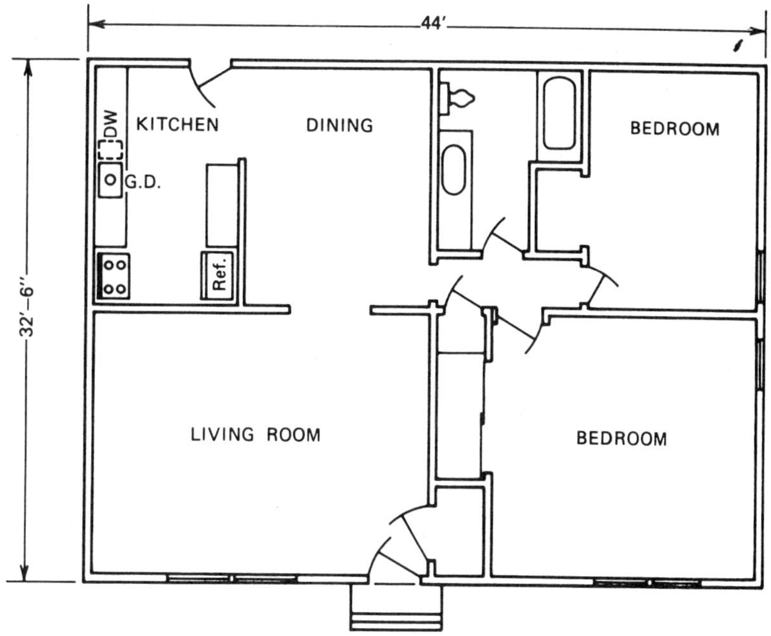

Figure 8.24 Floor plan for Problem 8-2.

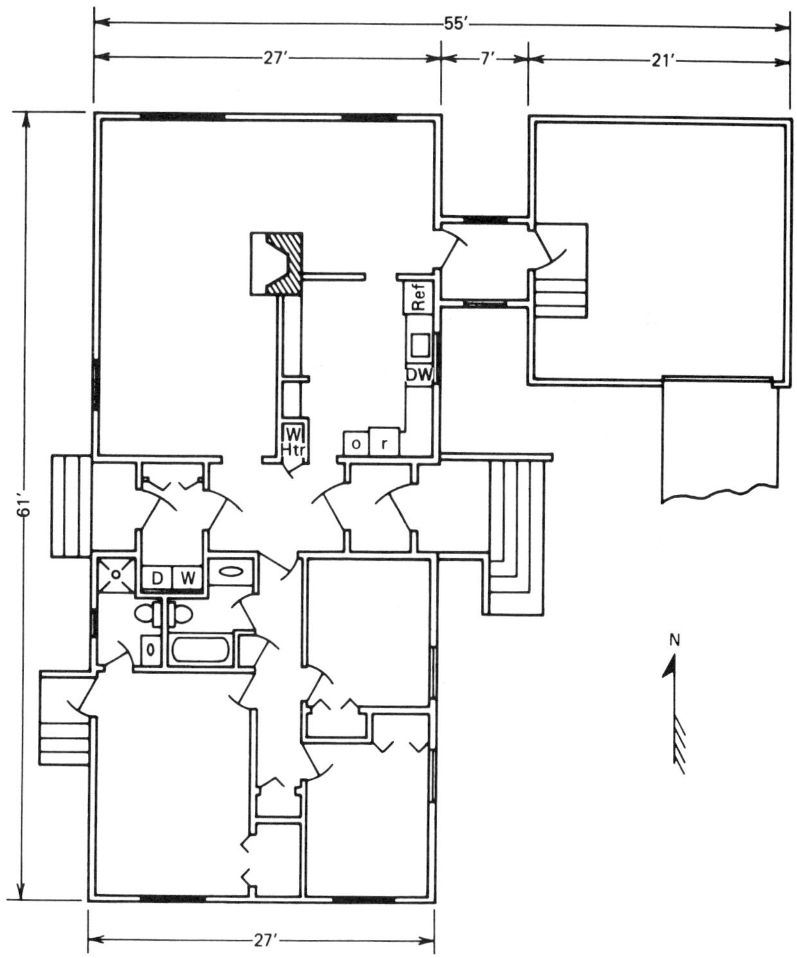

Figure 8.25 Floor plan for Problem 8-3.

8-5 Draw a sketch of a lighting fixture being controlled by two 3-way switches and one 4-way switch; the hot wire is fed to one of the three-way switches.

8-6 Draw a floor plan of a house and show the dimensions; from the dimensions calculate the minimum number of light circuits needed.

8-7 The NEC states in Article 220-2 (b), "A unit lighting load for dwelling units shall not be less than ________ watts per sq ft."

8-8 According to the NEC, Chapter 9, one 15-ampere, 115-volt branch circuit would be required for every ________ sq ft of occupied floor area.

8-9 The outlets in the kitchen must be supplied by at least ________ 20-ampere circuit(s).

8-10 According to the NEC, where are ground-fault circuit interrupters required in a house? ________________________________

____

____

Give Code section: ________________________________

chapter 9

SPECIAL-PURPOSE OUTLETS

Instructional Objectives

1. To learn how to interpret symbols on the electrical floor plan for ranges, counter cooking units, ovens, and other major appliances.
2. To develop the ability to determine any special installation requirements for special-purpose outlets.
3. To understand the grounding of all appliances according to the Code regardless of the wiring method used.
4. To learn how to select the proper conductor size for wiring special-purpose outlets.
5. To understand the need for selecting the proper overcurrent device based on the amperage rating of the device.
6. To develop the ability to calculate the dryer circuit.
7. To understand the National Electrical Code requirements for all special-purpose outlets.
8. To learn how to use service entrance cable to connect large appliances.

Self-Evaluation Questions

Test your prior knowledge of the information in this chapter by answering the following questions. Watch for the answers as you read the chapter. Your final evaluation of whether you understand the material is measured by your ability to answer these questions. When you have completed the chapter, return to this section and answer the questions again.

1. What determines the voltage of an appliance?
2. Would you calculate the load of a range the same as you would a counter-mounted cooking top and wall oven?
3. Are most hot water heaters considered a continuous duty load?
4. How many outlets are permitted on the laundry circuit?
5. Are clothes dryers basically a 240-volt appliance?

6. How are clothes dryers grounded?
7. Is it possible to use a split receptacle for the garbage disposer and dishwasher?
8. Do the receptacles in the garage have to be a ground-fault circuit interrupter?
9. Is it permissible to install an electric wall heater in the bathroom?
10. Will the NEC permit a cord on the garbage disposal?

9-1 TYPES OF BRANCH CIRCUITS

The NEC recognizes two types of branch circuits. The first is the ordinary circuit serving two or more outlets consisting of permanently connected appliances, light fixtures, or receptacles for portable loads such as mixers, lamps, vacuum cleaners, and other small hand appliances, as already discussed in Chapter 8. The second type is a circuit serving a single current-consuming appliance or similar load, such as a counter-mounted cooking unit, oven, range, water heater, dishwasher, garbage disposer, or trash compactor using a single circuit.

Some local codes require the furnace to be on a separate circuit regardless of the horsepower rating of the blower motor.

Circuits for appliances may be either 120 or 240 volts, depending on the nameplate of the appliance.

The NEC is not clear as to when an appliance requires a branch circuit. In general, you will not only meet the requirements but will also provide the homeowner with a better installation if you provide a separate circuit for the following appliances.

9-2 COUNTER-MOUNTED COOKING UNIT

The electric range in many areas has been replaced by the counter-mounted cooking top and built-in wall oven or by the microwave oven and radar range. The cooking top and oven are like separate units of a complete range divided for the sake of kitchen convenience, ease of operation, or kitchen appearance.

The two units are connected to individual branch circuits of lower rating than a single range unit. Normally, the range circuit terminates in a heavy-duty flush- or surface-mounted 50-ampere receptacle. The range should be equipped with a three-wire cord and plug (Figure 9.1).

A range is actually a multiunit load assembled as a single appliance. The watts rating of the range is the total of the oven load plus the total load of all of the heating elements turned to their highest setting. However, it isn't likely that all of the elements will be used at the same time. Therefore, the NEC Table 220-19 allows a demand factor for electric ranges, wall-mounted ovens, counter-mounted cooking units, and other cooking appliances over 1¾ kW rating.

A load of only 8 kW (8000 watts) may be used for any range rated at not more than 12 kW (12,000). If the range is rated at over 12kW, start with 8

Figure 9.1 50-ampere range receptacle, surface mounted. (Courtesy of Pass and Seymour, Inc.)

kW. For each kilowatt or fraction thereof above the 12 kW, add 400 watts (or 5 percent of 8000 watts).

EXAMPLE 1

A range might have a rating of 16 kW, which exceeds 12 kW by 4 kW. Add 400 watts for each of the extra kilowatts, or 1600 watts together.

8000 watts + 1600 watts = 9600 watts
9600 watts ÷ 240 volts = 40 amperes

The range should be wired for 40-ampere rating.

In other cases, the nameplate rating of each item, oven, and cook top must be used separately, and no demand factor may be used. It is much simpler and better to install a separate branch circuit for each cooking top and each wall oven. Rough-in according to the size of load of each unit, leaving enough wire stubbed out of the wall to make connections direct to the top and oven when installed at the finish.

Microwave ovens usually draw a load of 1500 watts. A separate circuit should be roughed-in when the owner intends to use this appliance.

9-3 ROOM AIR CONDITIONERS

An air conditioner is considered a room air conditioner if it is installed in the room it cools (in a window or in an opening through a wall), if it is single-phase, and if it operates at not over 250 volts.

Air conditioners are evaluated in two ways:

(a) "Cooling capacity" is the amount of heat (measured in BTU's) that an air conditioning unit can remove from the air in an hour. One BTU is the amount of heat needed to raise the temperature of 1 pound of water 1°F. This is about the amount of heat generated by a wooden match burned completely to ashes. A 12,000-BTU unit will therefore remove 12,000 BTU's of heat from an area every hour.

The capacity required for a home depends on many things, including area of the country, size of the space to be cooled, and the number of occupants in the home. However, as a general rule, 18 BTU's removed per hour will cool about 1 sq ft in a normal home.

(b) "Energy efficiency ratio" (EER) is the number of BTU's of heat that 1 watt of electrical energy will remove from the air in one hour. The EER is determined by dividing the capacity in BTU's per hour of a unit by its power required in watts. This information is usually found on the nameplate of the unit. The EER will be a number ranging from 4.7 to 12.2.

EXAMPLE 2

A 10,000-BTU unit requiring 2000 watts has an EER of 5. Purchase units with high EER's; 6 to 7 is fair, but you can do better; 8 to 9 is good; and 10 or over is great.

Installation Requirements

These are outlined in NEC Section 440. The air conditioner may be connected by a cord and plug. A unit switch and overload protection are built into the unit. The disconnecting means may be the plug on the cord or the manual control on the unit if it is readily accessible and not more than 6 feet from the floor, or a manually controlled switch installed where readily accessible and in sight of the unit. If the unit is installed on a circuit supplying no other load, the ampere rating on the nameplate of the unit must not exceed 80 percent of the circuit rating. Should it be installed on a circuit also supplying lighting or other loads, it may not exceed 50 percent of the circuit rating. Cords must not be longer than 10 feet if the unit operates at 120 volts and not over 6 feet if it operates at 240 volts.

9-4 HOT WATER HEATER

The hot water heater is the second largest consumer of energy in the home. Therefore, much thought must be given to this appliance.

The consumption of hot water is linked to the number of people in the family, their ages, and individual habits. Oversizing wastes energy, since you are maintaining a supply of unnecessary hot water. For the average four-member household, an 80-gallon low-power heater is recommended.

Locate the heater close to the point where it is needed. Some contractors suggest several smaller heaters in larger homes, with each placed near a location needing hot water.

The power consumed by a water heater varies greatly, from 700 watts to 5000 watts. A No. 12 wire with an ampacity of 20 amperes will carry 4800 watts and is suitable for most heaters. However, some inspectors might require a No. 10 AWG.

If your service equipment consists of circuit breakers, provide a 20-ampere two-pole breaker for the heater; provide a 30-ampere breaker if you use No. 10 wire.

NEC 422-14(b) states that "all fixed water heaters having a capacity of 120 gallons or less shall be considered a continuous duty load."

NEC 422-5(a) Exception 2 states that for a continuous loaded appliance, "the branch circuit rating shall not be less than 125 percent of the marked rating" of the appliance.

Figure 9.2 30-ampere dryer receptacle, surface mounted. (Courtesy of Pass and Seymour, Inc.)

9-5 CLOTHES DRYERS

Dryers are basically 240-volt, single-phase appliances, although most have a 120-volt motor-operated drum that tumbles the clothes as the heat evaporates the moisture.

A separate circuit must be provided for the electric clothes dryer because this appliance demands a comparatively large amount of power.

NEC Section 250-45(c) states "clothes dryers must be grounded," but may be grounded to the neutral of the circuit wire if the grounded conductor is No. 10 AWG or larger.

NEC 220-18 requires a circuit with a minimum of 5000 watts capacity or a capacity based on the rating of the dryer, whichever is higher. Dryers are usually installed by using a cord and plug and a receptacle, which must be 30-amperes, three-wire. The 120/240 volt type is shown in Figure 9.2.

Use any wiring method you choose. The NEC permits you to use service-entrance cable with a bare neutral as in the case of the electric range. However, it must start from the branch circuit overcurrent protection in the service equipment. The plug and receptacle will serve as the disconnecting means.

The nameplate on the dryer specifies the minimum circuit size and the maximum rating of the circuit overcurrent device.

The washer requires a separate laundry circuit [NEC 220-16(b)].

9-6 DISHWASHER

A built-in dishwasher is a motor-driven appliance and should be connected to a separate 20-ampere circuit. Usually the dishwasher is located under the drainboard near the sink. Should this be the location, the NEC now states that a built-in dishwasher and trash compactor intended for use in a dwelling, if it is provided with a three-conductor cord terminated with a grounding-

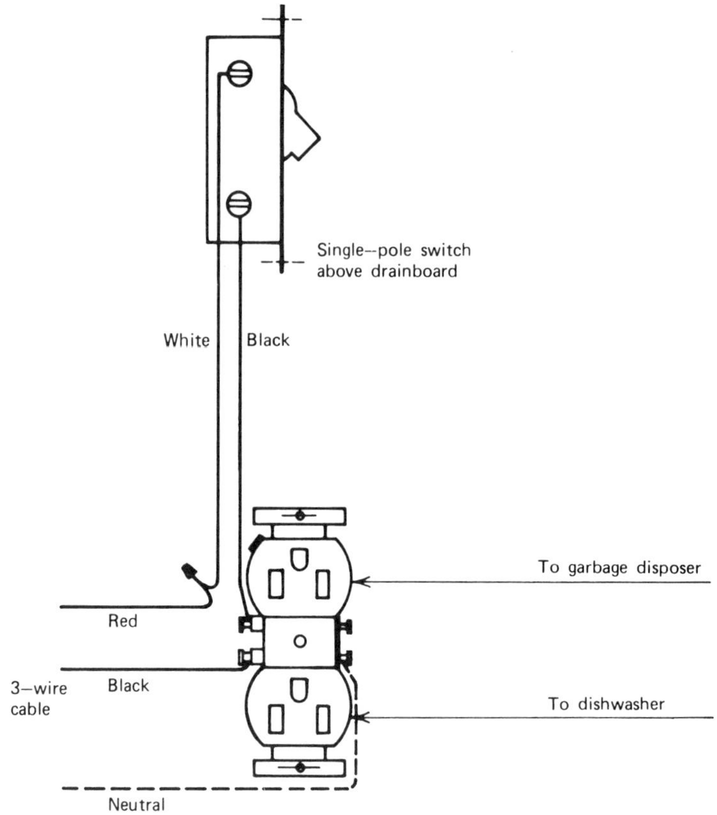

Figure 9.3 Split wiring of duplex receptacle. Upper half is switch controlled, lower half is hot all the time.

type attachment plug, shall be permitted if the cord is at least 3 feet to 4 feet in length and if the receptacle is located to avoid physical damage to the cord and located in an accessible area in the space occupied by the appliance or adjacent thereto. A split receptacle could be used for both the garbage disposer and the dishwasher when located under or near the sink (Figure 9.3). If the dishwasher is installed more than 4 feet from the disposer, another duplex receptacle could be roughed-in as the above directions state.

9-7 GARBAGE DISPOSER

Many different types of garbage disposal units are available. However, most units are furnished with a junction box or wiring space for wiring conductors to terminate.

Since the garbage disposer is a motor-operated appliance, the NEC requires running overcurrent protection not to exceed 125 percent of the full-load current rating of the motor.

When running overcurrent protection that is not an integral part of the disposer unit, the electrician must install separate protection; however, most disposals manufactured today have a built-in thermal protector to comply with the NEC requirements.

NEC Section 422-8(d) now permits a three-conductor cord not less than 18 inches and not over 36 inches with a grounding-type attachment plug to be used, providing the receptacle is accessible and located to avoid physical damage to the flexible cord. All of the above conditions must be met. A new exception to the rule is that if the disposal is protected by a system of double insulation or its equivalent and is marked, it shall not be required to be grounded. Usually a 20-ampere circuit will be sufficient for a 120-volt disposer.

9-8 BATHROOM HEATER

Because of lack of wall space, ceiling heaters are often installed in bathrooms. Three types of ceiling heaters are available; one with a resistance element, one with a resistance heater with a motor driven fan circulator, and one with infrared lamps. The resistance type heater is sometimes combined with a fluorescent lamp or with an incandescent lamp to provide light and heat at the same time (called a Heat-A-Vent light unit). See Figure 9.4.

The infrared lamp type, which also performs both functions, could be installed with one, two, or three lamps (Figure 9.4).

All of the above could be controlled by a timer located on the wall near the light switch. By turning the timer, the heater would run the required length of time, then automatically shut itself off.

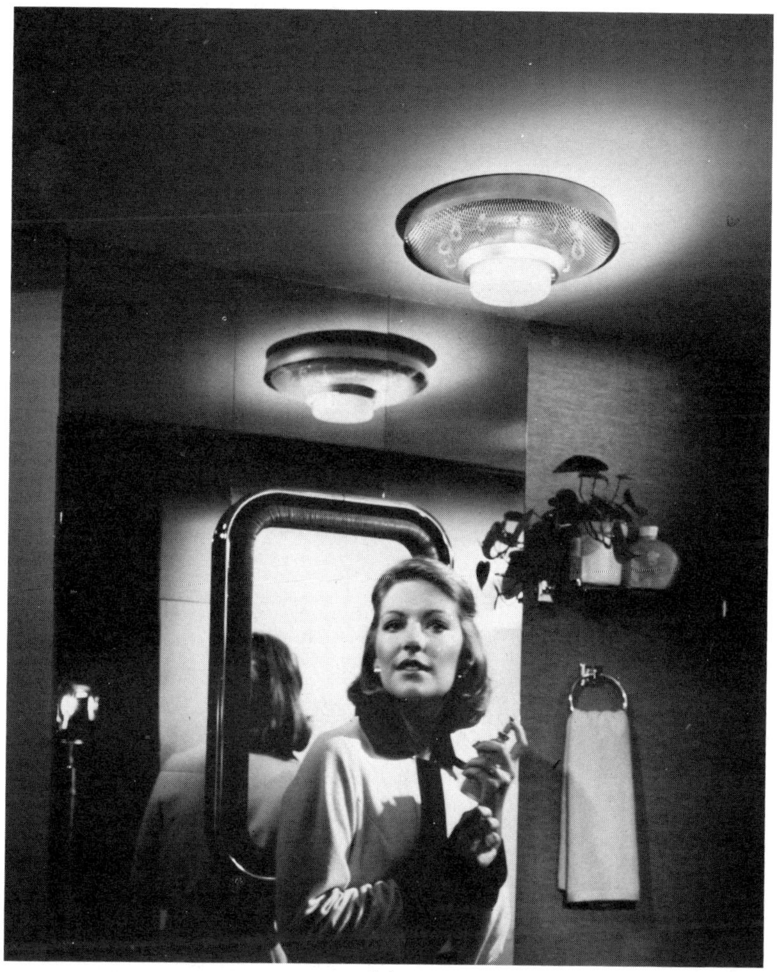

(*a*)

Figure 9.4 (*a*) Bathroom heat-a-ventlite, provides heat and ventilation together; ventilation and light together; ventilation only or light only. (*b*) Two-lamp infrared. (*c*) Three-lamp infrared ceiling heater. (Courtesy of Nutone Division, Scovill Manufacturing Company)

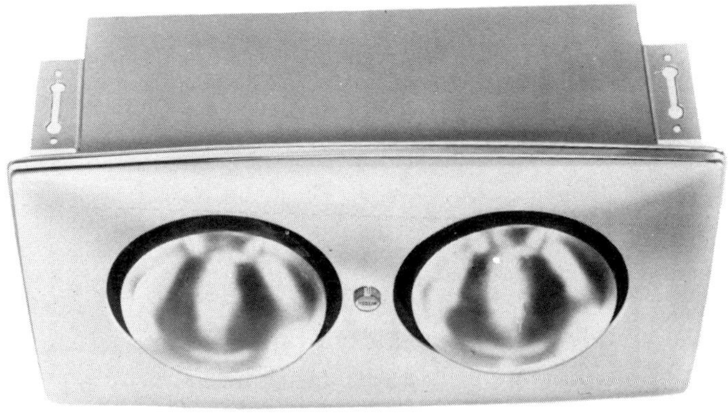

(b)

(c)

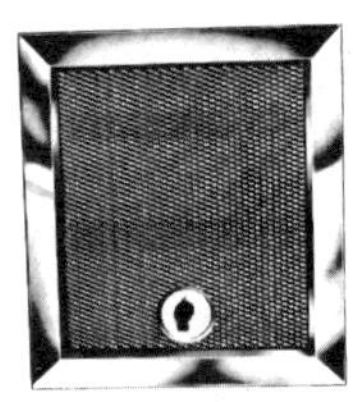

Figure 9.5 Bathroom wall heater. (Courtesy of Martin Industries)

If wall space allows, a resistance 1250- or 1500-watt wall heater could be installed (Figure 9.5).

Here again the heater name plate capacity rating dictates wire sizes. Usually an independent circuit should be provided.

The bathroom ground-fault receptacle was discussed in Chapter 8.

Figure 9.6 Automatic garage door opener. Power unit completely enclosed in steel housing to guard mechanism against moisture and dirt. Safety features include automatic reverse cutoff, time/delay light, overload clutch, automatic stop, manual release ring, and security locks. (Courtesy of Nutone Division, Scovill Manufacturing Company)

9-9 GARAGE DOOR OPENER OUTLET (FIGURE 9.6)

Automatic garage doors are designed for convenience, safety, and security. Several model variations are manufactured to accommodate different size and type doors and door hardware. An outlet must be installed overhead, centered on the door, usually 6 to 8 feet from the door opening.

9-10 ATTIC VENTILATOR FAN (FIGURE 9.7)

Intense summer heat often builds up in attics to 140°F or more. It's like a hot blanket over the entire house. An attic ventilator can reduce this attic heat to a temperature approaching that outside, permitting your air conditioning to work more efficiently, and in some homes, eliminate the need for air conditioning.

A minimum of one square foot of inlet area per every 300 cubic feet per minute (CFM) is required for proper fan operation. Inlet area should be located as far from the fan location as possible. More efficiency is achieved when air-intakes are installed in eaves.

(a)

Figure 9.7 Attic ventilation fan. (a) Shown installed in pitched room between 14½-inch rafters, can be used with standard on-off switch for manual control or with automatic thermostat. (Courtesy of Nutone Division, Scovill Manufacturing Company)

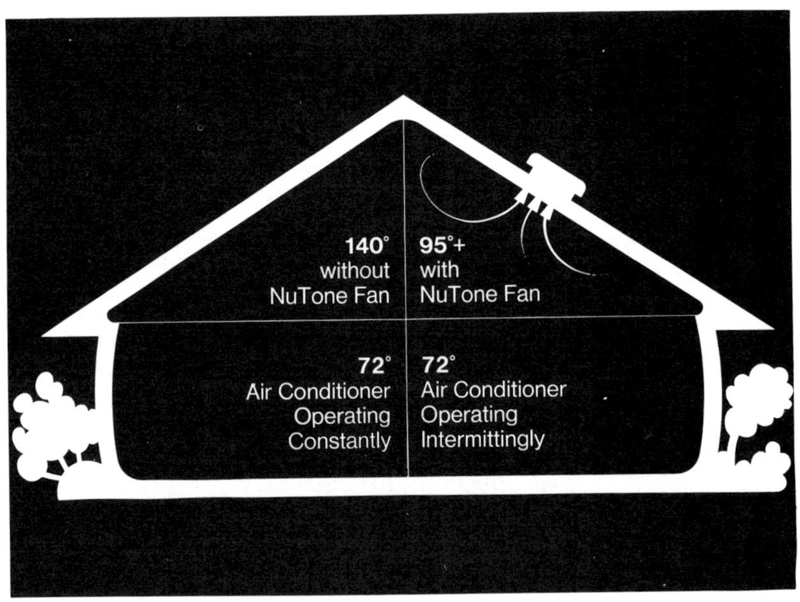

(b)

Figure 9.7 Attic ventilation fan. (b) Shows attic temperatures with and without use of attic ventilation fan.

How to Compute Size Needed

To determine the proper size of an attic ventilator fan needed for the average home, multiply the square feet of attic area by 0.7 CFM. This will give you the minimum air delivery needed to ventilate the attic.

EXAMPLE 3

1500 sq ft attic area × 0.7 CFM = 1050 CFM.
For dark roofs, add 15 percent to the required CFM.
For exceptionally large attic areas or split-level homes, use two or more ventilators to provide the needed air delivery.

SUMMARY

1. Circuits for appliances may be either 120 or 240 volts.
2. The NEC allows a demand factor for electric ranges and other cooking appliances over 1¾ kW.
3. A separate circuit should be installed for microwave ovens.
4. The power consumed by a water heater varies from 700 watts to 5000 watts.
5. The NEC applies to the installation of most special-purpose outlets.
6. A separate circuit must be provided for a clothes dryer.
7. The dishwasher should be connected to a separate 20-ampere circuit.
8. The garbage disposer requires running overcurrent protection not to exceed 125 percent of the full-load current rating of the motor.
9. Infrared lamps could be installed for both light and heat in the bathroom.
10. An attic ventilator fan can reduce attic heat temperature, permitting the air conditioner to work more efficiently.

PROBLEMS

9-1 A 6-kW counter-mounted cooking unit and a 4-kW wall-mounted oven are to be installed in the kitchen of a dwelling. Calculate the maximum demand according to Table 220-19 of the NEC.

9-2 A wall-mounted oven is rated at 7.5 kW. How many watts is this equal to?

9-3 The NEC states that water heaters that have a capacity of 120 gallons or less shall be considered continuous-duty, and as such, the circuit must have a rating of not less than what percent of the rating of the water heater? Circle one.

a. 80 percent
b. 70 percent
c. 125 percent

9-4 A small window air conditioner draws 13 amperes. To what size circuit must it be connected?

9-5 A free-standing range is rated at 11.8 kW, 240 volts. According to Table 220-19 of the NEC, what is the maximum demand?

9-6 A clothes dryer must be ______________________________.

9-7 Clothes dryers may be grounded to the ______________ of the circuit wire if the grounded conductor is ______________________________.

9-8 A built-in dishwasher is a motor-driven appliance; it should be connected to a separate ______________ circuit.

9-9 Usually a ______-ampere circuit will be sufficient for a 120-volt garbage disposer.

9-10 The power consumed by an electric hot water heater varies greatly from ______________ watts to ______________ watts.

PART 3

SIGNAL CIRCUITS

chapter 10

SIGNAL SYSTEMS

Instructional Objectives

1. To provide the basic instructions for roughing-in telephone/television outlets.
2. To make you aware of the need for residential fire/intruder alarm systems.
3. To become familiar with the location and installation of smoke detectors.
4. To learn how to plan for and lay out a radio/intercom system.
5. To become more familiar with components for residential security alarm systems.
6. To understand the operation of a two-note/two-door door chime.
7. To learn what the National Electrical Code requirements are for low-voltage conductors.

Self-Evaluation Questions

Test your prior knowledge of the information in this chapter by answering the following questions. Watch for the answers as you read the chapter. Your final evaluation of whether you understand the material is measured by your ability to answer these questions. When you have completed the chapter, return to this section and answer the questions again.

1. What requirements cover the installation of telephone outlets?
2. What must the electrician provide in the conduit run for the telephone installer?
3. Who should be consulted before work is started on television outlets?
4. Is it necessary to nail up an outlet for all television outlets?
5. Explain how the tone is made in a two-note/two-door door chime.
6. How many solenoids are contained in a two-note chime?
7. What is a signal circuit?
8. Where are smoke detectors located?
9. What is the advantage of installing a fire/intruder alarm system?
10. What are the NEC requirements for low-voltage signal circuits?

10-1 SIGNAL EQUIPMENT

Every residence is equipped with some signal equipment. The most simply constructed house has a doorbell and usually a telephone. A more expensive, modern, custom-built home will most likely have prewired telephone jacks, amplified television antenna outlets, a radio/intercom system, and a fire/intruder alarm system.

In Chapter 1, Article 100—Definitions, the National Electrical Code defines a signaling circuit as "any electric circuit that energizes signaling equipment."

10-2 TELEPHONES

Usually the installation of the residential telephone system is done according to the requirements of the local telephone company. In some areas the telephone company will have its own construction crews to rough-in a house; that is, the company furnishes and installs outlet boxes or plaster rings, cables, phone jacks, wall plates, connecting blocks, and other materials necessary to complete the installation.

In some areas, the electrician will rough-in the telephone outlets by nailing a raised plaster ring without the outlet box directly to the 2 × 4 stud at the same height as a duplex receptacle. A fish-wire is dropped from each outlet to the basement. When the telephone company installer is ready, cables are attached to the fish-wire, which is then withdrawn from the plaster ring outlet where the final connections are made. Some difficulty is to be expected in this procedure, since insulation in the walls tends to obstruct the fish-wire.

An alternate method would be to install the fish-wire in a run of ½-inch thin wall conduit from each phone outlet location to an accessible point in the basement. The installer is then able to pull the cables through the conduit and terminate them at the proper location (Figure 10.1).

Most telephone companies require at least one phone to be permanently installed. The remaining phones may be of the portable extension type that can be plugged into any of the phone jacks furnished by the telephone company.

10-3 TELEVISION

There are times when the electrician is called upon to install the television antenna and run the cable to the outlets. However, television is a highly technical field. It is recommended that a competent television installer be consulted before any work is started on the installation.

There are a number of ways in which television outlets may be installed. In general, a single-gang plaster ring may be used at each point on the system where an outlet is located. Nail the plaster ring to a 2 × 4 stud at the same height as the wall outlets and pull enough cable in so that a connection may be made to the TV plates on the finish.

Both indoor and outdoor antennas are available for black-and-white and color sets. Antennas may be installed in the attic or on the roof.

Figure 10.2 shows TV outlet on stud by outlet box.

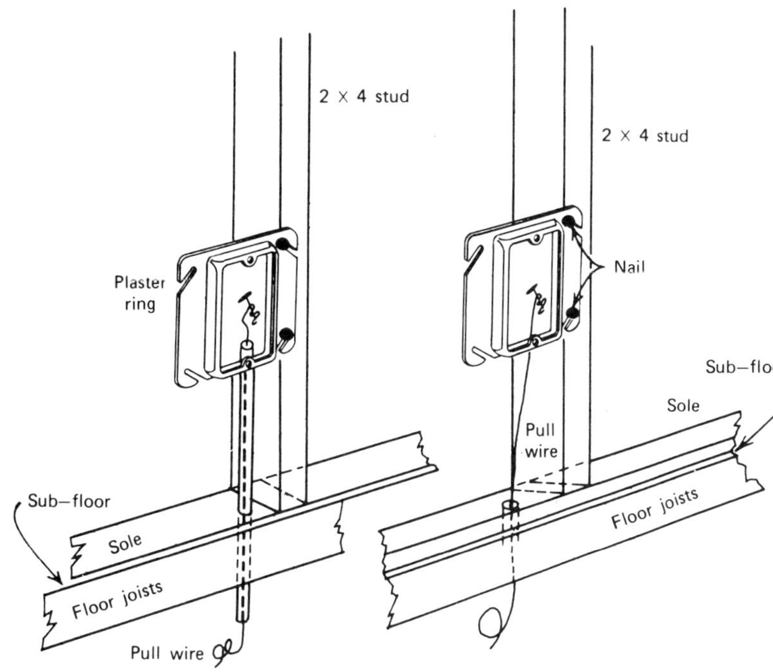

Figure 10.1 (*a*) Installation with pull wire run in ½-inch conduit for telephone installer. (*b*) Pull wires to accessible area in basement or attic.

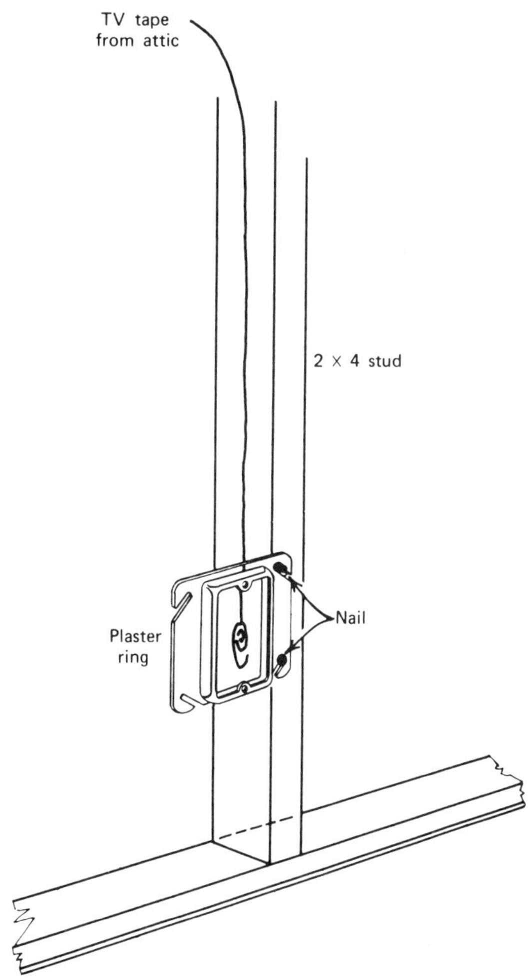

Figure 10.2 Television outlet complete with pull wire or TV tape pulled in from attic.

10-4 DOOR CHIMES

Most modern residential occupancies use chimes to announce the presence of someone at the door. Chimes are available in many different designs and styles. The simple two-note types are provided with two solenoids and two iron plungers. When one solenoid is energized, the iron plunger is drawn into the opening of the solenoid where a peg in the end of the plunger strikes one chime tone bar (usually the back door). When the solenoid is de-energized, the plunger returns by spring action and comes to rest against a soft felt pad so that it is prevented from striking the other chime tone bar. As a result, a single chime tone is heard. When the second solenoid is energized, one chime tone bar is struck. When this solenoid is de-energized, the plunger returns by spring action and strikes the second tone bar. In this manner a two-tone signal is produced (usually for front door). The plunger then comes to rest between the two tone boxes.

Other types and styles are available in up to eight-note and repeater-tone styles. In the repeater-tone model, both notes continue to sound as long as the pushbutton is depressed.

Electronic chimes that relay their chime tones through various speakers of an intercom system are also available.

Wiring the Chime Installation

Regardless of the type of style of the chime, the instructions of the manufacturer must be followed when installing the unit.

The wire used for the chime circuit is low-voltage wire, commonly called bell wire or thermostat wire. Most thermostat wire used in today's construction is insulated with a thermo-plastic compound identified as type "T." Because of the low voltage involved and the small current requirement, No. 18 AWG conductors are usually used.

Color-coded multicolor cables of two or three single wires are contained within a single protective covering, which may be fastened by insulated staples directly to the wall studs or run along the sides of floor joists in the basement.

When roughing-in for a chime, it is a good idea to install some backing between the wall studs at the required height. Should the chime be eight-note (four tube) and heavy, the backing will give support when installed.

National Electrical Code Requirements

Article 725, Sections 31–42

1. A general rule is: low-voltage wire with low-voltage insulation must not be installed in the same enclosure with higher-voltage light and power conductors.
2. Low-voltage wire must not be installed closer than two inches to open light and power conductors unless the low-voltage wires are permanently separated by some approved type of insulation.
3. Low-voltage wire may not enter an outlet box containing a higher-voltage light and power conductors unless the two voltages are separated by a metal barrier.

Figure 10.3 illustrates wiring diagrams for door chimes (Nutone).

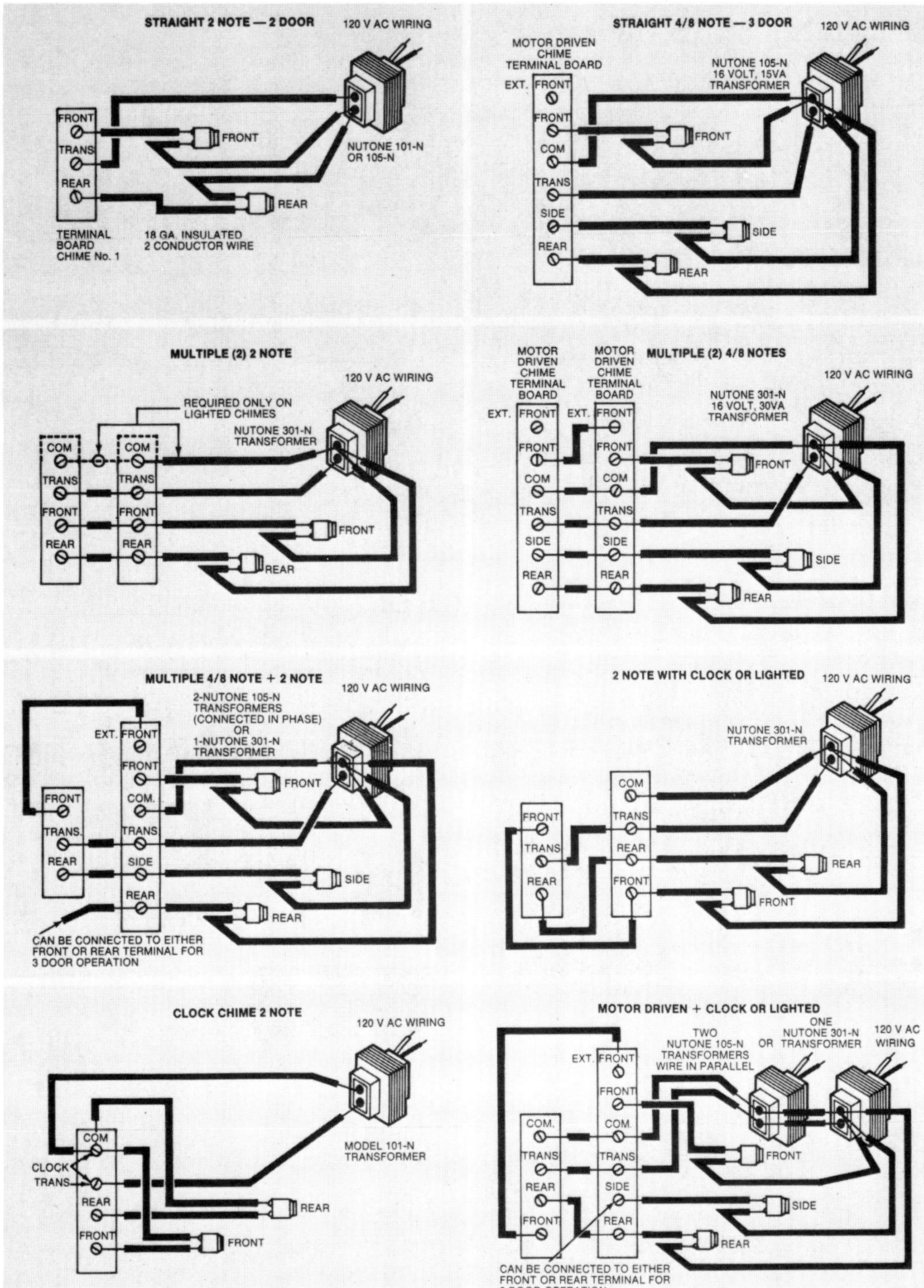

Figure 10.3 Wiring diagram for door chimes. (Courtesy of Nutone Division, Scovill Manufacturing Company)

10-5 RADIO-INTERCOM SYSTEMS

Combination units such as that in Figure 10.4 have become very popular. The system not only consists of a master station and a number of remote stations in various rooms but is also equipped with a cassette tape player/

(a) (b) (c) (d)

Figure 10.4 Radio intercom communicenter. (*a*) Radio intercom equipped with a cassette tape player/recorder as a family message center. Accommodates up to 12 remote stations. (*b*) Surface-mounted door speaker with push button control. (*c*) Eight-inch inside speaker. (*d*) Wiring diagram. (Courtesy of Nutone Division, Scovill Manufacturing Company)

recorder as a family message center. This centralized system accommodates up to 12 remote stations. Even door speakers have "hands-free" answering. Calls originating from remote stations are answered by pressing the "talk" button. Auxiliary input jacks allow addition of a foldaway record changer and eight-track tape player.

Connection of these devices by the electrician is relatively simple because the manufacturer furnishes color-coded wire, numbered terminal blocks, and detailed installation instructions.

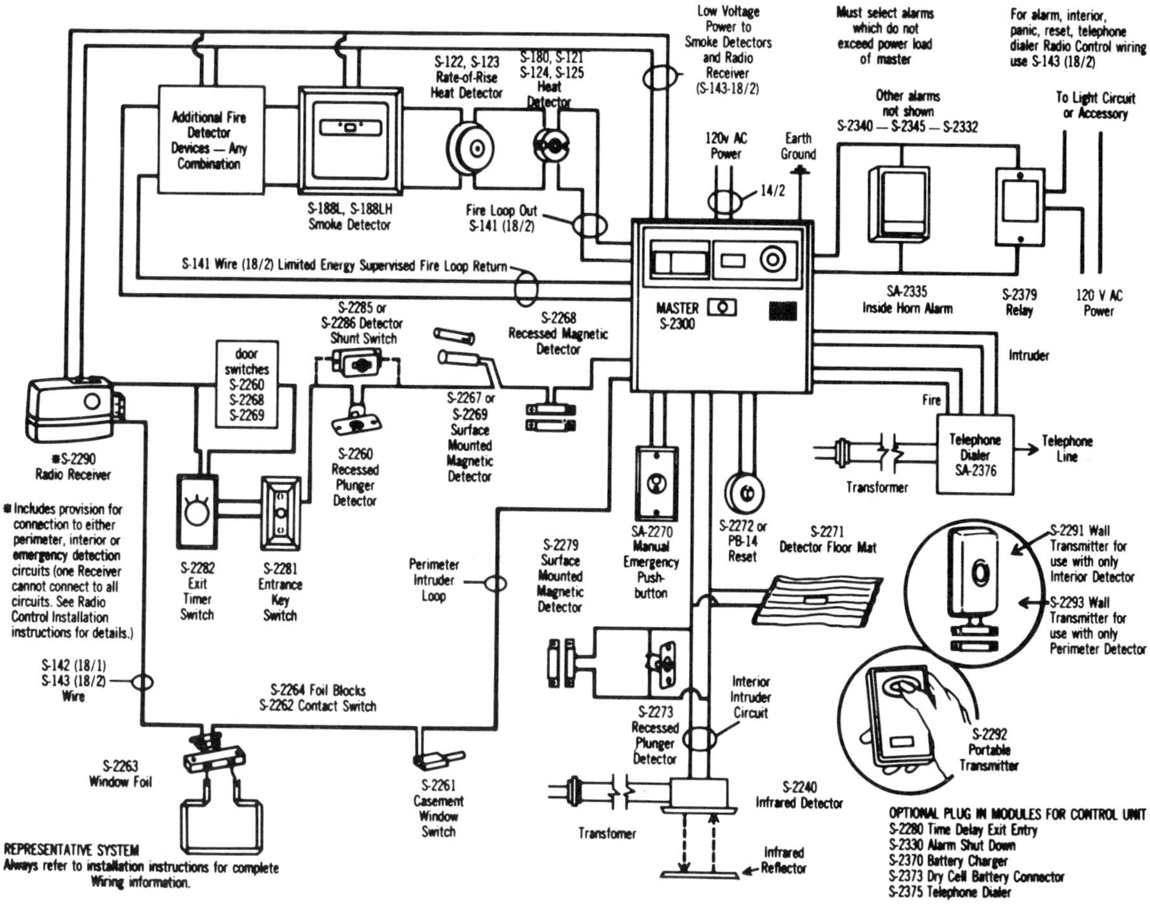

Figure 10.5 Intruder-fire alarm wiring diagram. (Courtesy of Nutone Division, Scovill Manufacturing Company)

10-6 RESIDENTIAL INTRUDER—FIRE ALARM SYSTEMS

There are several varieties of dependable electronic protection devices on the market to give early warning of fire, lethal smoke, or forced entry. They are designed to protect the home and family from fire and intruders. Most of these are sensing devices—they sense through sound and motion and announce the presence of fire or of an intruder. Some electronic systems signal an intruder's entry by alarms, by switching on lights, or by direct phone lines to the police station.

Figure 10.5 shows a fire/intruder alarm system designed for residential use. It has a master control unit, a series of intruder and fire detectors, and indoor/outdoor alarms. The control unit incorporates its own alarm horn, optional battery stand-by, and easy-to-use controls. Modular in design, the unit is solid-state with fuse protection circuitry. The control monitors heat and smoke detectors and supervises a comprehensive variety of fail-safe features, component options, and alarm accessories. It is customized to fit the size of the home and the requirements of the family.

Figure 10.6 shows the master control unit and Figure 10.7 shows fire/intruder alarm system components.

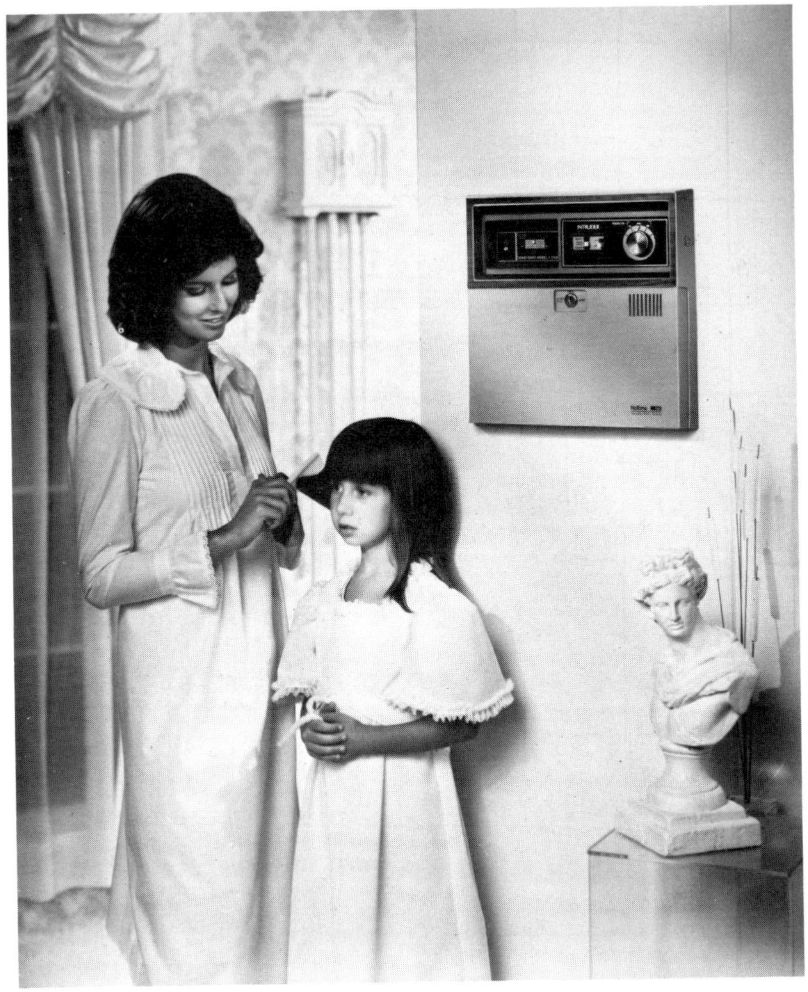

Figure 10.6 Intruder-fire alarm master control. (Courtesy of Nutone Division, Scovill Manufacturing Company)

10-7 SMOKE DETECTORS

A fire is unpredictable. A well-planned detection system should include both smoke and heat detectors in critical areas and in numbers sufficient to assure full coverage of the home.

There are two types of smoke detectors commonly used in homes: ionization detectors and photoelectric detectors.

Ionization detectors use a radioactive source to transform the air inside them into a conductor of electric current. A small current passes through this "ionized" air. When smoke particles enter the detector, they impede the flow of current. Electronic circuitry monitors the current reduction and sets off an alarm when the current gets too low.

Photoelectric detectors has a lamp that directs a light beam into a chamber. The chamber contains a light-sensitive photocell, which is normally tucked out of the way of the lamp's direct beam. But when smoke enters the chamber, the smoke particles scatter the light beam. The photocell now "sees" the light and, at a preset point, sets off an alarm.

(a)

(b)

(c)

(d)

Figure 10.7 Intruder-fire alarm system components. (*a*) Detector floor mat for easy concealment under rugs at doors, windows, or stairs. (*b*) Infrared face plate detector, protects entries, hallways, rooms, any large area; covers a span from 3 to 75 feet. (*c*) Outside electronic siren alarm. (*d*) Heat detector. It activates alarm when present temperature limit is exceeded, causing the contacts to close. (Courtesy of Nutone Division, Scovill Manufacturing Company)

For maximum protection: install a smoke detector (Figure 10.8) outside each bedroom area and at the top of stair wells; and heat detectors (Figure 10.9) in each enclosed living area, including bathrooms, closets, attic, and basement. It is recommended that at least one smoke detector be installed in the living level of multistory homes.

Install smoke alarms on the ceiling in the center of the selected area or mount on the wall with the top of the alarm not less than 6 inches and not more than 12 inches from the ceiling/wall junction.

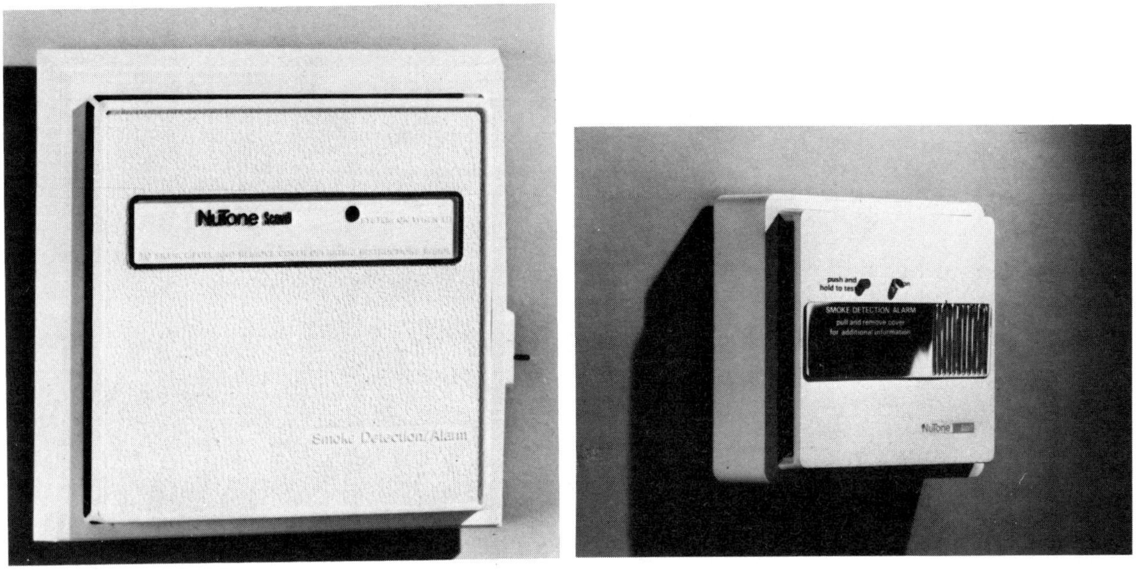

(a)

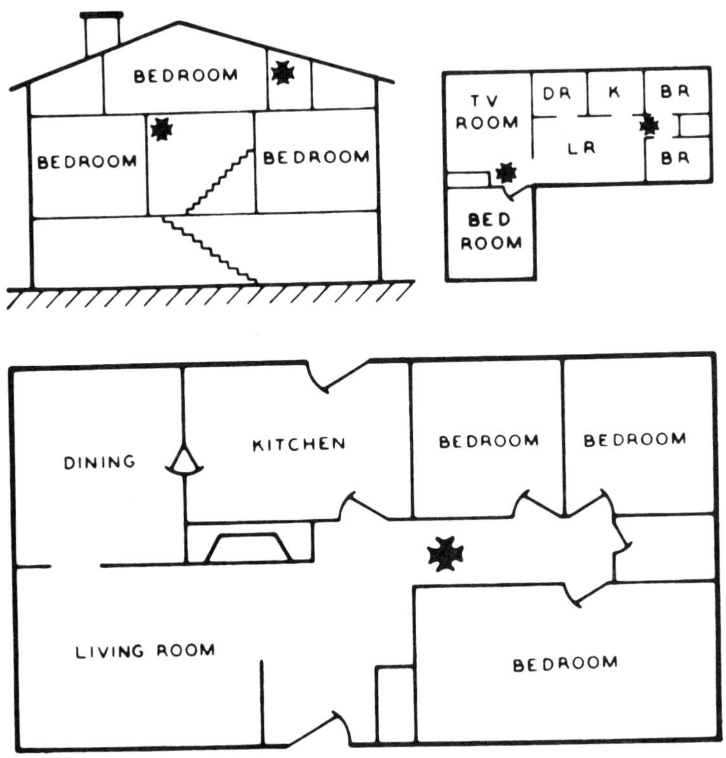

(b)

Figure 10.8 Smoke detectors. (*a*) Surface mounted on the wall or ceiling in hallways or other access to bedrooms. (Courtesy of Nutone Division, Scoville Manufacturing Company) (*b*) Shows location of smoke detectors. Detector should be located outside the sleeping area. In addition, a detector should be placed at the head of the basement stairs.

Figure 10.9 Heat detector. Rate-of-rise 135° fixed temperature. Surface mounted on ceiling or in ordinary living areas with normal room temperatures. (Courtesy of Nutone Division, Scovill Manufacturing Company)

SUMMARY

1. Telephone company requirements must be followed when installing telephone outlets.
2. When roughing-in a television outlet, a single-gang plaster ring is usually nailed to the wall stud.
3. Low-voltage door chime installations are covered by rules of the National Electrical Code.
4. A radio/intercom system consists of a master station and a number of remote stations in various rooms.
5. For convenience the radio/intercom master station is usually installed in the kitchen area.
6. Residential intruder and fire alarm installations are dependable protection devices.
7. For protection, smoke detectors should be installed in or near each bedroom.

PROBLEMS

10-1 According to the National Electrical Code's definition, how would you define a signaling circuit?

10-2 Make a list of the signaling outlets and equipment usually installed in a more expensive custom-built home.

10-3 Explain the procedure for roughing-in a telephone outlet

____

____

____

____

____

____

10-4 Draw a wiring diagram for a two-note/two-door door chime; show transformer, front-rear door buttons, and terminal board of one chime.

10-5 Draw a simple sketch of a hallway wall. Show where you would install a smoke detector for maximum protection.

10-6 Figure 10.5 shows a wiring diagram of an intruder/fire alarm system. From this diagram, locate and list at least six components to indicate an intruder setting off the alarm.

____

______________ ______________

____

____

____

____

10-7 Usually the installation of the residential telephone system is done according to the requirements of ______________ .

10-8 A general rule is: low-voltage wire with low-voltage insulation can or can not (circle one) be installed in the same enclosure with higher-voltage light and power conductors.

10-9 List the two types of smoke detectors commonly used in a house.

1. ______________________ 2. ______________________

10-10 A photoelectric detector has a lamp that directs a light beam into a chamber. This lamp is a ______________________

chapter **11**

LOW-VOLTAGE SYSTEMS

Instructional Objectives

1. To identify the various components of a low-voltage, remote control system for lighting circuits.
2. To explain the operation of various components of a low-voltage remote control system.
3. To install a low-voltage, remote control system using the proper components and conductors.
4. To understand the requirements of the National Electrical Code.
5. To explain the principles of basic remote control systems.

Self-Evaluation Questions

Test your prior knowledge of the information in this chapter by answering the following questions. Watch for the answers as you read the chapter. Your final evaluation of whether you understand the material is measured by your ability to answer the questions. When you have completed the chapter, return to this section and answer the questions again.

1. What is the advantage of remote control wiring?
2. Does the NEC require low-voltage signal wires to be supported?
3. What precautions must be taken when several relays are operated simultaneously?
4. What precautions must be taken when mounting both low-voltage wiring and wiring for power and/or light in the same outlet box?
5. What would be a likely cause for a relay to burn out?
6. How are the relays energized?
7. How is a low-voltage system checked?
8. Are switch boxes recommended at all low-voltage switch locations?
9. Is it permitted to run low-voltage conductors and 115-volt line conductors in the same raceway?
10. What is a master control?

11-1 REMOTE CONTROL WIRING

Remote control switching offers a safe, quiet means of switching electricity that is not possible to obtain with conventional switching methods; more important, it has provided a new freedom in the design and use of circuit control (Figure 11.1).

With low-voltage control it is practical, and desirable, for the designer to provide sufficient multiple switch control of individual electrical circuits to create a new level of convenience for the user. Also, the master control centers provide for remote control in the operation of multiple circuits from one or more locations and provide visual indication of the use of each circuit.

Remote control lighting systems have found wide acceptance in the modern construction of banks, churches, clinics, office buildings, schools, hospitals, warehouses, and communication systems.

Because of the low-voltage, low-current requirements of relay switching circuits, small, flexible wires may be fished in thin wall conduit or steel partitions without the protection of metal raceways (except where local codes prohibit such installations). Wires can be dropped down through hung ceilings into movable partitions as easily as wiring for a door bell.

Low-voltage wiring is one of the most advanced methods yet devised for the switching control of lights and other electrical loads. Because this system employs inexpensive 24–28 volt momentary-contact switches and lightweight conductor wire, low-voltage wiring permits multipoint switching in a practical and economical manner. It also provides for future flexibility for changing needs, since this type system allows adaption—either for expanding or altering switch locations—far more easily and economically than conventional wiring systems.

11-2 THE ECONOMICS OF MULTIPLE REMOTE CONTROL

Low-voltage wiring permits unusual flexibility in designing a multiple control layout. It is possible to remotely control lights from several convenient locations, which in many instances could prove economically prohibitive if wired in the conventional manner using three-way and four-way switches.

EXAMPLE 1

A motor-driven master control can actuate as many as 23 different lights or loads. Placed in the master bedroom, this controller could allow all lights in the house to either be turned OFF or ON with the touch of one switch.

In a utility room of a commercial or institutional building, this method would provide convenience for controlling the lights of whole floors or sections of buildings. See Figure 11.2.

11-3 AUTOMATIC LIGHT CONTROL

Low-voltage relay switching works simply by automatically turning OFF unneeded lighting whenever and wherever you want at the proper time. Circuits are controlled by timer, photocell, or computer. However, late office

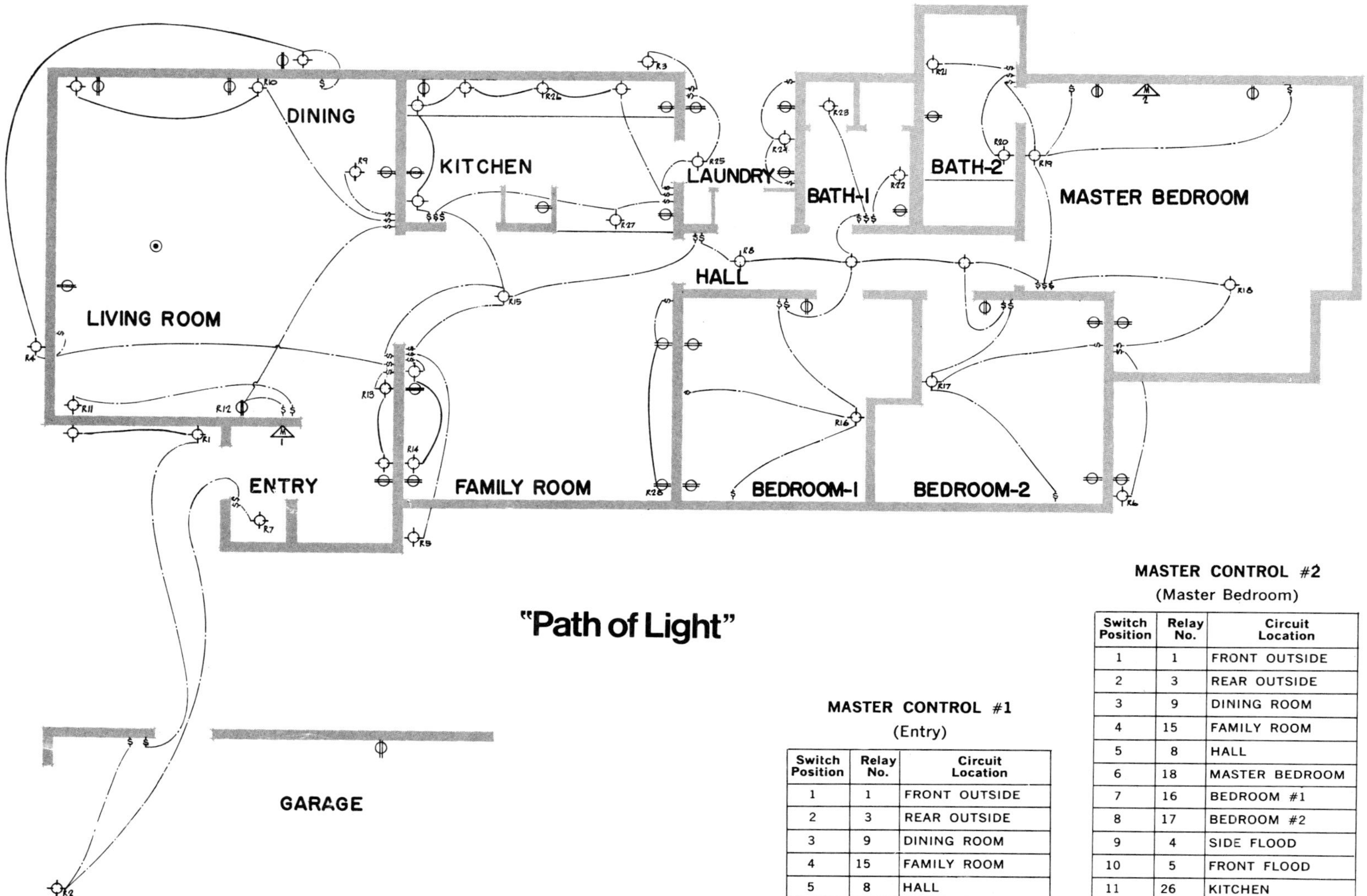

MASTER CONTROL #1
(Entry)

Switch Position	Relay No.	Circuit Location
1	1	FRONT OUTSIDE
2	3	REAR OUTSIDE
3	9	DINING ROOM
4	15	FAMILY ROOM
5	8	HALL
6	18	MASTER BEDROOM

MASTER CONTROL #2
(Master Bedroom)

Switch Position	Relay No.	Circuit Location
1	1	FRONT OUTSIDE
2	3	REAR OUTSIDE
3	9	DINING ROOM
4	15	FAMILY ROOM
5	8	HALL
6	18	MASTER BEDROOM
7	16	BEDROOM #1
8	17	BEDROOM #2
9	4	SIDE FLOOD
10	5	FRONT FLOOD
11	26	KITCHEN
12	27	KITCHEN

Figure 11.1 Typical layout showing flexibility in designing a multiple control system. (Courtesy of Touch-Plate Electro-Systems, Inc.)

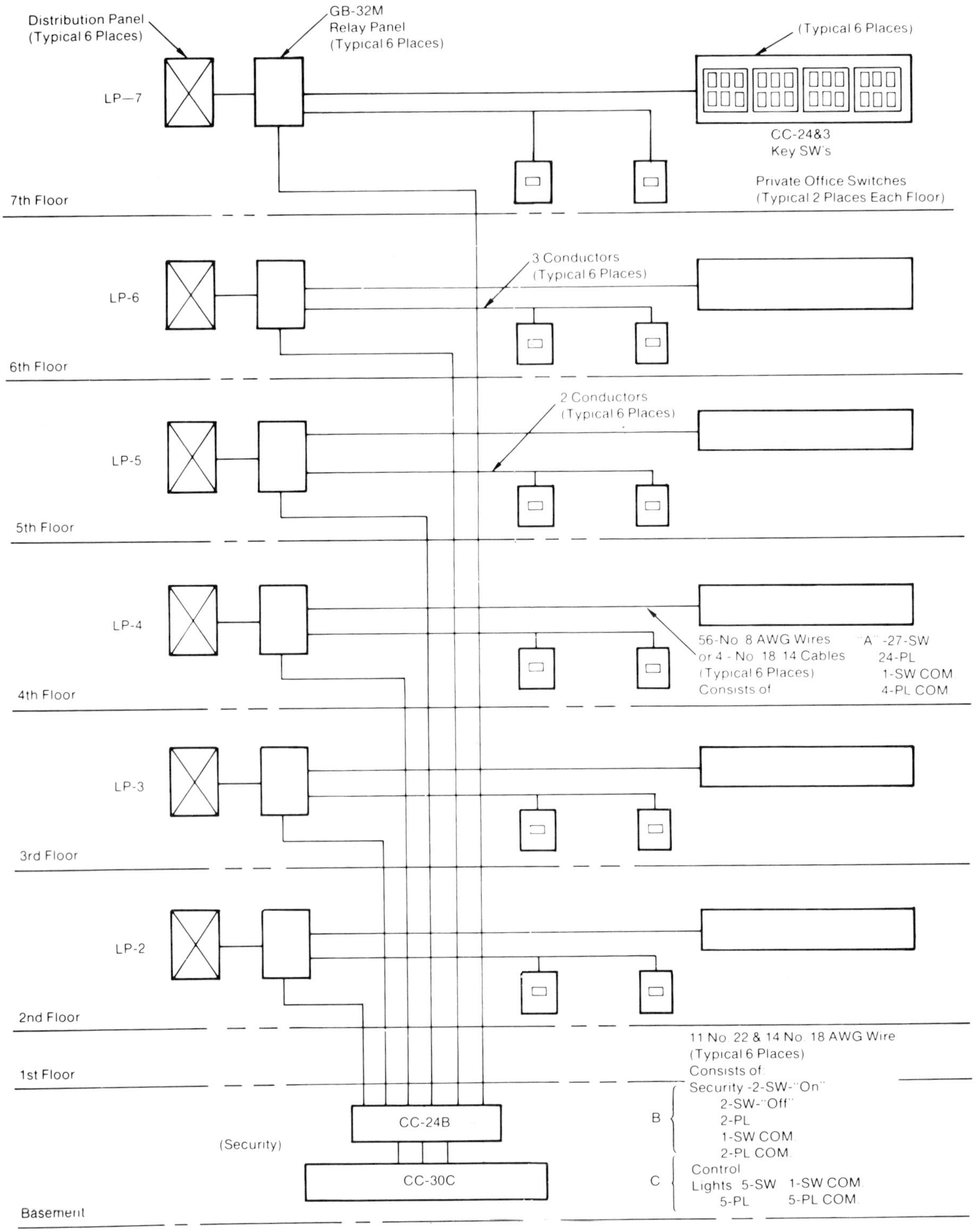

Riser diagram for a seven-story building.

(a)

Figure 11.2 (*a*) Riser diagram showing light control for a seven story building. (*b*) Typical prewired relay panel for automatic light control. (*c*) Wiring diagram using clock timer. (Courtesy of Touch-Plate Electro-Systems, Inc.)

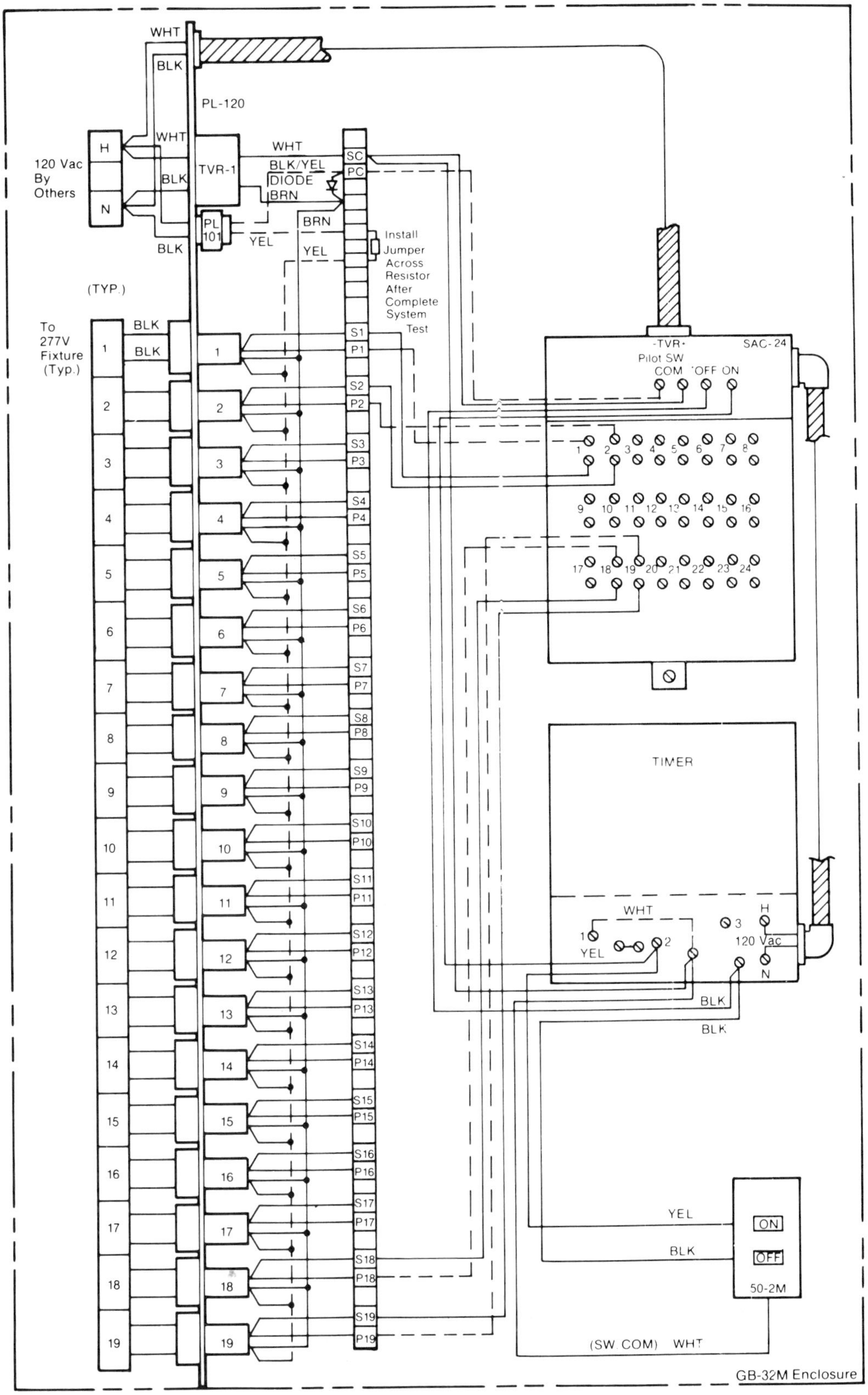

Typical relay panel for automatic light control.

(b)

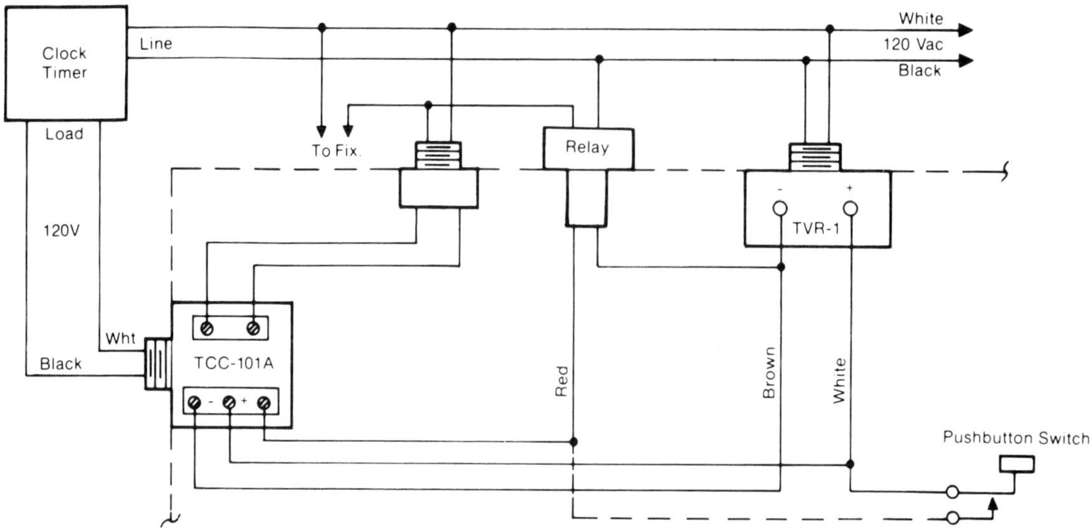

(c)

workers or cleanup crews could have a ready access to local control switches that override the system in designated areas. Low-voltage wiring is ideal for the modern energy saving home, office building, school, hospital and other heavy electrical users.

11-4 SYSTEM COMPONENTS

There are a number of component parts familiar to the apprentice.

Power Supply

One main power unit provides the voltage to operate all of the relays in the system. This power supply, or transverter, consists of a transformer, full-wave rectifier, and an electrolytic capacitor. It can easily be mounted in a relay cabinet or separately in a box (Figure 11.3).

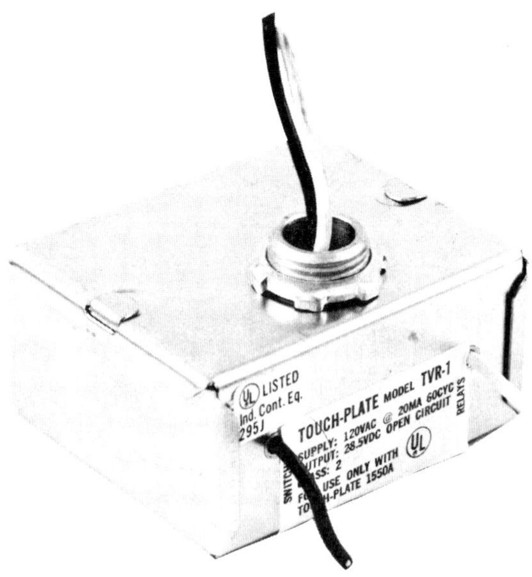

Figure 11.3 Power supply. (Courtesy of Touch-Plate Electro-Systems, Inc.)

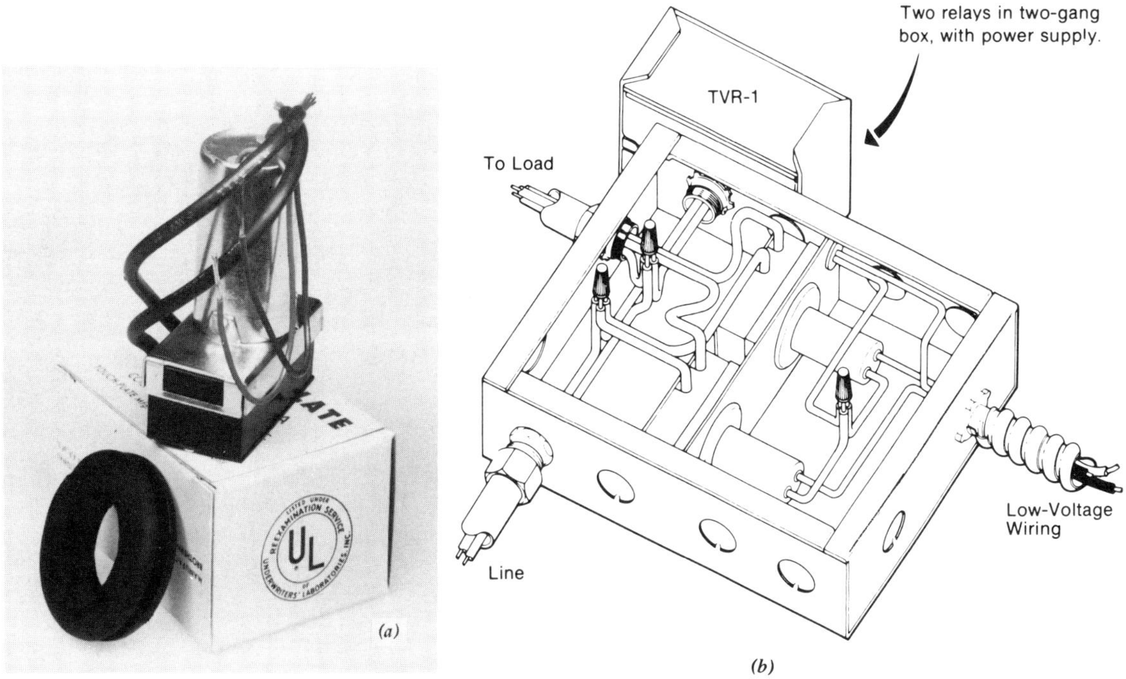

Figure 11.4 (*a*) Relay. (*b*) Relay wiring method using junction box where conduit is required for low-voltage wiring. Notice barrier separating low-voltage and line voltage. (Courtesy of Touch-Plate Electro-Systems, Inc.)

Input AC line voltage standards are 120, 208, 220 and 277 volts. Output voltage is a momentary 28 volts DC, occurring only when the switch is depressed. In the event of a short or stuck switch, closed circuit voltage on the output or load side of the power supply is a mere 3 volts.

Relays

Successive pulses from the transverter alternately open and close the relay. The cam-operated latching relay used requires only a single coil. Each relay may control any number of devices up to its rating. Although components can easily be assembled at the job site, relays are normally factory-installed and prewired in a metal cabinet or gang box (Figure 11.4). As required by the NEC, metal barriers separate the low-voltage and line voltage wiring in the gang boxes.

Switches

All switches are momentary contact, with no "on-off" positions (Figure 11.5). Pressing the switch momentarily energizes the relay and closes the load contacts. When the switch is pressed again, the contacts open.

As many switches as desired can control a single relay; they are merely wired in parallel. Switches are connected by No. 18 AWG or larger wire. Up to six switches can be supplied in a single gang, with either plastic wall plate or a decorative flush plate. Switches can also be combined in groups of up to ten gangs. All switches are mounted on plaster rings or standard outlet boxes. In most areas, local codes do not require a box.

Pilot lights are available with all switches and are normally used where the load is out of sight of the switch. When the pilot light is used, two additional No. 18 AWG conductors must be run. A transformer installed in the gang box supplies 6 volts AC for the pilot light.

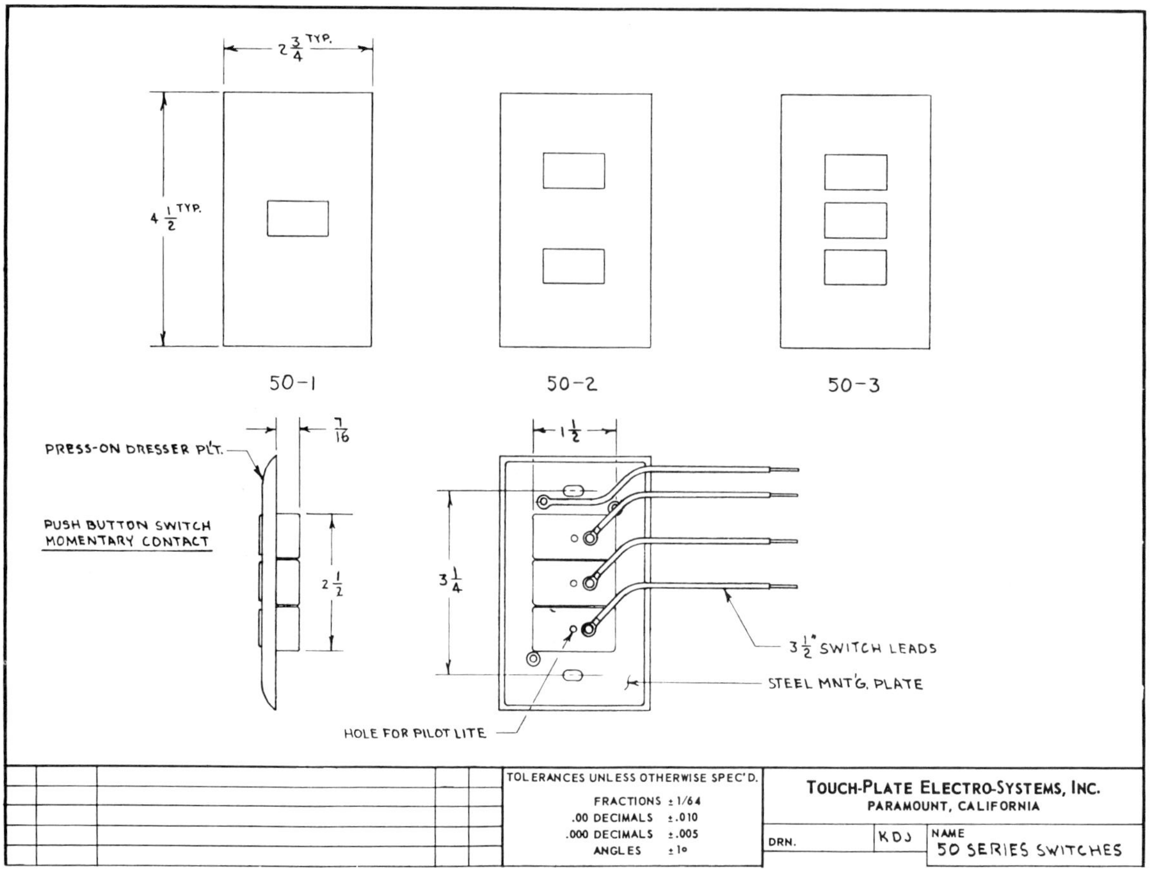

Figure 11.5 Switches. All switches are monentary contact, with no on-off positions. (Courtesy of Touch-Plate Electro-Systems, Inc.)

Master Control Panels

These units can be used to control any number of relays from a central point. Master control panels combine one or more modules of six pilot-lighted pushbuttons per gang (Figure 11.6). Special color-coded 14-conductor cable simplifies the wiring of each six-button unit.

11-5 DESIGNING A REMOTE CONTROL SYSTEM

The design and layout of a system is not difficult. The following step-by-step procedure is suggested to facilitate the reading of the plans and to provide easy reference to the circuits. The layout itself should be made directly on a plan drawing and a copy of the layout attached to the electrical specifications. See Figure 11.1.

1. Specify the location and type of lighting fixtures on the plan drawing. A residence is used for this example because it usually requires the most relays. Other applications are designed in a similar manner.
2. Specify the various outlets, other than switching outlets, in the conventional manner.

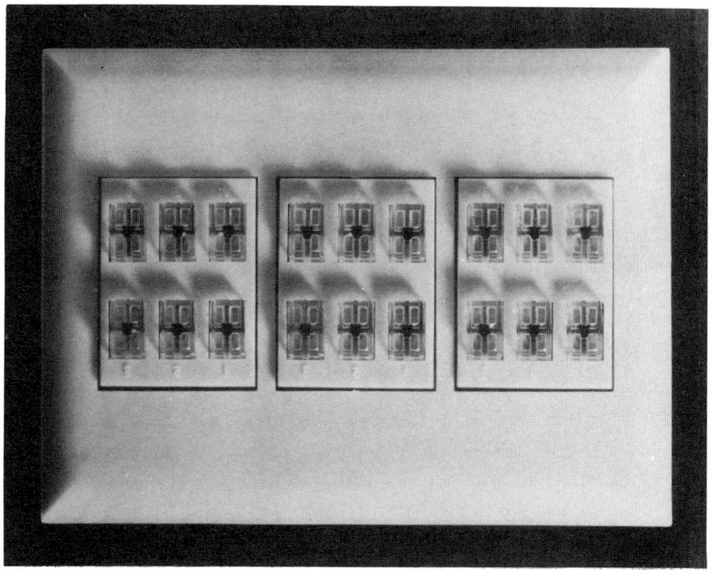

Figure 11.6 Master control panel. (Courtesy of Touch-Plate Electro-Systems, Inc.)

3. Use a solid line to connect any fixtures to be controlled together, as in a hallway or large kitchen. Do not assign more than 2400 watts to one controlled circuit.
4. Determine which convenience outlets should be controlled from switch locations. Normally, split-type outlets are specified for switch control. Where this type of outlet is used, the upper plug opening is switch controlled and the remainder left live for radios, clocks, and so forth. Whenever more than one outlet is to be switch controlled, connect all outlets with a solid line as was done in step 3.
5. Put the symbol for relay, "R," followed by a number, at each fixture or controlled outlet location throughout the house and grounds. Where a group of fixtures or outlets are to be controlled together (joined by solid line, steps 3 and 4), put the "R" and number by only one of the fixtures or outlets in the group. The numbers following relay symbols should run consecutively from 1 (one), until all circuits have been designated. This will provide a count of the number of relays on the job, and will enable you to refer to any specific circuit by its "R" number. In low-voltage switching, one relay is required for each outlet, or group of outlets. The same requirement applies to fixtures. For example, only one relay would be required for two front porch lights that are to be controlled from the same switch. The lamp post light, which is to be controlled from a separate switch, would require its own relay.
6. This step establishes the switching locations. It is at this point that the designer can provide the conveniences of multiple switching so often omitted in conventional wiring because of its high cost.

   Examine each room individually. Determine entrance and exit points and decide which lighting circuits should be controlled from these points to provide maximum convenience. Indicate a small "S" for each switch required. Connect the switch "S," with a dash line, to the "R" numbered circuit it is to control, or indicate the "R" number adjacent to the switch

(S15, to control relay 15). Use as many switches as desired to control a relay.

Although most switches are located at entry and exit points, there are exceptions. For example, the kitchen sink light is usually controlled from a point near the sink. In hallways, the lights may be controlled from either end and/or from inside each room adjacent to the hallway.

All of the desired switch points should be indicated on the plan.

7. The next step is to determine which circuits, if any, are to be remote controlled through the use of pilot-lighted master control centers. These control centers provide the ultimate in convenience. They may be located in any area without regard to the location of the circuit in the house. For example, a 12-button control center could be located in the garage, which would allow you to light an entire path of light through the house. A duplicate of this control center could be located in the master bedroom, allowing you to turn off all these lights when reaching the room. The master control center would be an addition to the individual switches located in each area (step 6). In planning the layout of master control centers, it is advisable to use a tabular form (Figure 11.1) and include the table on the plan drawing. With such a table it is possible to show:
   (a) which relay is to be master controlled from each position.
   (b) the area location of each circuit to be controlled by the master.
   (c) the location of the control centers on the drawing.

   Where identical relays or circuits are to be master controlled from more than one control center (i.e., one in the garage and one in the master bedroom), it is advisable to use the same switch positions on each master control center. This provides simplicity for the home owner in using the control center (the same button on both master panels will control the identical circuit). Also it allows for economy in wiring because both centers can be wired in parallel, using color coded cable.

8. Low-voltage factory prewired units are designed for economy, ease of installation, and service as well as for silent switching. If the location of these units is to be specified and shown on the plan drawing, note that circuits being controlled by master control centers use PMRU (prewired master relay units). Circuits not being master controlled use RU (relay units) prewired units.

11-6 INSTALLATION PROCEDURE

In a commercial building, the 120-volt lighting system is roughed-in as usual, except all switch legs are omitted. The 120-volt supply conductors are brought directly to the outlet that will have a switched lighting fixture or switched outlet connected to it. Throughout the United States, each manufacturer may number or color code their devices in a different manner; however, the electrician need only remember that three low-voltage conductors are required between the low-voltage side of the relay and the switch. Also, two low-voltage conductors carrying the low-voltage supply from the transformer must be run to either the relay or the switch. The wires are then connected as color coded or as certain colors are assigned to correspond to

Figure 11.7 Gang box. Mounted out of sight usually next to the circuit breaker. They contain the low- and full-voltage terminal strips, relays, pushbutton override switch, master controller, and transverter power supply. (Courtesy of Touch-Plate Electro-Systems, Inc.)

the numbers on the relays and switches. The black leads of the relay are connected to the 120-volt circuit in the conventional manner.

In wood frame construction the low-voltage cables are stapled in place, using insulated staples. For steel buildings cables are secured to the structure with plastic tape. Should the building be of poured concrete, then conduit must be installed and conductors fished through when roughed-in.

It is recommended that all of the relays be placed in one location. Some manufacturers gang relays in a box having a metal barrier; the line-voltage connections are made on one side of the barrier, while the low-voltage connections are made on the other (Figure 11.7).

11-7 ROUGHING-IN THE INSTALLATION

1. Mount prewired relay units in a central and accessible location (closet, basement, or in attic space). Run branch circuits to these units.
2. Mount a transverter to one of the prewired units and make the 115-volt connections. Only one transverter is required for the complete system.

3. Mount boxes at fixtures and device locations, and run 115-volt wiring to these boxes. With Romex or B-X installations, no pigtailing is necessary. Only two wires go to the fixture location.

4. Regardless of the type of 115-volt wiring required, practically all low-voltage wiring (check local requirements for minor variations) will be as follows: For switch locations in frame construction, a box ordinarily is not required. Mark the switch layout but instead of installing switch boxes, use switch plaster rings. Use a single gang ring for each switch location. Behind each switch ring drive an 8-D nail partially into the stud to hold the wire while pulling.

   Drill for low-voltage wiring the same as for conventional wiring.

5. Pull low-voltage supply lead. Using the loop method as illustrated, feed a white wire through the attic space from the relay gang box location to the farthest switch point and secure to nail behind the switch ring. "Pick up" each switch by looping wire down through the wall to hook on nail at each switch location. Work back toward the relay box without cutting the wire. To hold the loops in place, wind them around the nail behind the switch ring. This wire will supply all switches and connects to the transverter lead marked "switches."

6. From the relay gang box a colored wire to the farthest switch in any multiple switch circuit, loop the wire down to the other switch locations in that circuit in the same manner that the low-voltage supply lead was pulled in. This will be a continuous wire from the farthest switch to the relay gang box. It eliminates pulling separate switch lead to each switch of a multipoint controlled circuit.

   Remember, two leads go to each switch, but one of these is common to all switches (white colored wire). Simplify your work by using wire of different color for each run.

7. A simple wire rack, holding the different colors of wire, can be used for easy pulling. Where any individual switches are to be pilot-lighted (to indicate ON or OFF of the controlled outlet), two additional No. 18 wires must be run to those switches and connected to a pilot light transformer in the relay box (already installed in the master relay units).

   Pilot lights are appropriate for switches controlling remote lights, where it is not quickly apparent at the switch location whether outlet is ON or OFF.

8. Staple low-voltage wire loosely. If for any reason the wires must be pulled tight, use insulated staples.

9. The "makeup" at the prewired relay units is a simple matter of matching colors. The 115-volt feed connects to the prewired circuit which feeds the transverter and each relay. All 115-volt neutrals are jointed with wire nut or split-bolt type connector. Each 115-volt run from a switched outlet is connected to color-coded 115-volt pigtails on the relays. Identical colored low-voltage wires indicate the switch for that relay.

10. The colored wiring harnesses on the prewired units match the control cable which runs to the master control panels.

11. Wire master control centers by matching the colors of the wires on the control center with the colored cables. The end of the cable will also

match the harness in the factory prewired master relay unit. One run of 18-14 cable is used for every six circuits to be remotely controlled.

12. After walls are completed and decorated, switch units and snap-on wall plates (no exposed screws) are installed and appropriate connections are made at each switch.

Note. All wiring to service and branch circuits, including outlets which are not switch operated, is done in the conventional manner.

11-8 TROUBLE SHOOTING REMOTE CONTROL INSTALLATIONS

If a single light fails to light, be sure the bulb is good.

No Low-Voltage Control No power at any controlled outlet: Check your supply from the breaker to the relay gang box. You are not getting 110-volt power into the low-voltage controlled distribution system. If the breaker won't hold in, you are almost sure to have a short in the high-voltage makeup at some fixture.

No Control Some outlets on, others off: You are getting 110 volts into the system, but the low-voltage circuit is shorted or has an open. A shorted condition is the more probable, because it will not allow the transverter to charge. To find the cause of the short, disconnect the transverter pigtail labeled "switches" from the one or more wires connected to it. Touch these disconnected wires to the pigtail one at a time. When you hear one of the relays click, you have discovered the short. You can then isolate it to the proper relay by having someone watch the outlets to see which light is turning off and on. Then check your low-voltage makeup for all switches wired to this relay. Once in a while, wires pulled tightly around a ventilation or heating duct can break insulation and cause difficulty.

Some Low-Voltage Control But not all relays working: This may happen when you have relays in several gang boxes. Be sure you have a supply from the transverter pigtail labeled "relay" supplying the common relay connection in all gang boxes. This is occasionally overlooked, but is easily discovered because it will cause all relays in one box to be inoperative.

Low-Voltage Working Some pilot lights burn dim: If pilot light switches have dim or erratic pilot light operation, it frequently indicates an open in the common low-voltage AC lead to the pilot lights.

All Pilots Burn Dim When Any Button Is Operated 110-volt neutral to pilot light transformer is loose or disconnected.

Pilot Lights Burned Out Burned out pilot lamps on the new installations are generally caused by reversing the low-voltage common leads. Solid white wire from TVR-1 transverter is common to switches. White wire with black tracer from PL-6 is common to pilot lamps. Reversing these will cause difficulty.

Existing Older Installations Where One, Two, or More Relays Cease Operating This is not relay failure. It is due to a weakening low-voltage power supply. Check for stuck switches. A continuous drain reduces the life of the rectifier and is usually the cause of failure. If the installation does not have the transverter (TVR-1), but is a two separate-component power supply, it can be checked by taking the converter out of the system, hooking the low-voltage lines directly to the transformer, and testing the operation on AC. A TVR-1 transverter should be used to replace both the 78K transformer and 17C converter.

General Notes—Relays

1. Relays not overloaded will seldom fail. In trouble shooting a job you can count on the relays as being O.K. until you have checked all other parts and makeup in the system.
2. If a relay fails to operate, it may be due to the relay making contact against a dead short in the 110-volt wiring at the fixture. Such a dead short can pass enough current before the breaker drops out to fuse the contact points in the relay together. If you do find a relay failure, be sure and eliminate cause before you replace relay.
3. Two relays will not operate from one switch.
4. Use only one transverter in a single prewired unit.
5. Painted switches will cause difficulty. Broken switch plates that let the low-voltage wires free can also short out. Remember that any continuous drain on the low-voltage system will considerably reduce the normal expected life of the rectifier.
6. In low-voltage failure, always look for the cause. If you replace the low-voltage power unit and the cause of failure is not eliminated, you will naturally have a reoccurrence of the trouble.
7. Several relays that fail to operate or operate erratically, indicate a weakened power supply. Install a new transverter before changing relays.

11-9 NATIONAL ELECTRICAL CODE

The low-voltage, remote control system is not subject to the same code restrictions as the standard 120-volt lighting system.

All wiring on the supply side of the transformer must conform to the wiring methods of Chapter 3 of the NEC. All wiring on the low-voltage of the transformer must conform to Article 725-38 of the NEC.

SUMMARY

1. Remote control switching offers freedom of design in use of circuit control.
2. In most installations low-voltage wires are not required to be placed in conduits.
3. In a commercial building, remote switching control installation, the 120-volt lighting system is roughed-in as usual, and the switch legs are omitted. It is the switch leg that is controlled by low voltage.
4. Remote control systems are not subject to the same code restrictions as the standard 120-volt lighting system.

PROBLEMS

11-1 Draw a sketch of one switch and one relay controlling a load.

11-2 Draw a sketch showing three switches added to an existing switch (show only the switches).

11-3 Draw a sketch showing two relays added to an existing relay and switch (show one switch, three relays, and a transformer).

11-4 Low-voltage relay switching works simply by automatically turning OFF unneeded lighting whenever and wherever you want at the proper time. These circuits are controlled by ____________, ____________, or ____________.

11-5 Low-voltage wiring is ideal for modern energy saving homes, ____________, ____________, ____________, and other heavy electrical users.

11-6 In remote control systems, all switches are ____________ contact type.

11-7 Master control panels control any number of relays from a ________ ________.

11-8 How many power units are required for a low-voltage system?

____

11-9 Any number of switches can control a single relay; these switches are wired in ________.

11-10 Remote control lighting systems have four components. List them.

________, ________, ________, ________.

chapter 12

SIGNAL AND COMMUNICATION SYSTEMS

Instructional Objectives

1. To become familiar with power supply sources relating to fire alarm systems.
2. To understand the operation of automatic fire doors.
3. To identify the components of a fire alarm system.
4. To understand the need for end-of-line resistors.
5. To understand the operation of the three basic fire alarm systems.

Self-Evaluation Questions

Test your prior knowledge of the information in this chapter by answering the following questions. Watch for the answers as you read the chapter. Your final evaluation of whether you understand the material is measured by your ability to answer the questions. When you have completed the chapter, return to this section and answer the questions again.

1. What is an alarm initiating device?
2. What does the National Fire Protection Association (NFPA) Code require?
3. What is the purpose of the end-of-line resistor?
4. List three automatic alarm initiating devices.
5. What does a smoke detector do?
6. What are the three basic fire alarm initiating circuits?
7. Where is the resistor located in the parallel end of the line resistor circuit?
8. What are the symbols used for fire alarm circuits?
9. In a series circuit, are the alarm initiating devices contacts normally open or closed?
10. Of the three systems, which system is usually used in schools?

12-1 PURPOSE OF A FIRE ALARM SYSTEM

The modern fire alarm system is designed to save both lives and property. Most newly installed systems in commercial buildings do both with a remarkable record of efficiency. On the average, more than three and one-half times as many persons perished in unprotected properties, and physical damage was nearly three and one-half times greater. No value can be put on the loss of life; to physical damage you must add the much greater hidden losses: business wiped out, production and wages lost, the many hours, perhaps years required to reconstruct records and plans, as well as the higher insurance costs for unprotected buildings.

Early warning ensures immediate evacuation of all in danger, and it permits fire fighters to get to the scene and take prompt action that often spells the difference between a minor fire and a major conflagration. Aside from the physical protection it affords, the local integrated fire alarm system (frequently installed in conjunction with sprinklers) means freedom from fear of fire, which can mean a great deal where the occupants are vulnerable, or the material stored is expensive or highly combustible.

Fireproofing Limitations

Fireproofing isn't the answer; no building is truly fireproof. Fire resistant materials and such devices as fire walls, fire doors, and ventilation closures have done much to prevent or contain fires; however, every material has its combustion point, and nearly every building contains varying quantities of easily combustible material; no property is immune to destruction by a hot fire that gets out of hand.

Automatic Sprinkler System Limitations

Automatic sprinklers are not enough. In the first place, the use of sprinklers is confined to areas where water can be tolerated. They should not be installed where materials such as potassium, sodium, or sulfuric acid are stored or used (these materials react violently with water), nor should sprinklers be used around petroleum, since water will spread blazing petroleum. They should not be installed for use in areas containing valuable records, books, paintings, soft goods, hard goods with expensive finishes, and other items which might be damaged as much by water as by fire. Most of all, sprinklers should not be used near live electrical equipment.

In the second place, an automatic sprinkler may spray water for many unnoticed hours in an unoccupied area, if there is no means of simultaneous warning; and it can in some cases create large quantities of smoke, which can be more dangerous and damaging than fire, if other fire fighting methods are not brought to bear as soon as possible.

If sprinklers are a required part of a fire protection system, they should always be installed in conjunction with a modern fire alarm system which includes:

1. Automatic fire or smoke detectors, Figure 12.1, which must give warning of a fire before the sprinklers are actuated, providing fire fighters the opportunity to put out the fire without water damage.

Cat. No.	Description	Input Power	Pilot Lamp	Temp. Range	Alarm Horn
217P	Plug-in Installation–13′ (3.9 meters) lead	110/120V AC 1.050A	Light emitting diode	32°-120°F (0-49°C)	85 dB at 10 Ft.
217T	Permanent Connection	110/120V AC .050A	Light emitting diode	32°-120°F (0°-49°C)	85 dB at 10 Ft.

Figure 12.1 Automatic smoke and fire detector. (Courtesy of Edwards Company)

2. Water flow switches, which will automatically signal the flow of water in a sprinkler system.

12-2 A FIRE ALARM SYSTEM

A good local fire alarm system combines all the elements required for early detection and warning of fires in a single, reliable system that can be diverse and extensive enough to fully protect the most complex property, or simple and inexpensive, to meet the requirements of most commercial buildings.

The modern fire alarm system has three basic elements:

1. *Alarm Initiating Devices.* These may be any combination of the following:

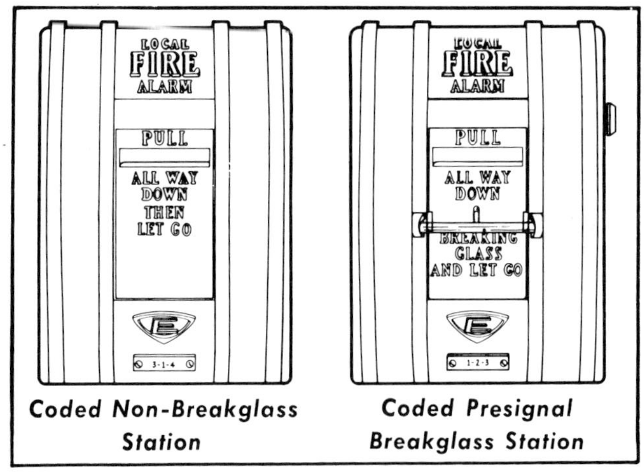

Figure 12.2 Manual fire alarm station. (Courtesy of Edwards Company)

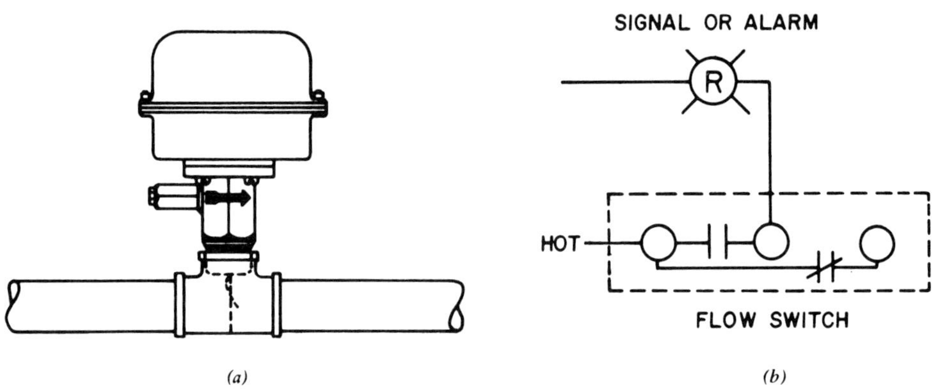

Figure 12.3 Automatic device. (*a*) Flow switch. (*b*) Wiring diagram.

(a) *Manual fire alarm stations.* These stations are of two basic types. The simplest is a switch which is closed, manually, by pulling a handle. This is called a noncoded station. See Figure 12.1. The second and more complicated station contains a spring-driven mechanism which, when operated by the pull of a handle, closes and opens an internal switch in some predetermined time sequence. This type is called a selective-coded station (Figure 12.2).

(b) *Automatic Devices.* These include (1) thermal fire detectors, (2) smoke dectors, and (3) water flow switches (in automatic sprinkler systems), each of which is designed to automatically operate upon a condition indicative of fire. See Figure 12.3.

2. *Alarm Indicating Devices and Accessories.* These devices and accessories include:
   (a) Audible alarm indicating devices such as horns, single-stroke bells, vibrating bells, and sirens. See Figure 12.4.
   (b) An automatic connection to close fire doors. See Figure 12.5.
   (c) Lamp or drop annunciation to indicate the specific location or the actuated initiating device or group of devices. See Figure 12.6.

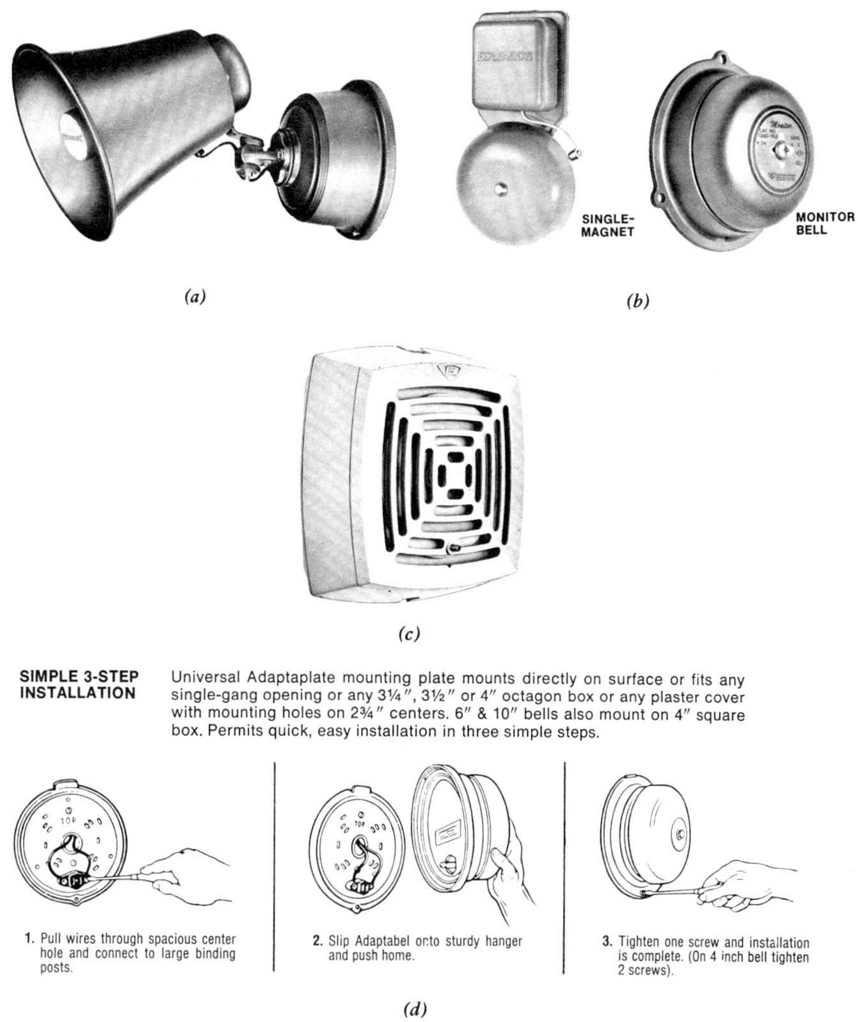

Figure 12.4 Alarm indicating devices. (*a*) Electronic horn/siren. (*b*) General purpose bells. (*c*) Vibrating bell. (*d*) Shows ease of installation. (Courtesy of Edwards Company)

3. *The Control Unit.* This contains all the components necessary to receive the signals from alarm initiating devices and transmit them to the correct alarm indicating devices and accessories.

Power Supply Sources

There are various types of power supply sources for fire alarm systems, depending upon the type of system involved. Of course, these power sources are subject to the acceptance by the authority having jurisdiction. The power sources could be:

1. Light and power services
2. Engine generator set
3. Storage batteries
4. Rectifiers

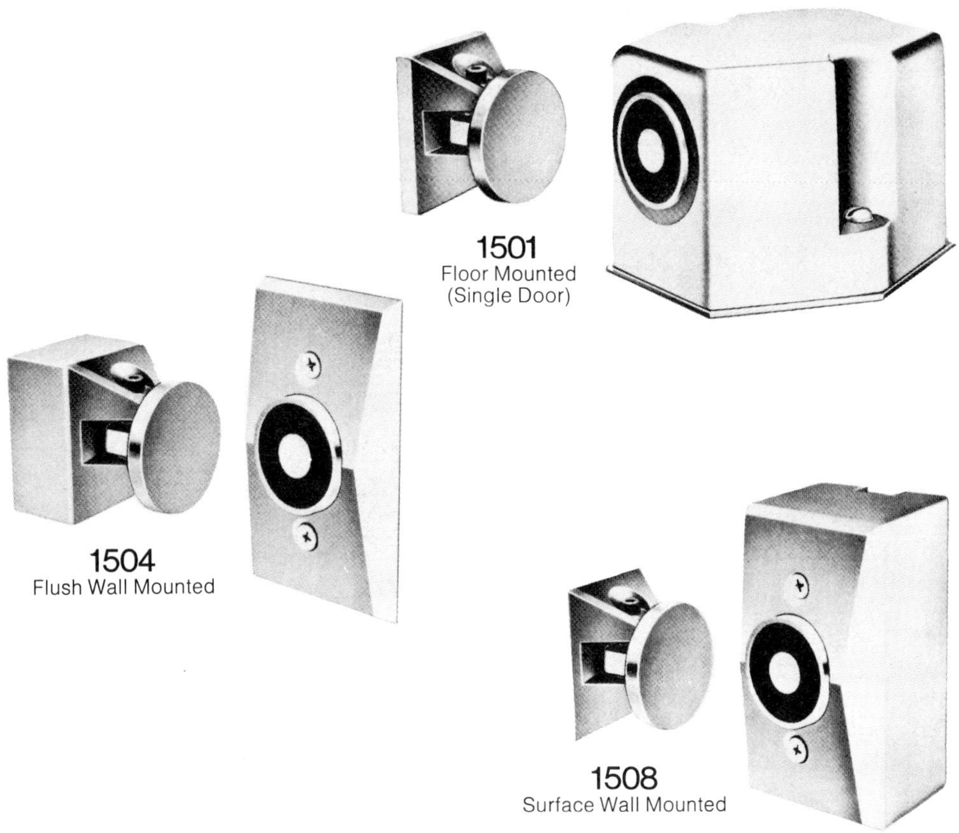

(a)

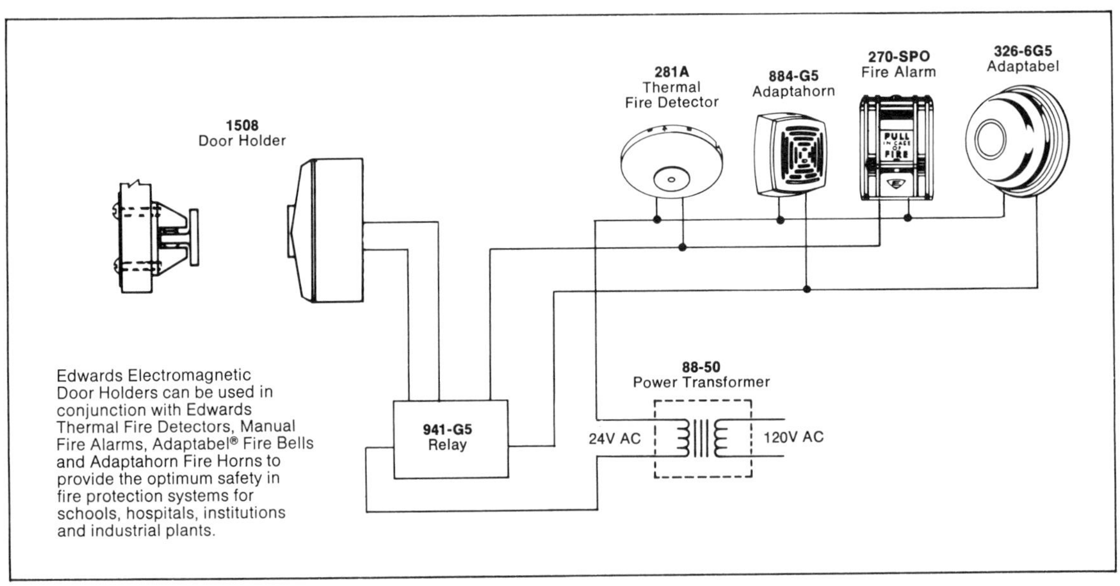

(b)

Figure 12.5 Electromagnetic door holder. (*a*) Door holders are designed to be used on self-closing, swinging doors to automatically isolate an area in conjunction with fire alarm or smoke detection systems. The swivel contact plate that mounts on the door adjusts to any door angle. It has a shock absorbing nylon swivel ball. (*b*) Typical application of door holder used in conjunction with thermal fire detectors, manual fire alarms, and fire horns. (Courtesy of Edwards Company)

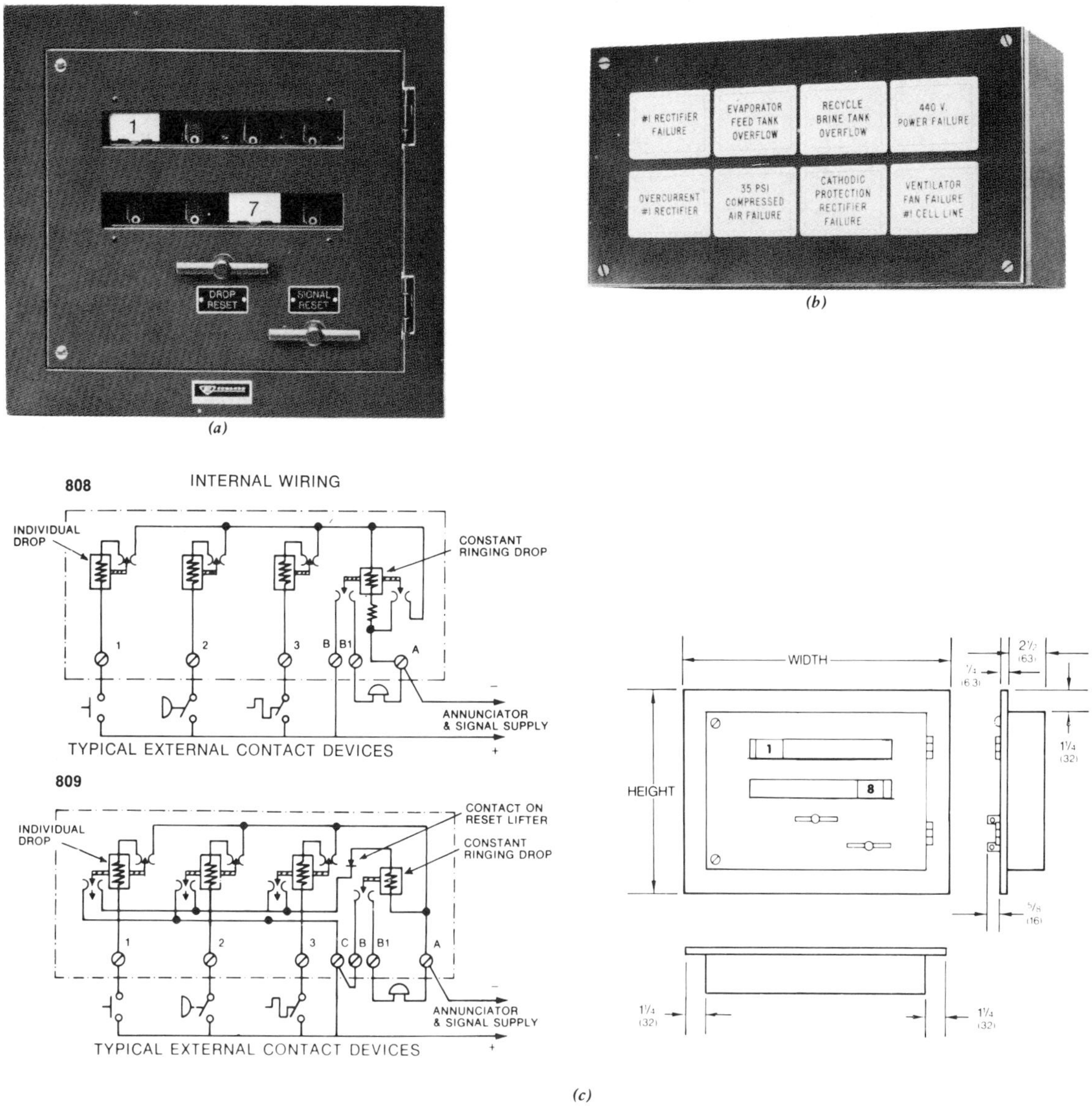

Figure 12.6 Annunciator systems. (*a*) Drop annunciator. When contact is activated, an electromagnet causes an indicator to fall into the appropriate window. At the same time an audible signal can be sounded calling attention to the event. (*b*) Lamp annunciator. In the lamp annunciator, the electromagnetic indicator is replaced by an illuminated window. Lamp annunciators use a locking relay or maintained switch to keep the lamp illuminated until reset action has been taken. (*c*) Typical wiring diagram. (Courtesy of Edwards Company)

For the purpose of this section, light and power supply sources will be covered, because they are the most popular ones used for the average fire alarm system in commercial construction.

The source which is commonly used and for which a majority of fire alarm systems are designed is three-wire 120/240 volt single-phase 60 cycle. The main operating power to operate the system is connected to one leg and neutral and the trouble bell is connected to another leg and neutral.

12-3 THE NATIONAL FIRE PROTECTION ASSOCIATION CODE

The NFPA Code requires the following:

1. The conductors of the signaling system power supply circuit shall be connected on the line side of the main service of a commercial light or power supply circuit or to the main bus bars of an isolated power plant located on the premises. A circuit disconnecting means shall be so installed that it will be accessible only by authorized personnel. [No. 2222]
2. An overcurrent protective device of suitable current-carrying capacity and capable of interrupting the maximum short-circuit current to which it may be subjected shall be provided in each ungrounded conductor. The overcurrent protective device shall be enclosed in a locked or sealed cabinet located immediately adjacent to the point of connection to the light and power conductors. [No. 2223]

Electrical Supervision of Fire Alarm System

Because fire alarm systems must, by nature of application and use, be highly reliable, electrical supervision is used as a required means to assure that the system is in an operable condition at all times.

An electrically supervised system may be defined as a system in which a break or ground in the wiring, which prevents the transmission of an alarm signal, will activate a trouble signal, which shall be operated by a separate source of power. From these requirements, it should be observed by the apprentice that the integrity of the installation wiring of fire alarm systems is of such importance that it is necessary to do more than simply interconnect alarm initiating devices, as is done in connecting light switches in multiple to operate lights.

12-4 ALARM INITIATING CIRCUITS

The function of the alarm initiating circuit under normal conditions, when there are no faults in the supervised circuit, is to energize supervisory relay by a small current in the main circuit; the relay armature is held in the attracted position with contacts open. Under the off-normal condition (an open or ground), the supervisory relay is de-energized and the relay contacts close the circuit of a separate supervisory signal power supply, thus actuating a trouble bell or any other trouble indicating device.

Three Basic Systems

1. *The Series Type Closed Contact Circuit System.* It should be noted that this type of circuit should be used for coded evacuation systems only and with single-stroke bells for the alarm signals. See Figure 12.7. The reason for this, as will be seen when studying circuits, is that an open or ground in the wiring will give the same circuit operation as the opening of an alarm initiating device and would cause vibrating alarm signals such as bells or horns to sound an alarm, which is in violation of fire codes. An open or ground when using single-stroke devices such as single-stroke

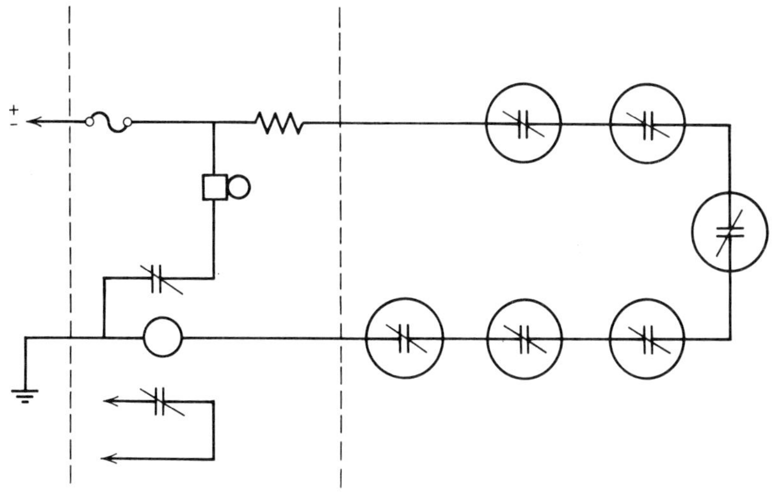

Figure 12.7 Basic series closed contact devices fire alarm system. System is shown with power disconnected.

bells, on the other hand, will cause only a single tap on these devices and this is not considered a fire alarm signal by fire authorities.

Referring to Figure 12.7, notice that the alarm initiating devices have normally closed contacts and, therefore, may be placed in one series circuit. The current flowing through this circuit is sufficient to hold the armature of the relay in the energized position with the contacts open. When the contacts of the initiating devices open, the relay R becomes de-energized, closing contacts R-1 and sounding a signal on the alarm bell AB. At the same time, relay contacts R-2 close, sounding the trouble circuit. Should the wiring to the alarm initiating devices become open or grounded, relay R will drop out, sounding the trouble signal and striking one signal on the alarm bell. If the fault condition is a ground, a short circuit across the alarm initiating devices contacts would exist, depending upon the location of the ground. To prevent an overload of this circuit during a ground-fault condition, a fuse and limiting resistor are placed in the circuit.

Figure 12.8 shows the correct wiring methods of the basic fire alarm system. Also illustrated is a riser diagram showing the alarm initiating devices placed in the circuit along with junction boxes such as may be found on a typical architect's riser diagram.

2. *Parallel Return Loop Circuit System.* Figure 12.9 illustrates one of the noncode fire alarm circuits that utilizes normally open contacts on the initiating devices. They may be manual or automatic devices. In this circuit, a small amount of current flows through the alarm relay (AR), trouble relay (TR), field wiring, end-of-line resistor (ELR), limited resistor (LR), and fuse. The small amount of current is sufficient to energize the trouble relay and keeps its contacts, TR_1, open but insufficient to energize the alarm relay. This system will cause the trouble signal to sound as in the basic system should one of the wires to the alarm initiating devices open. Ground-fault protection is also provided. Should a ground occur in any of the field wires, the current for the circuit would then travel through the fuse (F_1), the limiting resistor (LR), the field wiring, and the

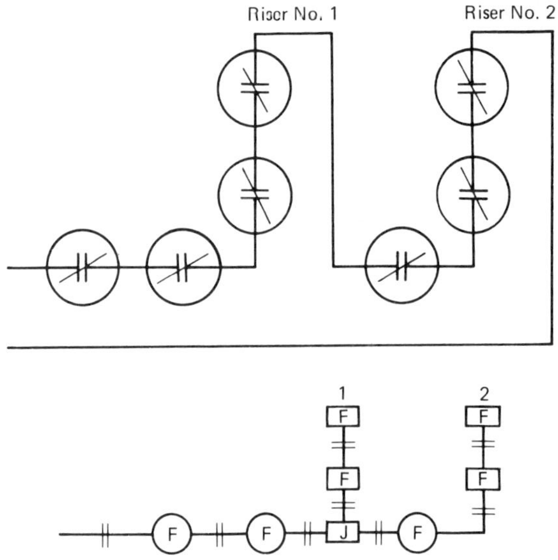

Figure 12.8 Riser diagram of series circuit.

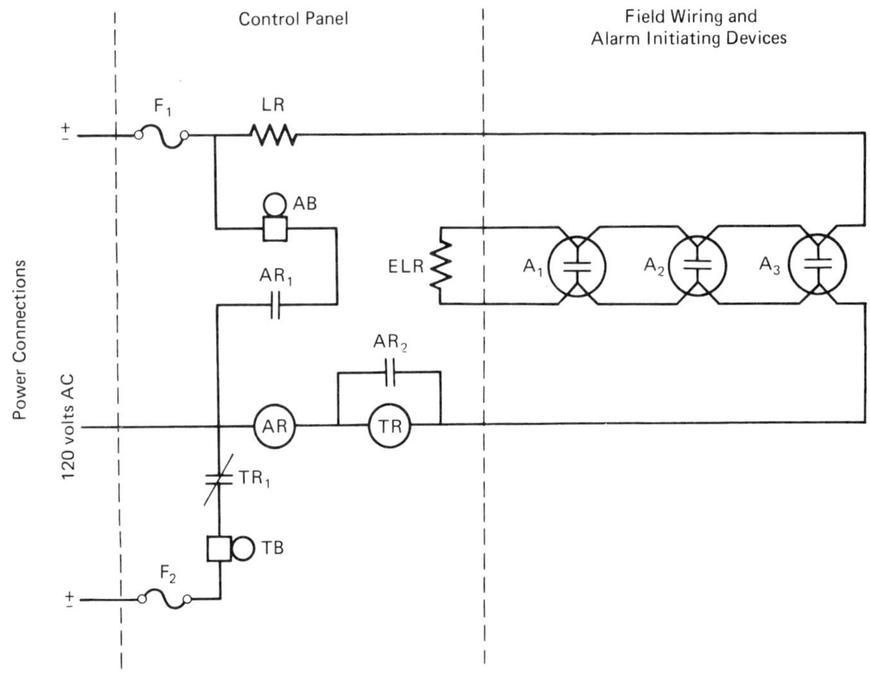

Figure 12.9 Basic parallel loop circuit.

grounded point of wiring to ground. If grounded wire is beyond the end-of-line resistor (ELR), the current would also pass through the ELR. This will shunt the relays causing the trouble relay to drop out and sound the trouble signal. In this, as in the series circuit, the limited resistor and fuse prevents excessive current from overloading the circuit in a ground-fault condition.

In an alarm condition, one or more of the contacts of the alarm initiating devices will close. This will shunt ELR and increase the current through the alarm relay sufficiently to energize it and close the alarm relay contacts

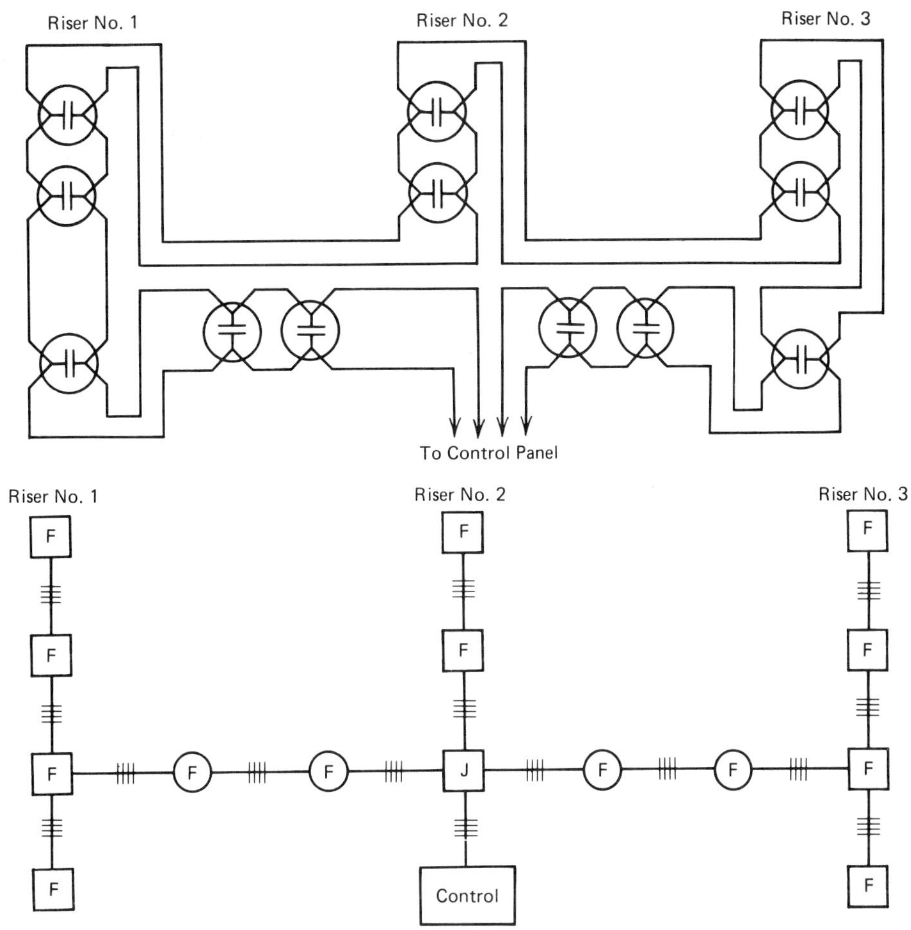

Figure 12.10 Wiring diagram of return loop circuit with riser.

AR_1. In this, as in the series type, the alarm and trouble signals sound during the alarm condition.

Figure 12.10 shows a typical return loop circuit and riser diagram correctly wired. Basically, it is just a parallel pair of wires going from the control panel to each of the alarm initiating devices. The devices are wired across the conductors, so that when they operate they shunt the pair of wires. Notice that each initiating device is wired in parallel with the two conductors and then, when the last device is reached, the wires return to the control panel.

3. *Parallel End-of-Line Resistor Circuit.* This system is used quite commonly in many fire alarm systems in schools and other commercial buildings. See Figure 12.11.

   Here as in the return loop circuit, a small amount of current flows through the alarm and trouble relays field wiring to a resistor placed across the last alarm initiating device and back through the limiting resistor to the other side of the supply line. In this arrangement, the current flowing through the circuit is limited by the end-of-line resistor to a point that is sufficient to attract the armature of the trouble relay but insufficient to attract the armature of the alarm relay. Either a ground- or an open-fault condition in any of the field wiring will cause the trouble

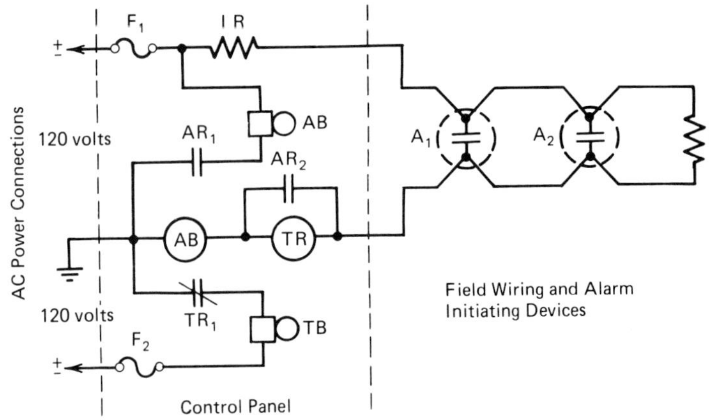

Figure 12.11 Basic end-of-line resistor circuit.

relay to become de-energized and close contacts TR_1 which will sound the trouble signal. In an alarm condition, the contacts of any of the alarm indicating devices will shunt the end-of-line resistor, increasing the current flow through the circuit sufficiently to energize the alarm relay AR and close its contacts (AR_1), sounding the alarm signal. In this circuit, the coil of the trouble relay must be low enough in resistance to pass sufficient current to the alarm relay to energize it.

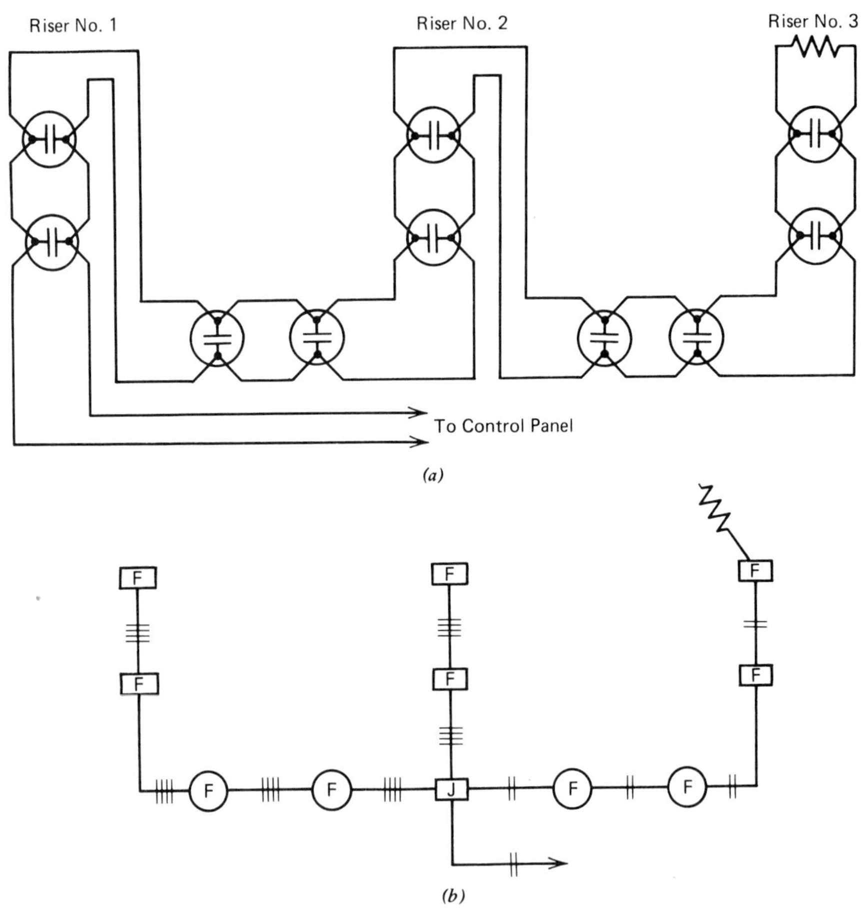

Figure 12.12 (*a*) End-of-line resistor circuit. (*b*) Wiring of riser for end-of-line resistor circuit.

One of the advantages of this system is shown in Figure 12.12. Notice, four wires are required in riser 1 and 2 to maintain supervision of the circuit; only two wires are required on riser 3.

12-5 NURSE-CALL SYSTEM

The nurse-call signal system is usually classed as an open-circuit type. Its main components are an annunciator, bedside pushbutton or switch, and over the door light located in the hallway (Figure 12.13).

When the patient uses the bedside pushbutton or toggle switch, one of the annunciator drops will fall to uncover a white number. At the same time a buzzer will sound, and the hallway light outside the patient's room will also come on. The nurse at the nurses' station will observe the exposed number. The nurse then presses the station reset button to restore the drop and disconnect the buzzer. The light outside the patient's door will remain on until the nurse enters the room to reset the patient's bedside pushbutton.

The electrician should be concerned about the supply conduit to the system. Usually the supply feeds the annunciator box. If the signal wiring is of the low-voltage type, a step-down transformer will usually be located here, and low-voltage wire will be run to the signal lights and bedside pushbutton stations. However, if the system operates at line voltage, the signal wire will be pulled in a conduit.

12-6 ANNUNCIATOR SYSTEMS

Annunciator systems have application virtually wherever people communicate, but yet they need a quiet restrained atmosphere. Schools, banks, medical and legal offices, and restaurants are typical instances. Today, most successful users of annunciator systems hired the system around one or as many as three central command positions. Key people can be aware of room status and the locations of each key person in the building. Status is observed through the use of color-coded lights on centrally located panels or local displays near the door to each room.

In medical clinics a lighted annunciator system would show the status of each of the examining rooms and its relationship to key staff members. Typically, it could be set up like this:

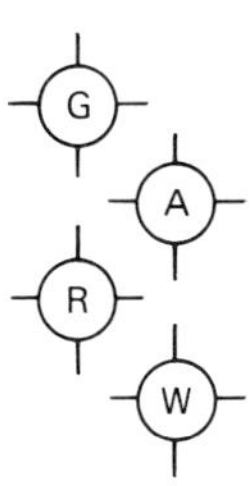

Green light — Room vacant
Amber light — Patient waiting for doctor
Red light — Doctor with patient
White light — Doctor's call for nurse

How it works: At the receptionist's master control center, a room is set as green when it is empty and ready for use. As the patient is admitted to an examining room, the room is set as amber, thus letting the doctor know the patient is waiting. A local control in the room is then activated to red by the doctor as she enters. Should the doctor desire assistance, she would simply

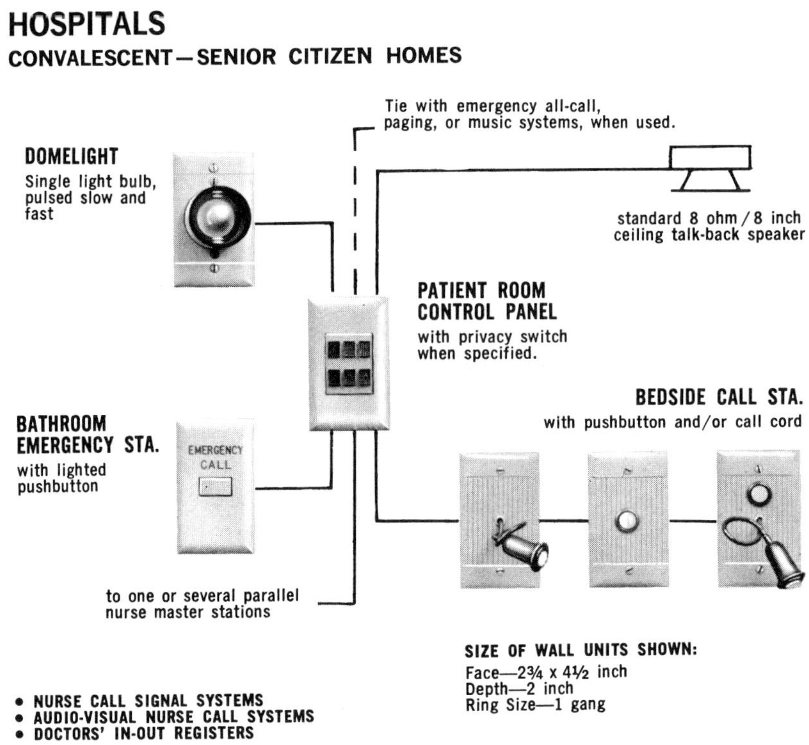

Figure 12.13 Typical nurse call system. (Courtesy of Touch-Plate Electro-Systems, Inc.)

push her local "white" button which would call the nurse to the room. Chimes may also be tied into the system to audibly notify key people of a status change. As the doctor leaves she shuts off the red light, which indicates that the room needs cleaning. When cleaning is complete, the green light is activated, which indicates that the room is ready for the next patient.

The status of each room is shown by a local one gang display in the corridor above the door to that room. In addition, the status of all rooms and all doctors' calls can be shown on master display panels at the nurses' station, the receptionist's desk, and at other convenient locations.

SUMMARY

1. A good local fire alarm system combines all the elements required for early detection and warning of fires in a single, reliable system that can be diverse and extensive enough to fully protect the most complex property, or simple and inexpensive, to meet the requirements of most commercial buildings.
2. Because fire alarm systems must, by nature of application and use, be highly reliable, electrical supervision is used as a required means to assure that the system is in an operable condition at all times.
3. The system most commonly used in many fire alarm systems in schools and other commercial buildings is the parallel end-of-line resistor circuit.
4. The nurse-call signal system is classed as an open-circuit type.
5. The electrician should be concerned with the installation of the supply circuit to the nurse-call system.

PROBLEMS

12-1 An automatic sprinkler system is installed in a commercial building. Is this system enough for fire protection? Explain your answer.

____

____

____

____

12-2 If sprinklers are installed in a commercial building for fire protection, what should be installed in conjunction to the sprinkler system? Explain.

____

____

____

____

12-3 List four alarm indicating devices

1. ______________ 3. ______________

2. ______________ 4. ______________

12-4 There are various types of power supply sources for fire alarm systems. Name four types of systems.

1. ______________ 3. ______________

2. ______________ 4. ______________

12-5 Explain the operation of a nurse-call system.

____

____

____

____

12-6 A modern fire alarm system is designed to save ______________

______________________________ .

PART 4

HEATING SYSTEMS

chapter 13

ELECTRIC SPACE HEATING

Instructional Objectives

1. To learn the advantages of installing electric space heating in a residence.
2. To become more familiar with the methods of heating.
3. To understand the different types of electric heating systems.
4. To learn how to install electric space heating with appropriate temperature control.
5. To become familiar with the difference between baseboard and ceiling heating cables and the installation of each.
6. To learn how to calculate heat loss.
7. To learn the requirements of the NEC for the installation of electric space heating.

Self-Evaluation Questions

Test your prior knowledge of the information of this chapter by answering the following questions. Watch for the answers as you read the chapter. Your final evaluation of whether you understand the material is measured by your ability to answer the questions. When you have completed the chapter, return to this section and answer the questions again.

1. What are the advantages of installing electric space heating units?
2. List the different types of electric heating systems.
3. What are the three methods of heating?
4. Normally, on which walls should thermostats be located?
5. What is the recommended elevation above the floor that thermostats should be located?
6. How is a heating cable installed in the ceiling of a residence?
7. According to the NEC, what code limitations exist for line thermostat controllers for fixed electric space heating?
8. What code requirements exist for splicing heating cables?

9. Why is closer spacing of radiant heating cables recommended near outer walls and windows?
10. What is heat loss?

13-1 ELECTRIC HEATING

Unlike combustion heating, electric heating leaves no residue. "Pure" energy in the form of electricity flows into the house through wires. There is no waste; all of the energy expended is converted into heat, and this is why we call electricity "pure" energy. Combustion of fuel may take place, but the fire is far away in the boiler of the power plant.

Heating an entire house requires a lot of heat, many thousand BTU's per hour, regardless of the method of heating. Electrically, the load during the heating season may be continuously 10 to 20 kw. This is not an unusual or especially large load, but when contrasted with that of a typical home having only lighting and normal household appliances, there is quite a difference.

This means several things. To homeowners just becoming acquainted with electric heating, their electric bills are higher than they are used to. More of the cost of the home is in the electrical system. And, there may be other economic considerations, but these have to be evaluated in terms of the benefits. However, they will not be discussed here.

To the electrical contractor, it means checking the available electrical service carefully before selling the homeowner a job; knowing additional building codes and regulations; and installing bigger, more complex, heavier electrical services as well as more equipment and controls.

To the utility company, it means having to provide completely satisfactory electrical service. To do this may involve such changes as a considerable increase in generating capacity, new regulations allowing higher voltages, and different electrical rates.

Surely, more people need to know more about the use of electricity in heating installations.

13-2 ELECTRIC HEAT IN PERSPECTIVE

In the past year, electrical usage has soared as never before. Everywhere we look, electrical energy is being used in more and bigger ways—light, heat, and power. It now seems clear that in the early 1980s electrical energy will assert itself as clearly superior to all other forms of energy both now and in future generations.

It is interesting to note that the shocking worldwide energy crisis—with its oil and gas shortages hitting all segments of the economy—has been a major stimulus of the growth of electrical application. Even though the push to nuclear generation of electricity has been slowed down by an assortment of technical, ecological, and economic problems, it will surely be one of the dominant power sources of the future. Another dominant source will be the harnessing of the sun's radiant waves—the use of "solar energy." The pace of developing these power sources that lead to the total electrical energy age will be even faster than anyone might estimate, since research, development, economics, and ecology are all on the side of electricity.

The growing availability of electricity, its ease of installation, its flexibility and cleanliness at the point of use, and many other factors, coupled with the steadily dwindling availability of fossil fuels, have all played significant roles in causing the remarkable growth of electrical energy usage, especially in the areas of electric space heating and comfort conditioning.

When economic conditions are favorable, heating by electricity is found to have many advantages.

13-3 ADVANTAGES

1. *Flexible.* Just as electricity is flexible in terms of generation methods, so is electric heat flexible at its point of use. It can be used for completed heating or just partial heating. For example, an electric heating unit can be installed to supply heat to just one room—such as an addition to a house—without affecting whatever other system may be involved. Electric heating units can be used to heat just one area without having to start and operate the main heating system. Unlike other heating systems, electric heating units are available for almost all types of applications, including single and multifamily housing units.
2. *Clean.* The process of converting electricity into heat involves: no dirt, no dust, no vapors, no fumes or odors. Because electric heating units typically include equipment that limits the amount and velocity of air, there is less likely a discoloration of walls; and because electrically heated homes usually involve better insulation and weatherproofing, there is less chance of infiltration of dust and dirt from outside air. In fact, electricity is the cleanest source of energy available for producing heat.
3. *Efficient.* Electric heat is 100 percent efficient so far as conversion of electric energy into heat is concerned. It requires no air for combustion; therefore there is no heat loss up a chimney.
4. *Durable.* In many of today's homes, fuel-fired systems will require eventual replacement due to overall deterioration caused by the combustion process. Electric heating units can be expected to last the life of the heating element, usually comparable in cost to a single cleaning of a fuel-fired system.
5. *Quiet.* Electric heating units—most of those now being manufactured—are silent. This can prove very valuable in helping to control noise pollution, especially when a well-insulated and weatherproofed dwelling is involved.
6. *Controllable.* Unitary electric heating equipment usually has a thermostat of its own. As a result, a homeowner can adjust the temperature of the room to meet the requirements of comfort for that area. On the other hand, single-thermostat installations require that the system provide heating for the entire space served by the system, which can be wasteful unless motorized dampers and other expensive, intricate mechanical controls are utilized. In fact, sophisticated new solid-state controls for electric heating units are being developed, resulting in more selective and energy-efficient use of heat.
7. *Completely Comfortable.* There are no drafts, no cold spots, no uneven irregular heat.

8. *Good Environmental Impact.* Electric space heating emits no air pollution at the home site. The pollution control at the utilities' generating plants are so effectively controlled that pollution there is only a small percentage of what it would be if fossil fuel were burned at each home.

13-4 HEAT TRANSFER

It must be understood that heat is a form of energy and therefore cannot be created or destroyed. However, it can be moved or transported from one place to another through varied mediums.

In order to understand how a heating system works, it is necessary to understand the ways in which heat transfer can occur. Water always flows downhill, never uphill, always from a higher level to a lower level. In a similar way, one might think of heat as always flowing in one direction, from a position of higher temperature to one of lower temperature.

When water flows downhill, the steeper the hill the faster the water travels. Likewise in the transfer of heat, the greater the temperature difference, the greater the quantity of heat will flow in a unit of time. There are three main ways the transfer of heat takes place: conduction, radiation, and convection.

Conduction

This is the flow of heat through a substance due to the transfer of heat energy from particle to particle, from a warmer region to a colder region. For example, if a rod is heated over an open flame, heat travels by conduction from the hot end to the cooler end. Conduction heat transfer occurs not only within an object or substance but also between different substances that are in contact with one another.

A good example of this is house construction. In house construction, there is a combination of wood, insulation, sheetrock, plaster, brick, and even concrete. These materials are often touching each other. If it's hotter in the house than outdoors, heat by conduction will pass through these materials. If it's hotter outdoors than it is in the house, heat will flow into the house (Figure 13.1).

(a) (b)

Figure 13.1 Conduction. (*a*) Heat flows from hot to cold. (*b*) Conduction takes place between different materials.

Figure 13.2 Radiant energy travels in straight lines until it is interrupted or absorbed by some object or body. (Courtesy of The Singer Company, Climate Control Division)

Radiation

Radiation is the transfer of heat through space just as light travels through space. Radiant heat passing through the air does not warm the air through which it travels. Radiant energy travels in straight lines until it is interrupted or absorbed by some object or body. When this energy strikes an object, it causes the molecules of the absorbing object to vibrate. This vibration converts the energy into heat. Radiant energy will pass through certain materials such as glass without heating them, yet it is reflected by various other materials. It will pass through air regardless of its temperature without heating it to any appreciable extent (Figure 13.2).

A good example of the effect of radiant heat is when you are lying on the beach with your face in the sand. You are being warmed by energy waves from the sun. Suddenly, you feel cool. You know immediately, without turning to look, that a cloud has come between you and the sun. The air temperature remains the same, yet you feel cool because your source of "radiant heat" has been blocked.

Convection

Convection heating is where the natural upward flow of heated air results from contact with convective surfaces that have been heated by an electric current (Figure 13.3).

13-5 TYPES OF HEATING UNITS

There are many styles and designs of electric space heating units manufactured in this space age. However, the type unit to be installed will depend

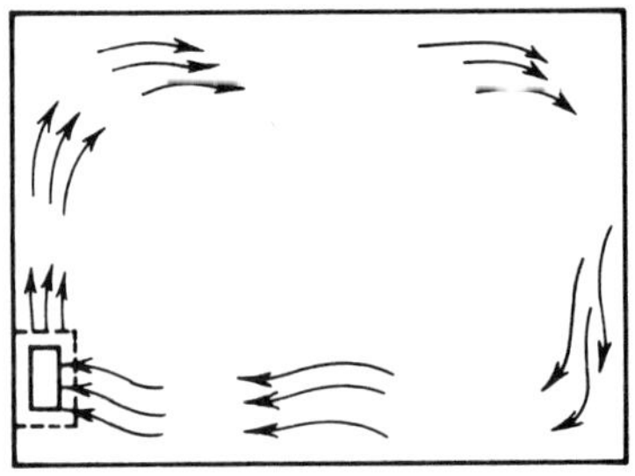

Figure 13.3 Convection heating. Heat transfer depends on air movement over the heated surface in the heating equipment and then into the room. The circulation of air over the convector surface may be produced by a fan or blower or by natural convection.

(*a*)

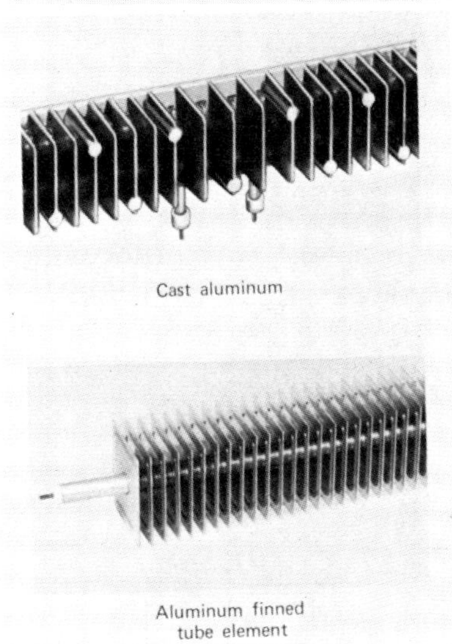

(*b*)

Figure 13.4 (*a*) Basement heating unit. (*b*) Metal heating elements through which the resistance wiring runs. Cast-aluminum grid retains warmth somewhat longer than the aluminum finned-tube element. (*c*) Cutaway view of a heater employing a cast-aluminum grid. End sections with removable covers may be used for wiring convenience outlets or for built-in thermostat to control the baseboard. (Courtesy of The Singer Company, Climate Control Division)

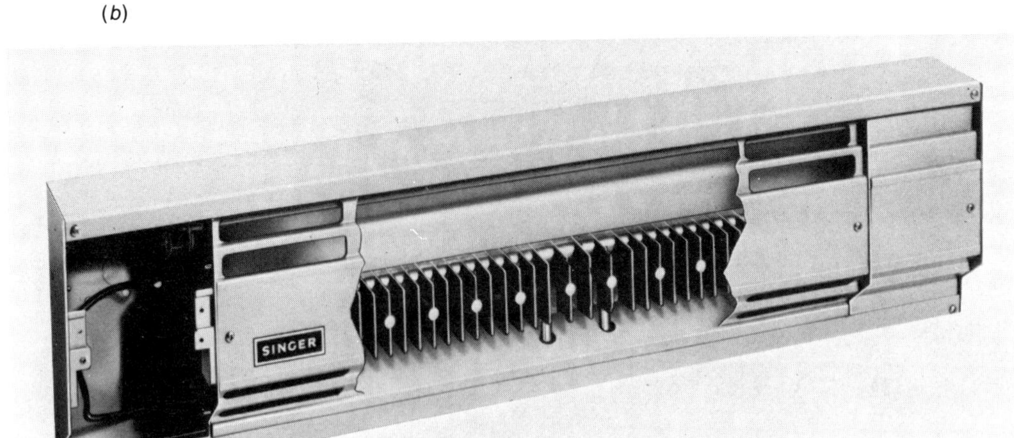

(*c*)

upon several factors: economic conditions, structural conditions, the kind of room and the purpose for which the room is to be used, and whether the homeowner should plan for the future installation of a "heat pump" or for converting the unit to "solar energy". All of the above will be considered as we look at the different types.

Baseboard Heaters

The baseboard heater is used for general comfort heating of a residence. Baseboard sections are available in a wide range of heat output ratings. However, standard density heaters (250 watts per foot) are the most commonly used. They may be used in single sections or joined to supply the heat needed for any given area. These sections can be employed as a sole heating source or as an addition to a central plant, heat pump, or solar energy unit. This type of heater is installed at the floor line, usually on an outside wall under the windows (Figure 13.4).

Baseboard heaters have the advantage of taking a minimum of space in the room and of delivering the heat along a broad area at the outer perimeter of the room where it is needed. In this location, the cold air will be warmed before it circulates through the room. However, the heater should be located so that an adequate circulation of warm air will be provided to the area to be heated.

With convectors, room air flows in through an opening at or near the bottom. The air then flows around the heating elements, out through the opening at the top, and around the room.

The baseboard heater in any of its forms is suited to control by a wall mounted thermostat, though some are manufactured with built-in thermostats. Many now use a high-limit safety control along the full length of the heater to provide against hot spots in case of blocked circulation (Figure 13.5).

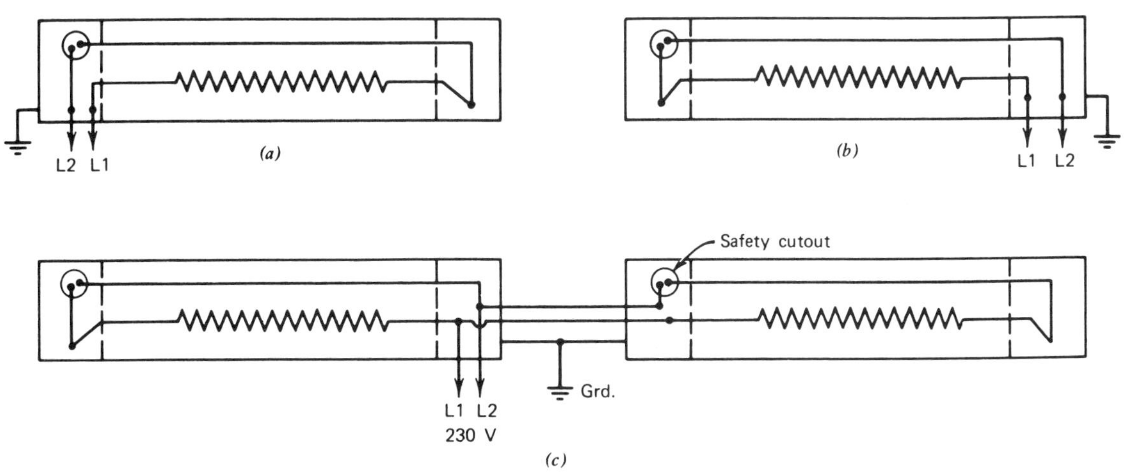

Figure 13.5 Wiring diagram for standard baseboard electric heaters. (*a*) Left-hand line connection. (*b*) Right-hand line connection with safety cutout. (*c*) Power supply to two baseboard heaters each having its own safety cutout.

ELEKTRA-COVE heating effectively and efficiently gives instant satisfying comfort to people of all ages...providing a gentle warmth that permeates every corner of the room...with a uniform heat distribution from floor to ceiling unmatched by any other single system. It is accomplished in a most unique manner, through Primary and Induced radiant wave energy beamed into the space being conditioned.

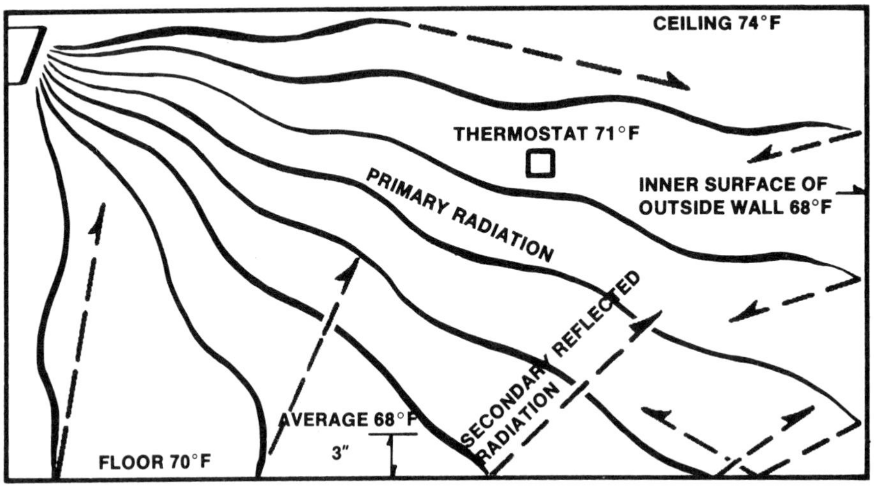

Figure 1 shows Primary Radiation of wave energy being emitted from the face of the unit. This wave energy has the property of directly heating all walls, objects and people. Depending on the type of surface it strikes a percentage of wave energy is absorbed and the balance is reflected and again comes in contact with other surfaces. The energy which is reflected is referred to as secondary radiation and it is this energy, coming from all directions that contributes to uniform comfort heating.

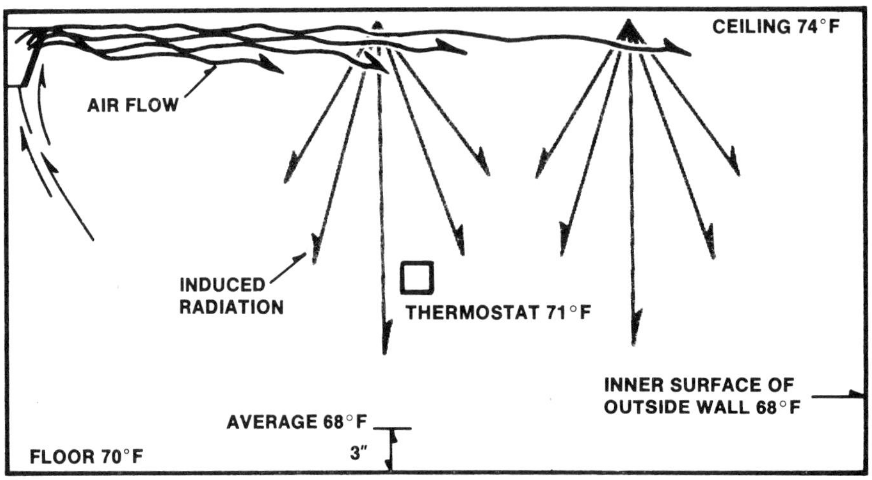

Figure 2 shows how induced radiation is caused. Since the ceiling has a considerable mass it acts as a heat bank which holds the heat and thus radiation continues to distribute warmth throughout the entire room even after the thermostat is satisfied, resulting in more uniform temperatures.

Because of its design features the **Elektra-Cove** enables the induced heat from the ceiling together with the highly efficient (75 to 80%) direct radiation from the face plate to provide satisfying comfort within moments after the system is turned on.

Figure 13.6 Cove radiant heating. (Courtesy of Elektra Systems, Inc.)

Cove Heater Systems

This is a radiant heating unit installed on the wall near the ceiling as a valance. The unit works on the principle that uniform heat emissions of the radiant type will blanket the entire room, directly heating objects in the room (Figure 13.6).

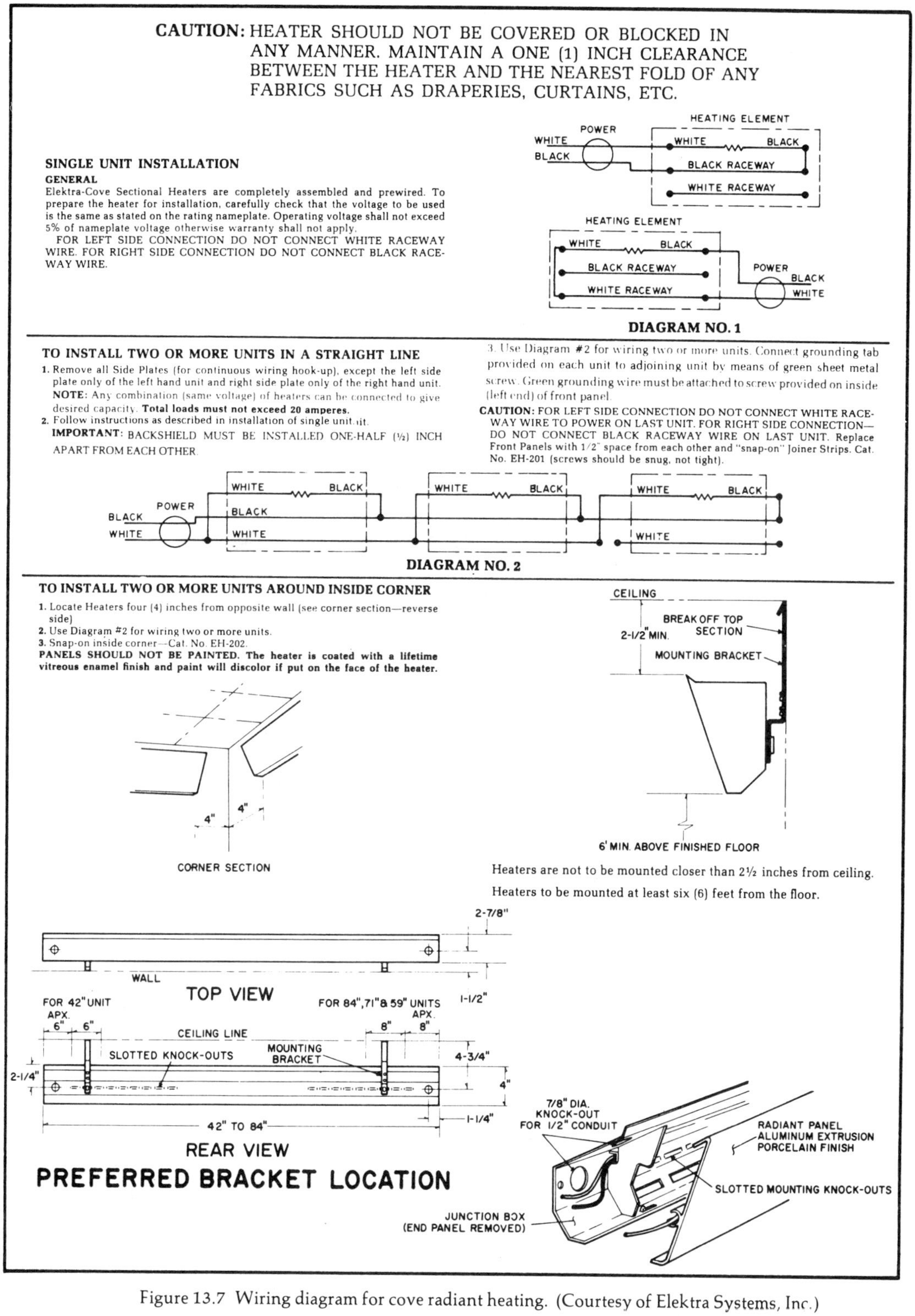

Figure 13.7 Wiring diagram for cove radiant heating. (Courtesy of Elektra Systems, Inc.)

It is the same kind of heat that warms you instantly when the sun strikes a window pane. You are warm as long as the sun is shining and cooler when the sun no longer shines on the window.

Cove units usually come in standard lengths of 42, 60, and 84 inches to accommodate the proper wattage for heating in any room. See Figure 13.7 for ease of installation.

Units are thermostatically controlled so a homeowner will get zoned heat in each room. Temperature requirements are met quickly, quietly, and silently without chilly drafts, blasts of hot, dry air, soot, or dust. Since there are no moving parts to the heater, it is foolproof and maintenance-free (Figure 13.8).

Electric Furnace

Electric furnaces are basically similar to fossil fuel furnaces except that electric resistance heating elements replace the burner (Figure 13.9). Generally, a number of electric heating elements are used. The furnace cabinet also encloses the air filter and a circulating fan that forces the air through the heating section and ductwork to the various rooms. Temperature-limit controls prevent the furnace's overheating. Comfort control of the system is

Figure 13.8 Cove heating unit installed in kitchen. (Courtesy of Elektra Systems, Inc.)

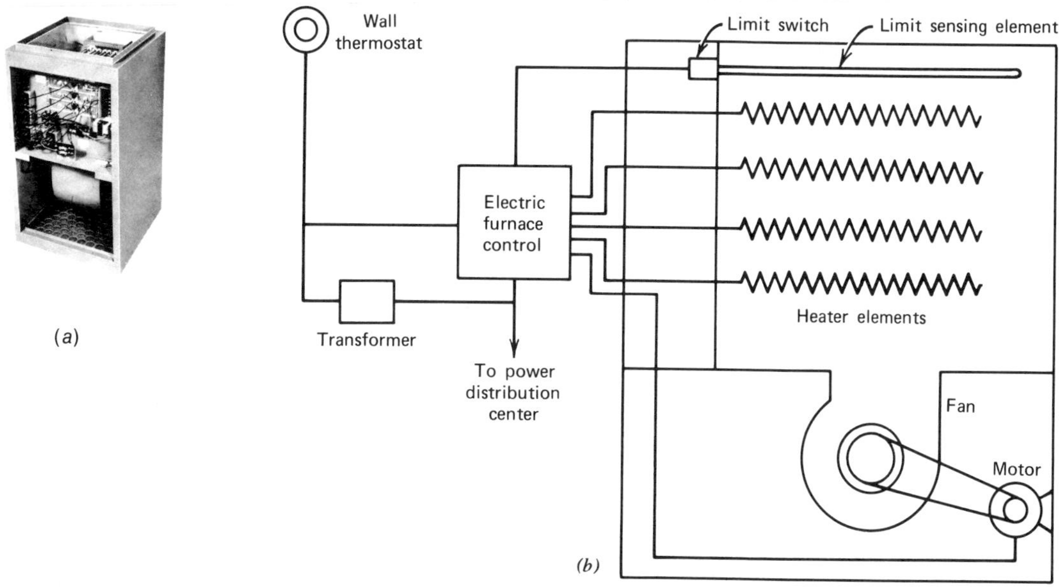

Figure 13.9 (*a*) Warm air electric furnace. (*b*) Wiring diagram showing controls and heating elements. (Courtesy of Martin Industries)

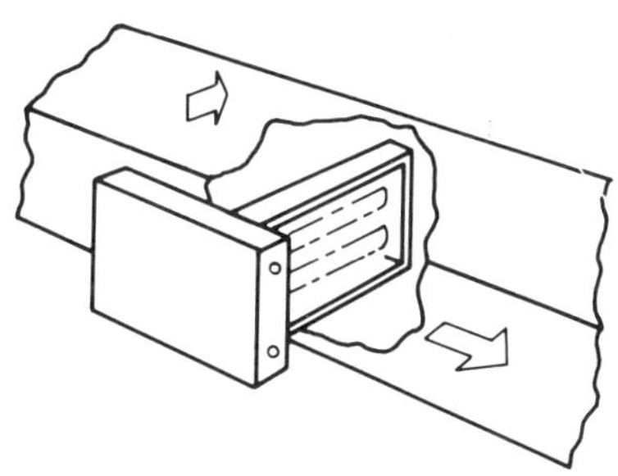

Figure 13.10 Duct heater. Basic unit is complete with heater frame, heating elements, control compartment with hinged or removable cover, primary silent controls, backup contactors, automatic reset thermal cutout, manual reset thermal cutout, fan interlock, fan motor terminal board, factory wired.

provided by a low-voltage thermostat and a relay or sequencer and fan control.

The electric furnace, with a properly designed duct system, may be combined readily with central cooling systems and an electronic air cleaner to provide year-round air conditioning.

Central Fan with Duct or Register Heaters

Warm-air duct or register-heater systems do not employ a large central furnace as the major heat source. They include a central circulating fan or blower and air cleaning system with trunk and lateral ducts to each room. The main heat supply for each room is from a heater in the duct (Figure 13.10) or in the diffusing register in the room.

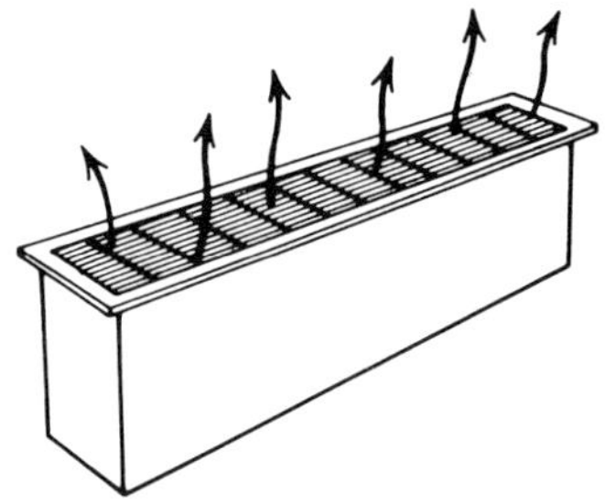

Figure 13.11 Floor furnace.

The central fan system offers advantages of individual room or zone control and is adaptable to central year-round air conditioning.

A duct heater may supply one room or several rooms and be controlled by a zone thermostat. Register heaters installed in each room are controlled by individual room thermostats.

Direct and register units can be used to augment a "heat pump" or "solar energy" system in extreme weather. They may be used similarly to supplement any ducted heating plant having hard-to-heat rooms or additions.

Radiant Heating Panels

Fabricated panels rely on heat-producing wires embedded in plasterboard, tempered glass, or aluminum circuitry sandwiched between two flat sheets of durable polyester to provide radiant heat. These panels are ideal for entryways, vestibules, and other places where a concentrated heat is needed.

Floor Furnaces

These are often used under large picture windows where there is no space for baseboard units and for numerous other situations where the floor location of the heater is preferred. They are designed to heat by gravity convection of warm air from the heater into the room. They may be made with a built-in thermostat; however, they perform better if controlled by a wall-mounted room thermostat (Figure 13.11).

Radiant Wall Heaters

Radiant wall heaters offer advantages in bathrooms, entryways, and other areas where warmth is desired. They should be installed so that the heat rays will not be blocked from the desired heating target or area. They must "see" the heating target in order to heat it (Figure 13.12).

Electric Radiant Heating Cables

For complete invisible heating, the cable can be embedded in plaster or between two sheets of wallboard. Installation of the cable is permanent and care should be exercised to avoid any abnormal condition that would cause damage to the cable or circuit overload. Ceiling electric heating cable radiates heat evenly to rooms below (Figure 13.13). Proper installation is of utmost importance in deriving full efficiency from the system.

Each room has temperature control permitting room-to-room variations for individual needs.

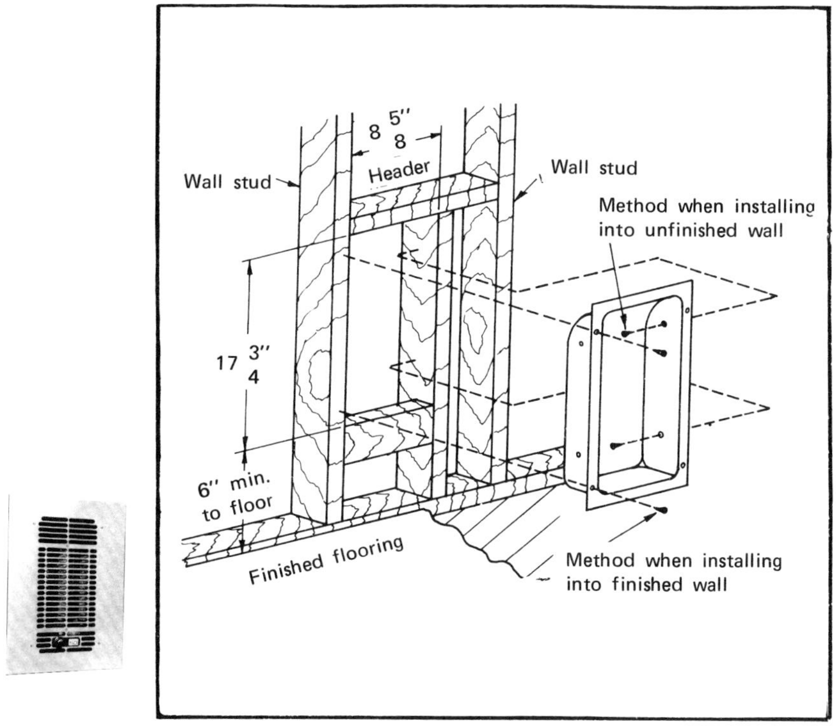

Figure 13.12 Radiant wall heater offers advantages of instant concentrated heat which makes them particularly useful in bathroom, entry ways, and other areas where quick warmth is desired. (Courtesy of Martin Industries)

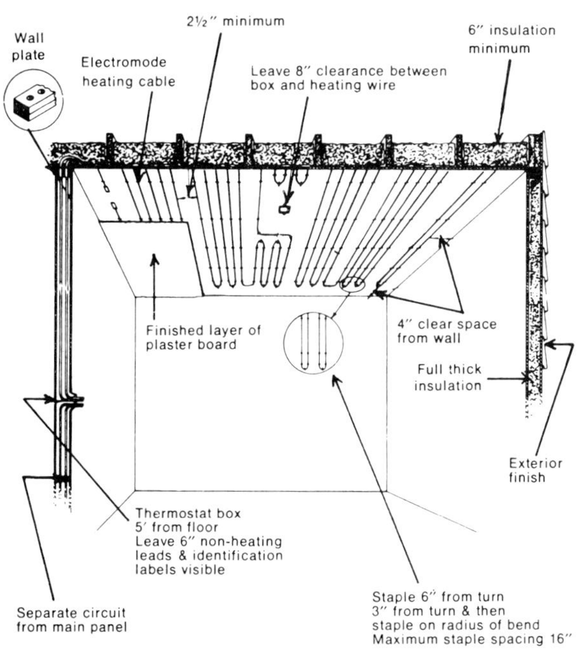

installation information

The diagram above depicts a typical installation for a dry wall ceiling. The cable is stapled directly to the ceiling lath, which should be of a non-metallic fire resistant type such as gypsum board or sheet rock. Under no circumstances should the cables be shortened. In the event that the factory labels have been inadvertently removed from the spool, the cable can be identified by referring to the identification tag on the non-heating leads, or by connecting suitable test meters into the circuit.

Full details on installation procedures can be found in the instruction manual which accompanies the cable.

Figure 13.13 Installation of electric radiant heating cables. (Courtesy of The Singer Company, Climate Control Division)

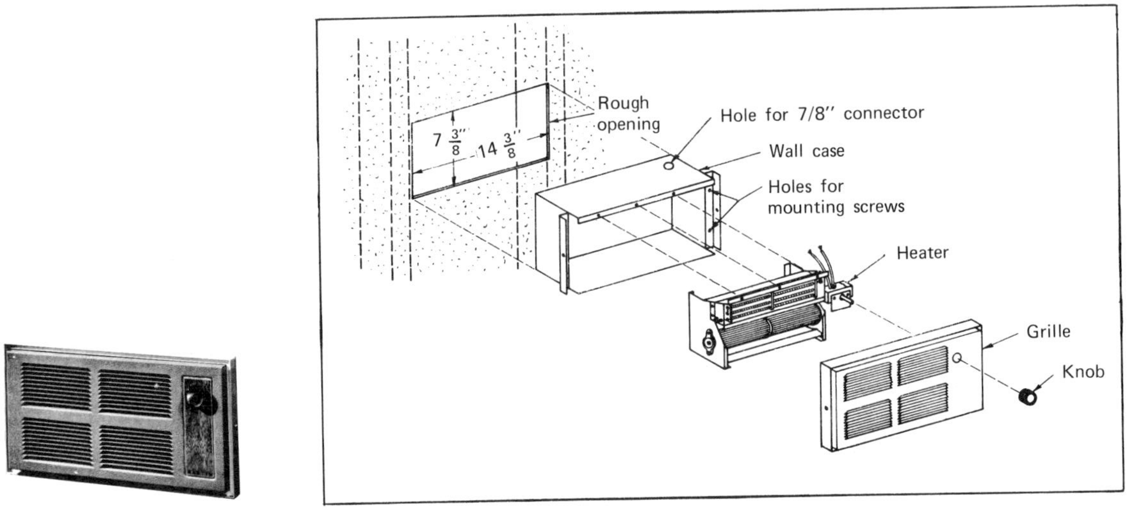

Figure 13.14 Wall-mounted convectors offer the advantage of room air circulation which distributes warmth throughout the space. Fan or blower models produce more positive circulation and more rapid distribution of heat. (Courtesy of Martin Industries)

Wall-Mounted Convectors

Wall-mounted warm air convectors offer the advantage of room air circulation that distributes warmth throughout the space. Fan or blower models produce more positive circulation and more rapid distribution of heat (Figure 13.14). For best results, wall convectors should be installed where circulation of air to the room will not be blocked.

Wall heaters are usually equipped with either wall-mounted or cabinet-mounted thermostats.

Usually approved listed heaters are equipped with a limit switch to shut down the heater and prevent overheating if the air flow should be stopped for any reason.

Ceiling Heaters

Ceiling heaters are often used in bathrooms. In general, two types are available: one having a resistance element, the other infrared lamps. The resistance type is usually combined with a circline fluorescent lamp or with an incandescent lamp to provide light and heat at the same time. The infrared unit, which also performs both functions, utilizes one, two, or three infrared lamps (Figure 13.15).

13-6 TEMPERATURE CONTROL

Thermostats

Years ago homeowners had only one heat control—themselves. When it was cold, they turned up the heat. When the house was too hot, they turned it down. Some of the first thermostats were crudely constructed and allowed temperature variations of several degrees. However, today the homeowner has a wide choice of controls to fit their exacting heating requirements.

This section is aimed at covering some of the basic types of controls for electric space heating.

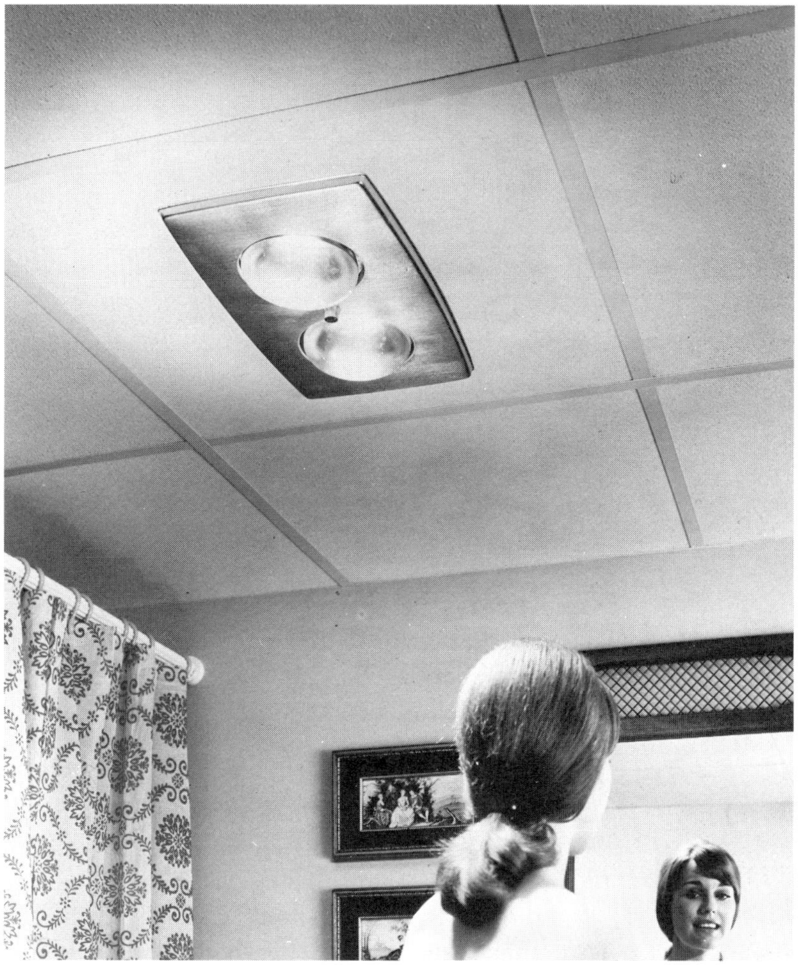

Figure 13.15 Two-bulb heat-a-lamp. Provides luxurious auxilary heat in bathrooms. May be wired for one- or two-bulb operation. Use 250-watt R40 infrared heat resistant glass lamps. Quiet cooling system regulates internal temperatures, has automatic reset thermal protector. (Courtesy of Nutone Division, Scovill Manufacturing Company)

1. *Single-Pole Line-Voltage Thermostat* (Figure 13.16). It has one set of contacts that act as an electric switch. They are connected in the service line, in series with the heating load. Since the single-pole thermostat breaks only one side of the line, it isn't considered to have an electrical "off" position.
2. *Double-Pole Line-Voltage Thermostat* (Figure 13.17). It has two sets of contacts connected to the circuit so that one set breaks one side of the line and the second set breaks the other side of the line. Since both sides are

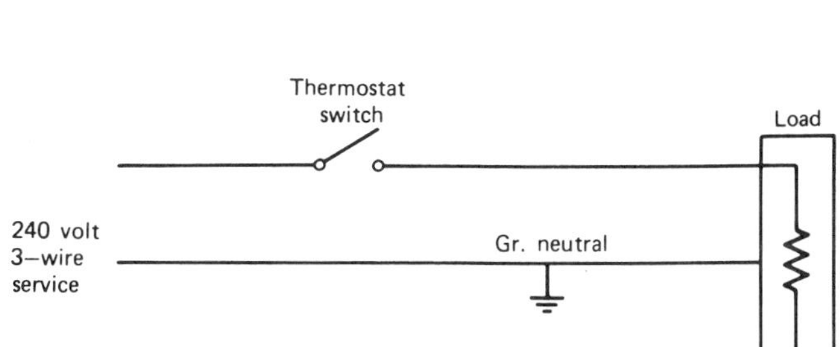

Figure 13.16 Single-pole line voltage thermostat.

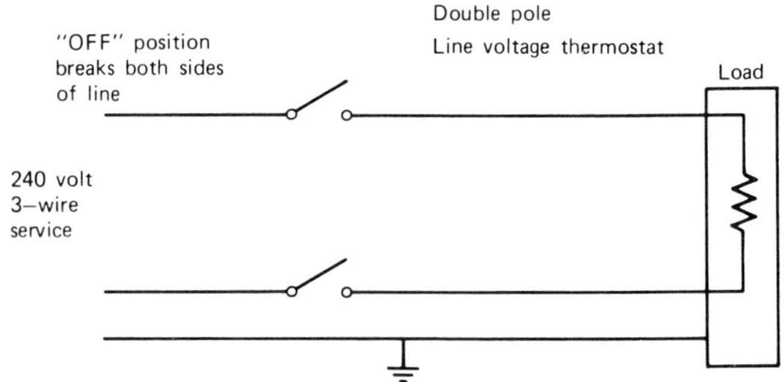

Figure 13.17 Double-pole line voltage thermostat.

broken when the thermostat contacts are locked in the "open" position, the double-pole unit should be used when it is desirable to break both sides of the line and when codes require such a break.

Thermostats built into electric heating equipment are, by their nature, line voltage. Built-in thermostats are either fixed or optional on most wall heaters and baseboard heaters. As a rule, wall-mounted line- voltage thermostats are most sensitive to ambient-surrounding temperatures than built-in types.

3. *Single-Pole Low-Voltage Thermostat* (Figure 13.18). The line or service voltage is applied across the primary winding of a transformer within the relay and less than 30 volts is impressed across its secondary winding. The thermostat is connected to this low-voltage circuit.

   When the thermostat contacts close, current flows through a resistance wire that activates a bimetallic element, which, in turn, operates a switch, and the switch closes. The switch is connected into one side of the service line. When its contacts close, the heater load is energized.

   Opening the thermostat contacts disrupts current through the resistance wire. The bimetallic element cools and operates the switch so that the contacts open to disrupt current to the heating load.
4. *Double-Pole Low-Voltage Thermostat* (Figure 13.19). Just as in the single-pole unit, a step-down transformer delivers about 30 volts to the thermostat.

When the thermostat contacts "make," the voltage energizes a solenoid coil that pulls the mercury switch over and closes their contacts. Opening

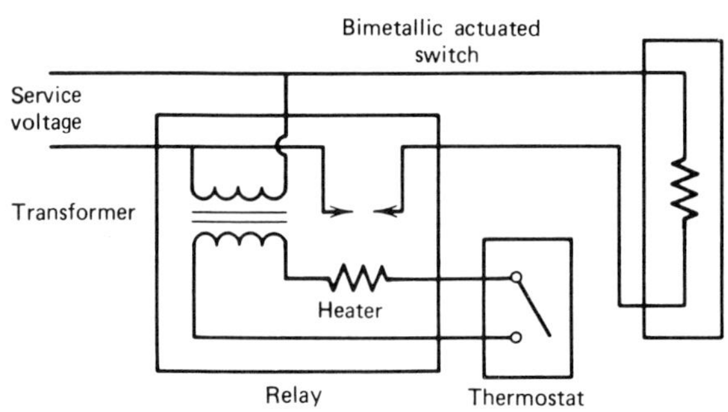

Figure 13.18 Low-voltage thermostat with relay.

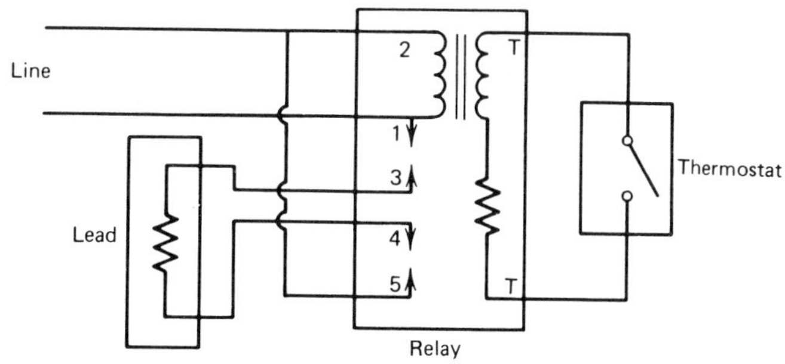

Figure 13.19 Low-voltage double-pole thermostat with relay.

the thermostat contacts de-energizes the coil, and a spring forces the mercury switches to assume their normal open-contacts position.

Low-voltage thermostats are less susceptible to "droop" than line- voltage thermostats. Since the heating load requirements are handled by the relay, the size of the load does not affect the operation of the thermostat. Moreover, the thermostat is less susceptible to contact arcing and welding, since the unit is required to "make" and "break" less voltage.

Code Standards

The following section from the NEC applies to thermostats.

(424-20 Controllers and Disconnecting Means.)

(a) Thermostats and thermostatically controlled switching devices that indicate an off position and interrupt line current shall open all ungrounded conductors when the control device is in this off position.

(b) Thermostats and thermostatically controlled switching devices that do not have an off position shall not be required to open all ungrounded conductors. Remote-control thermostats shall not be required to meet the requirements of (a) and (b) above. These devices shall not be considered as a disconnecting means.

Thermostat Installation

The location of the thermostat is an important consideration in any type of heating.

There are five basic rules:

1. Mount it on an inside wall approximately 5 feet from floor level.
2. Keep the thermostat away from all direct heat and direct light sources. That means don't expose it to the sun or a lamp, and don't put it near a fireplace or a TV set. Heat from such sources will activate the sensing element and cut off the heater when the ambient temperature in the room is below the control point.
3. Don't install the thermostat near ductwork or piping in the wall for the same reason. The unit won't accurately record ambient temperatures in the room.
4. When mounting the thermostat on the wall, use a spirit level or a plumb bob to ensure a vertical position. The calibration does not hold true unless the unit is perfectly level.

5. Finally, when location is desired, it is a good practice to drive a 16 penny spike into the nearest 2 × 4 wall stud. Wrap the thermostat wire around the spike several times. This will mark your location, as the sheetrock installers will have to cut a hole in the sheetrock at this point. Not only do you have your location marked out, you also have sufficient backing for mounting the thermostat.

13-7 DETERMINING HEATING REQUIREMENTS

To determine the heating requirements and design of the heating system, it is essential to calculate the rate of heat loss per hour so that a comfortable environment will be maintained for the activity inside the house.

The temperature to be maintained can vary to conform to the needs and activities of the occupants. For the healthy active child, 60° might be comfortable, while 80° might be more suitable for the older person. Naturally, the definition of comfort will vary from person to person and from one type of activity to another. We therefore have to make our analysis based on average types of usage.

Where Do We Lose Heat?

Look at a typical house, shown in Figure 13.20, and see where the heat escapes. (This is an example of contemporary houses with adequate insulation.)

What Determines Heat Loss?

The total heat loss from a house can be divided into two major parts:

1. The heat loss by conduction through the structure.
2. Heat loss through air infiltration. All homes have some air exchange—cold air entering the house and warm air leaving. This is called infiltration. This cold air must be heated and constitutes a significant load on the

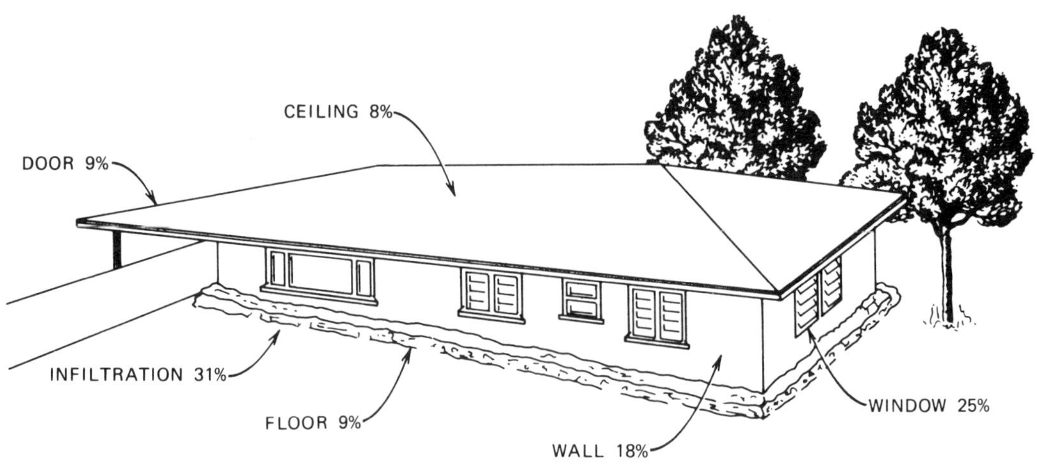

Figure 13.20 A typical house with adequate insulation, showing heat loss by conduction through the structure. (Courtesy of Martin Industries)

heating system. The type of structure and its insulating values can greatly affect the amount of heat loss. You can see from the house that glass areas lose more heat than insulated areas.

Insulation

Insulation is any material that reduces the rate of heat transfer from one area to another. All building materials have some insulation value, but the term insulation is generally applied to a group of products designed mainly to provide this one service.

Insulation performs several functions:

1. It helps conserve heat during cold weather and therefore reduces the energy required to heat the home.
2. It helps reduce the rate of heat gain in hot weather, thereby enabling a house to remain cooler inside in summer than it otherwise would.
3. It maintains higher inside surface temperatures. A person loses heat by radiation to warmer surfaces and will be more comfortable. Warmer inside surfaces also minimize condensation or sweating. As an example, double windows collect less condensate than single windows.

Insulation Values

One way to compare insulating materials is by thermal resistance (R) or ability to resist the passage of heat. This resistance, determined by standard testing procedures, may be expressed per inch of thickness as manufactured. The higher the R value, the better the insulation properties.

The difference in "R" value can greatly affect the heat loss. One example is in wall and ceiling insulation. A 2 × 4 has an "R" value of 1.25 per inch thickness, where fiberglass has an "R" value of 4.00 per inch thickness. Therefore, the part of a structure that is framing loses more heat.

Infiltration

Some air filtration is necessary in any home. Fuel-burning furnaces require air for combustion and to remove the products of combustion from the house. All homes require some air exchange to remove water vapor that is produced within the home. If the house is extremely tight and there is not enough air exchange, severe condensation may occur on windows and other cold surfaces during cold weather. Infiltration is commonly expressed in air changes per hour. Loosely constructed houses may have an infiltration rate of 2 air changes per hour, whereas tight houses may have an infiltration rate of ½ air change per hour or less. The average home will have an infiltration rate of about 1 air change per hour. Infiltration can be reduced by weather-stripping doors and windows, caulking sills and other openings, keeping fireplace dampers closed, and reducing the operating time of exhaust fans.

Design Temperature

The extremes in temperature from inside to outside, called the design temperature, will also determine the amount of heat required to make a space comfortable.

TABLE 13.1 Winter Outdoor Design Temperatures

Degrees Fahrenheit		Degrees Fahrenheit	
40	Miami Beach, Florida	−20	North Platte, Nebraska
30	Southern Texas	−20	Northern Maine
20	San Diego, California	−20	Helena, Montana
10	Arkansas	−20	Minneapolis, Minnesota
0	New York City	−25	LaCrosse, Wisconsin
−10	Kansas	−30	Bismarck, North Dakota
−10	Chicago, Illinois	−35	Miles City, Montana
−15	Des Moines, Iowa	−40	Regina, Saskatchewan

The design temperature varies from area to area. Table 13.1 gives a guideline in selecting the proper value.

Terminology

As in any analysis, there are certain terms that are used and should be understood:

BTU British Thermal Unit (the amount of heat required to raise 1 pound of water 1°F in 1 hour).

Watt 1 kWh = 3.413 BTU.

"U" The units of BTU per hour loss per square foot per degree Fahrenheit temperature differential.

"R" Value The amount of heat a material will hold back. The resistance to heat transfer expressed in total "R" or "R" per 1 inch thickness:

$$R = 1/U$$

13-8 CALCULATING HEAT LOSS

It has been shown that the amount of heat loss is determined by design temperature, "R" value of the surface, and area of the surface. Putting these into a formula, we have the following:

$$\text{Heat loss (watts)} = \frac{A \times DT}{\text{"R"} \times 3.413}$$

A = area of exposed surface
DT = design temperature (indoor to outdoor)
"R" = "R" value for surface (resistance to heat flow)
3.412 = conversion from BTU to watts per hour

To simplify it, we have worked out multipliers in watts per sq ft. This multiplier includes the design temperature, "R" value, and 3.412 conversion factor. All you have to do is multiply by the area of a surface. The multipliers in Table 13.2 are for typical construction. They naturally do not include every possible situation, and multipliers can be developed for any situation by using the "heat loss formula" above.

TABLE 13.2 Heat Loss Multiplier (In Watts per Sq Ft)

	"R" Value (Average)	Design Temperature Differential 70	80	90	100
Walls (Frame)					
3½-inch fiberglass (10 percent framing)	14.6	1.4	1.6	1.8	2.0
3½-inch fiberglass (10 percent framing) (w/1-inch rigid insulation overall)	18.6	1.1	1.3	1.4	1.6
(Block)					
8-inch no insulation	1.04	19.7	22.5	25.3	28.2
12-inch no insulation	1.56	13.1	15.0	16.9	18.8
w/2-inch insulation	9.56	2.1	2.5	2.8	3.1
Ceiling					
4-inch fiberglass (10 percent framing)	14.3	1.4	1.6	1.8	2.0
6-inch fiberglass (10 percent framing)	21.7	0.94	1.1	1.2	1.3
8-inch fiberglass (2 inches over framing)	29.1	0.7	0.8	0.9	1.0
10-inch fiberglass (4 inches over framing)	36.5	0.56	0.64	0.72	0.8
Windows					
Single glass	0.89	23.0	26.3	29.6	32.9
Single glass w/storm	1.79	11.4	13.1	14.7	16.4
Double pane, ½-inch air space	1.73	11.8	13.5	15.2	16.9
Triple pane, ¼-inch air space	2.12	9.7	11.1	12.4	13.8
Doors					
2-inch solid, no glass	2.5	8.2	9.4	10.5	11.7
2-inch solid, w/storm	4.6	4.5	5.1	5.7	6.4
Floors					
(over Crawl Space) (40° Min. Temp.)					
No insulation	1.25	16.4	18.7	21.1	23.4
2-inch fiberglass	9.25	2.2	2.5	2.8	3.2
4-inch fiberglass	17.25	1.2	1.4	1.5	1.7
(Concrete Slab)*					
No insulation	1.2	17.0	19.5	11.0	24.4
2-inch rigid insulation 4 feet down	9.2	2.2	2.5	2.9	3.1
Rim Joist Area†					
No insulation	3.0	6.8	7.8	8.8	9.8
4-inch fiberglass	19.0	1.1	1.2	1.4	1.6
Infiltration‡					
Tight Construction (½ air change per hour)		0.35	0.4	0.45	0.5
Loose Construction (1 air change per hour)		0.7	0.8	0.9	1.0
Fireplace§					
Tight Damper		410	470	525	586
Loose Damper		1022	1168	1314	1460
No Damper		3070	3512	3951	4390

*Heat loss on slabs is based on perimeter of slab exposed to the outside; therefore it is in lineal feet.
†Rim joint area is the framing space between floors and concrete block walls—if uninsulated, can be an area of high heat loss.
‡See discussion of infiltration on page 219.
§Fireplace losses are expressed in watts for each design temperature. Add these values to your normal calculation for each room with a fireplace.

Sample Calculation

Design temperature ________° difference

Room ____________________		Multiplier (10)	Loss watts per hour (11)	Notes—heating equipment
Size ______X______X ______				
Outside wall gross	______ sq ft			
Windows	______ sq ft			
Doors	______ sq ft			
Outside wall net	______ sq ft			
Ceiling	______ sq ft			
Floor (wood over air)	______ sq ft			
Floor (concrete slab)	______ lin ft			
Air change	______ cu ft			
TOTAL HEAT LOSS FOR THIS ROOM			(12)	

1. **Design temperature**—Select from chart. In Minneapolis 90°F is normally used.
2. **Room and size**—Identify the room or area.
3. **Outside wall gross**—This is total wall area exposed to the outside.
4. **Windows**—Total window area in the room. Remember, sliding glass doors are the same as windows in calculation.
5. **Doors**—Include any glass area in doors in (4) above.
6. **Outside wall net**—Subtract door and window areas from outside wall gross. This gives you the insulated wall area.
7. **Ceiling**—If there are any skylights, they must be considered separately.
8. **Floor**—Overheated basement, no loss. Unheated or crawl space depends upon insulation. If it is a concrete slab, it depends on lineal footing around the perimeter, rather than square footage of area.
9. **Air change**—Depends upon how tight the structure is and if it has a fireplace or not. Remember how tight the damper is on the fireplace could also affect this. *Note:* This will be based on cubic footage or volume, not area.
10. **Multiplier**—After you have put in the areas or volume, insert the multiplier from page 221 or from your own calculation.
11. **Loss in watts per hour**—Multiply them out and you can determine how much watt loss you have from each part of the structure.
12. **Total heat loss for this room**—Add these up and it gives you a minimum size of heating unit in watts that is required to make up for the heat loss in this room.

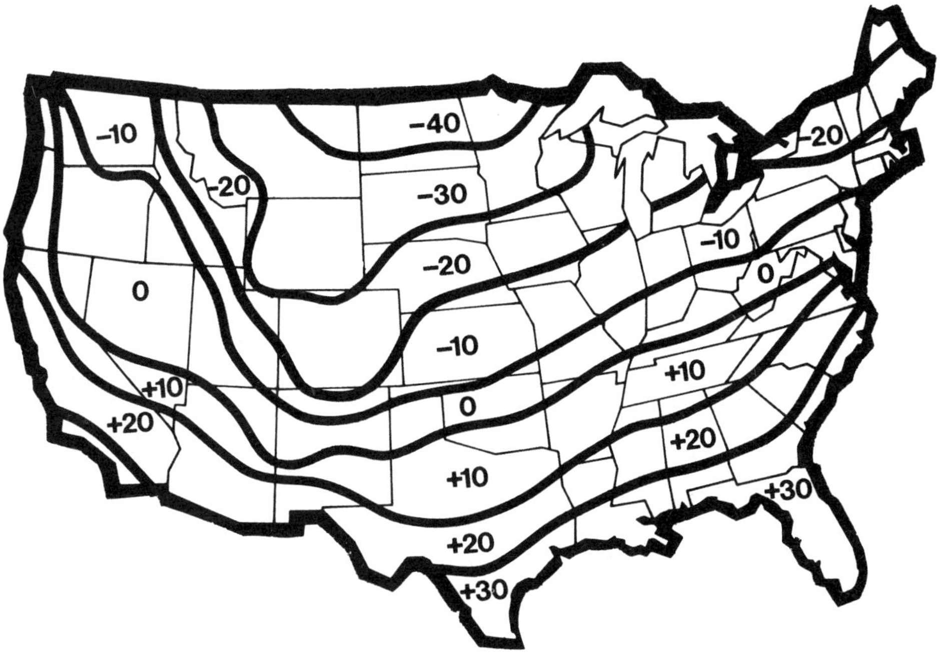

1. Figure the square footage of the room to be heated by multiplying length × width.	STEP 1: ____ × ____ = ___ sq ft
2. Find the outside wall area of the room by multiplying the horizontal wall length by the height.	STEP 2: ____ × ____ = ___ sq ft
3. Determine whether windows exceed 10 percent of exposed wall area. (a) Add up square footage of windows in room. (b) Multiply wall area (No. 2) × 0.10 (10 percent). (c) If (a) is *larger* than (b)—enter the overage in (b) of STEP 5.	STEP 3: (a) ___ + ___ + ___ = ______ sq ft (b) ______ × .10 = ______ sq ft (c) ____ − ____ = ______ sq ft (a) (b) Enter 5(b)
4. From the map, select the temperature zone in which heater is to be installed.	STEP 4: ____________ zone
5. (a) Multiply the area found in STEP 1 × MULTIPLIER I (below) from chart for proper zone. (b) Multiply window area in excess of 10 percent of wall area (3c) × WINDOW MULTIPLIER II (below) from chart.	STEP 5: (a) ____ × ____ = ______ watts (b) ____ × ____ = ______ watts ____ + ____ = ______ watts (a) (b) Total

TEMPERATURE ZONES—MULTIPLIERS I AND II:

Zone:	−40	−30	−20	−10	0	+10	+20
I—Floor Area	10	9	8	7	6	5	4
II—Window Excess	16	15	13	12	10	9	7

STEP 6: Select an electric heater with wattage equal to or greater than Total Watts in STEP 5.

Note: Unusual designs and vaulted ceilings may not be adequately determined by this method. Local utilities may be contacted for assistance.

SUMMARY

1. With electric space heating, temperature requirements are met quickly, quietly, and without chilly drafts.
2. A steadily dwindling availability of fossil fuels has caused a growth in the use of electrical energy.
3. Electric space heating is foolproof and maintenance-free.
4. Thermostats are installed on inside walls away from ductwork and piping.
5. Inside temperatures can vary to conform to the needs and activities of the occupants.
6. The type of structure and its insulating value will effect the amount of heat loss.
7. Insulation reduces heat transfer from one area to another.
8. Electric heat panels are ideal for an area where a concentrated heat is needed.
9. Infiltration is expressed in air changes per hour.

PROBLEMS

13-1 List six advantages that electric space heating has over other heating systems.

1. ____________ 4. ____________

2. ____________ 5. ____________

3. ____________ 6. ____________

13-2 List seven different types of heating units available for installation in a residence.

1. ____________ 5. ____________

2. ____________ 6. ____________

3. ____________ 7. ____________

4. ____________

13-3 Explain what is meant by convection heating.

____

____

____

____

13-4 Draw a wiring diagram of a baseboard heater unit being controlled by a double-pole line voltage thermostat.

13-5 Explain what is meant by the R designation of an insulation.

____

____

____

____

13-6 In conduction heat transfer, heat travels by conduction from ____________________ to ____________________

13-7 Radiant heating panels are ideal for ____________________ ____________, and other places where a ____________ heat is needed.

13-8 Thermostats are mounted on inside outside walls (circle one).

13-9 To determine the heating requirements and design of the heating system, it is essential to calculate the rate of ____________ per hour so that a comfortable environment will be maintained for the activity inside the house.

13-10 The total heat loss from a house can be divided into two major parts. Name the two parts.

____

____

chapter 14

SOLAR ENERGY

Instructional Objectives

1. To understand the need for harnessing the sun's power.
2. To make you aware of our environment.
3. To learn why we need other sources of energy.
4. To provide the basic components in the solar system.
5. To learn some advantages of the storage system.
6. To understand the operation of heat transfer.
7. To become familiar with the types of collectors.
8. To make you aware of the need for auxiliary energy.

Self-Evaluation Questions

Test your prior knowledge of the information in this chapter by answering the following questions. As you read the chapter, watch for the answers. When you have completed the chapter, return to this section and answer the questions again.

1. Why do we need to find ways to harness the sun's power?
2. What types of fuel are in short supply?
3. Will solar energy be used for more than space heating?
4. How is energy collected?
5. Why is a good storage system essential?
6. What will happen to solar energy on a cloudy day?
7. Is solar heat effective in a cold climate?
8. What is the purpose of adding auxiliary energy?
9. Where will the auxiliary system be located?
10. What is the purpose of the control component?

14-1 ENERGY FROM THE SUN

This section is about solar energy and how it is used for space heating, cooling, and domestic hot water heating. It has been prepared as an introduction for the newcomer, apprentice, and owner or builder interested in reducing heating costs by harnessing the sun's power.

The United States currently faces two interrelated problems. One, we are endangering our environment by polluting the atmosphere, ground, and water with the by-products of our technological society. Two, we are quickly running out of fuels that have enabled us to achieve a high standard of technological development. This perplexing dilemma has resulted in a nationwide investigation of energy choices.

Among the many energy alternatives being considered is solar energy. Harnessing the sun's power is considered an attractive alternative because it is a renewable resource that does not pollute. In contrast to conventional fuels, its use eliminates the need for refining, transportation, and conveying fuel and power over long distances. The use of solar energy for heating and cooling promises a more rapid payoff than other energy alternatives, because the basic technology already exists and needs only minor refinement. Considerable research, development, and demonstration activities have been initiated in the public and private sector to facilitate the widespread utilization of solar energy.

14-2 SOLAR SYSTEM COMPONENTS

Several characteristic properties apply to all solar heating, cooling, and domestic hot water systems, whether they are simple or relatively complex. Any solar system consists of three generic components: collector, storage, and distribution, and may include three additional components: transportation, auxiliary energy system, and controls (Figure 14.1). These components may vary widely in design and function. They may, in fact, be one and the same element (a masonry wall can be seen as a collector, although a relatively inefficient one, that stores and then radiates or "distributes" heat directly to the building interior). They may also be arranged in numerous combinations dependent on function, component compatibility, climatic conditions, required performance, and architectural requirements.

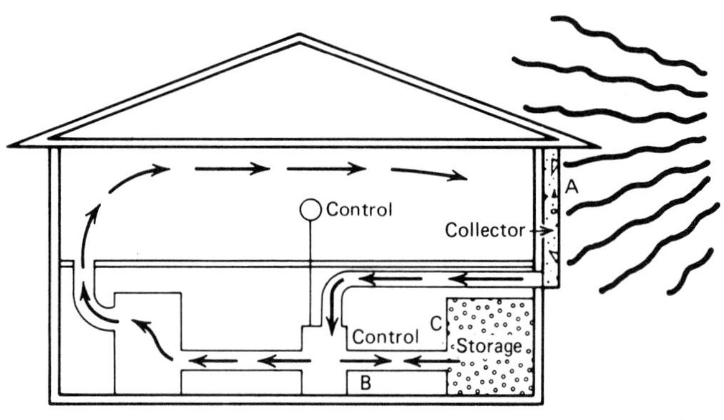

Figure 14.1 Solar system components. (*a*) Taking heat from collector. (*b*) Storing heat from collector. (*c*) Heating from storage.

The Collector

The collector converts incident solar radiation to usable thermal energy by absorption on a suitable surface. The thermal energy captured is transformed to a heat transfer medium, usually gas or liquid within the collector.

There are numerous concepts for the collector of solar radiation. These concepts range from the most simple—a window—to those that are quite complex and require advanced technology for their development.

Collectors are generally classified as focusing or nonfocusing, depending upon whether the sun's energy is concentrated prior to being absorbed or collected at the density received at the earth's surface.

Of the many solar heat collection concepts presently being developed, the relatively simple flat-plate collector has found the widest application. Its low fabrication, installation, and maintenance cost as compared to higher temperature heat collection shapes has been the primary reason for its widespread use.

Flat-plate collectors utilize direct as well as diffuse solar radiation. Temperatures to 250°F (121°C) can be attained by carefully designed flat-plate collectors.

A flat-plate collector generally consists of an absorbing plate, often metallic, which may be flat, corrugated, or grooved, painted black to increase absorption of the sun's heat, insulated on its backside to minimize heat loss from the plate, and covered with a transparent cover sheet to trap heat within the absorber. The captured solar heat is removed from the absorber by means of forced air circulating underneath the collectors. A fluid such as water or air passes through the collector, Figure 14.2, picks up the sun's heat from the hot absorber surface and transports that heat away from the collector. The heated air or liquid can be used directly or it can give up its heat to a storage container (Figure 14.3).

Storage

The storage component of a solar system is a reservoir capable of storing thermal energy. Storage is required since the largest portion of total heat usage will occur at night or on consecutive sunless days when collection is

Figure 14.2 Flat-plate solar collector, with a high-absorption, black chrome coated absorber plate, an acid-etched, tempered, low-iron glass cover for increased light transmission, and rugged, fully insulated construction. (Courtesy of Lennox Industries, Inc.)

Figure 14.3 A solar heat storage tank, heat exchanger, and control center all in one. A heat transfer fluid is circulated through the collector where it absorbs heat. This fluid is then piped to the built-in heat exchanger in the storage tank. The cooled transfer fluid is then pumped back to the collectors to continue the cycle. As hot water is drawn from the conventional water heater, hot water from the solar module storage tank replaces it. (Courtesy of Lennox Industries, Inc.)

not occurring. Storage acquires heat when the energy delivered by the sun and captured by the collector exceeds that demand at the point of use.

Air systems generally have a large insulated storage tank consisting of washed river rock about the size of a golfball. The rock storage is heated as the air from the collector is forced through the rock container by a blower. A decrease in rock size increases the air flow resistance through the storage and may affect blower and duct size and distribution efficiency. Rock storage does not have to be in close proximity to the collector. However, as the distance increases, the heat transfer losses between the heated air and the rocks also increase, and larger ducts and more electrical power are generally required for moving air between the collector, storage, and heated spaces.

Domestic hot water piping is run through the storage tank prior to passing through a conventional water heater. Storage heat is transferred to the hot water piping, thereby either eliminating the need for additional heating or substantially reducing the energy required to raise the water to the needed distribution temperature.

Distribution

The distribution component receives energy from the collector or storage component and disperses it at points of consumption—spaces within the dwelling. For example, comfort heat is usually distributed in the form of warm air or warm water by ducts or pipes within a building. Distribution of energy will depend upon the temperature available from storage space heating if baseboard convectors are increased in size or if used in conjunction with a heat pump or auxiliary heating system.

Because solar-produced temperatures in storage are normally in the low range (90 to 180°F), distribution ducts and radiating surfaces are normally larger than those used in conventional heating systems. Therefore, careful consideration is required in the design of heat distribution systems throughout the dwelling.

Domestic water heating is also a part of the distribution component. Its distribution system generally consists of a heat exchanger, backup heater, piping, and controls.

Transport

Most solar systems have an energy transport component that provides the means of moving a fluid carrying thermal energy to and from the collector and storage. The transport component also regulates the flow through the collector and storage. In liquid or air systems this component consists of pumps, valves, and pipes, or blowers, dampers, and ducts.

Auxiliary Energy

The auxiliary component provides a supply of energy for use during periods when the solar system is inoperable or during periods of extremely severe weather or extended cloudy weather when solar-produced temperatures from the collector and storage are not sufficient to satisfy the dwelling's heating or cooling load. Currently, the experimental nature of solar heating, cooling, and domestic hot water systems and the possibility of extended sunless days generally require that the auxiliary energy components be capable of providing for the total energy demand of the house if the solar system is inoperative.

Almost any type of auxiliary energy system may be used in conjunction with a solar system. The auxiliary system may be completely separate or fully integrated with the solar heating/cooling system. However, in most cases, it makes economic sense to integrate the backup system with the solar system. This means running the distribution component from heat storage to the occupied space through the auxiliary system where an energy boost may be supplied when storage temperatures are low. Heat from storage may also be used in conjunction with heat pumps. The heat pump, a device that transfers heat from one temperature level to another by means of an electrically driven compressor, utilizes the solar heat available from the storage to supply necessary heat to the occupied space. The advantage of the heat pumps/solar system integration is the reduction of electrical energy

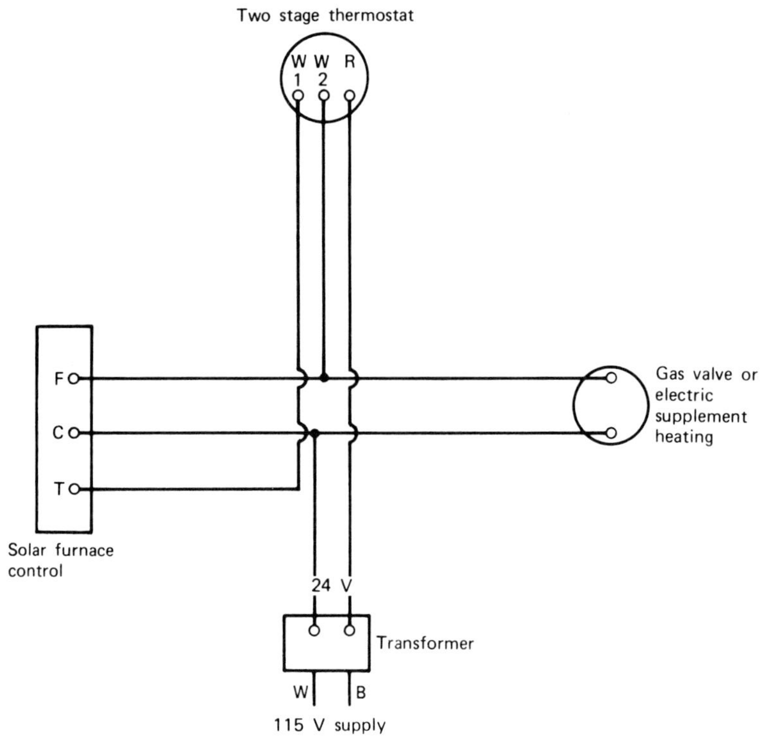

Figure 14.4 Wiring diagram showing two-stage thermostat control for solar furnace tied in with gas valve on furnace or electric supplement heating.

required by the heat pump because of heat supplied by solar storage. (Heat pumps are discussed in Chapter 15.)

Supplemental heat may also be supplied by installing electrical resistance duct heaters into the duct work of the system. (Refer back to Chapter 13 and types of heating units.)

Control

The control component performs the sensing, evaluation, and response functions required to operate the system in the desired mode. For example, the temperature in the house is sensed by a thermostat and relayed to the distribution component (pump or blower) when heat is required. The controls generally distribute information, including fail-safe instructions, throughout the system by means of electrical signals. However, the control function can be performed by automatic pneumatic controls or by the dwelling occupants who initiate manual adjustments to alter the system (Figure 14.4).

SUMMARY

1. A solar system consists of three generic components: collector, storage, and distribution.
2. The collector converts solar radiation to usable thermal energy by absorption.

3. The storage component of a solar system is a reservoir capable of storing thermal energy.
4. An auxiliary system can be powered by conventional fuels or electricity.
5. The control component performs the sensing, evaluation, and response functions required to operate the system.

PROBLEMS

14-1 List the six components required for a solar heating system.

1. ____________ 4. ____________

2. ____________ 5. ____________

3. ____________ 6. ____________

14-2 Explain how conventional fuels or electricity could be added to supplement heat on a cloudy day or cold night.

____

____

____

____

14-3 Draw a sketch of a solar system showing a collector, storage, and distribution of air to a single-story house.

14-4 Draw a line drawing of the control system, showing a two-stage thermostat hooked to a solar furnace.

14-5 Explain why a good storage system is essential.

____

____

____

____

14-6 What is the purpose of the solar collector?

chapter 15

THE HEAT PUMP

Instructional Objectives

1. To become familiar with the basic operation of the heat pump.
2. To make you aware of the alternatives.
3. To become familiar with the outdoor compressor.
4. To understand why the heat pump is sized for the cooling load of the house.
5. To learn the advantages of the heat pump.
6. To develop an understanding of efficiency loss.
7. To understand unit maintenance of the system.

Self-Evaluation Questions

Test your prior knowledge of the information in this chapter by answering the following questions. Watch for the answers as you read the chapter. Your final evaluation of whether you understand the material is measured by your ability to answer these questions.

1. What is a heat pump?
2. How does the heat pump transfer heat/cooling to the rooms?
3. How is the refrigerant reversed?
4. What is needed during extreme cold weather?
5. How is heat/cooling controlled?

15-1 THE HEAT PUMP

In these days of lower thermostat settings and revamped heating systems, the heat pump is gaining attention as a possible energy-saver, full-efficient alternative to the more traditional home heating systems.

Actually, the heat pump may sound new, but it's not. It was used experimentally in the 1930s and was adopted by the United States Army during World War II for use in some government buildings in the southern United States.

It wasn't until quite recently that the technology of heat pumps had advanced to the stage that they could be considerd as viable heating alternatives in more severe northern climates.

Today, manufacturers say, heat pumps are in use as far north as southern Canada.

15-2 THE HEAT PUMP DEFINED

What is a heat pump? To oversimplify, the heat pump is an electrically operated system that can both heat and cool the conditioned area (Figure 15.1).

During the heating cycle, heat is "pumped" from outside air through coils containing a pressurized liquid (there is heat in cool outside air). The heated liquid is then vaporized and, as the inside air circulates around the coils, the heat is transferred to the air and pumped into the house heating ducts.

In the summer months, this operation is reversed. The heat pump reverses the flow of heat by using a system of valves in the refrigerant that turn the

HOW A HEAT PUMP WORKS

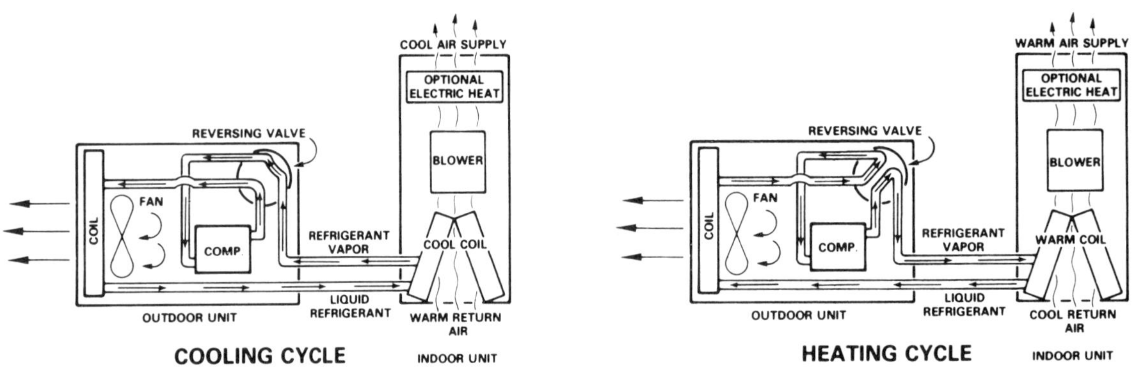

Cools in Summer

During cooling, a heat pump removes heat from indoor air by circulating liquid refrigerant from the outdoor coil to the indoor coil. Here, warm indoor air is forced over the coil surface by a blower. Liquid refrigerant in the coil changes to cold vapor and absorbs the heat. Air flowing over the finned coil is cooled, excess moisture is condensed. This cool, dry air is then circulated gently and quietly throughout your home.

The refrigerant vapor, having absorbed heat, is returned to the outdoor unit where the compressor and outdoor coil convert it back into a liquid and discharge the heat. The liquid refrigerant then travels back indoors to continue the cooling process as long as there is a need.

Heats in Winter

During heating, a reversing valve changes refrigerant flow so the heat pump can use solar energy present in outdoor air to heat your home. Refrigerant in the outdoor coil absorbs heat from air passing over it. Even at 0°F, air still contains over 82% of the heat that was available at 100°F. (That may seem strange but it's true.) The compressor pumps the refrigerant, now in a hot vapor form, to the indoor coil. The blower circulates indoor air over the hot coil, warming it for distribution throughout your home. As the hot vapor cools, it condenses and the resulting liquid refrigerant returns to the outdoor coil. There it once again absorbs heat and repeats the cycle as long as heating is needed.

Figure 15.1 Operating principles of the heat pump. (Courtesy of Lennox Industries, Inc.)

refrigerant around and make it flow in the opposite direction. The pressurized liquid is pumped into the indoor coils where warm air is blown over them. The coils absorb the heat from the air and pump it to the outside. The cooled indoor air is then circulated through the house.

Heat pumps distribute the heat or cooling to the rooms through a central duct system in the same manner as other central forced-air systems (Figure 15.2).

It should be noted, however, that the heat pump is usually sized for the cooling load of the house. If heat pumps were sized to handle total heating demands in homes during the coldest winter weather periods, they would be considerably oversized with relation to summer cooling needs.

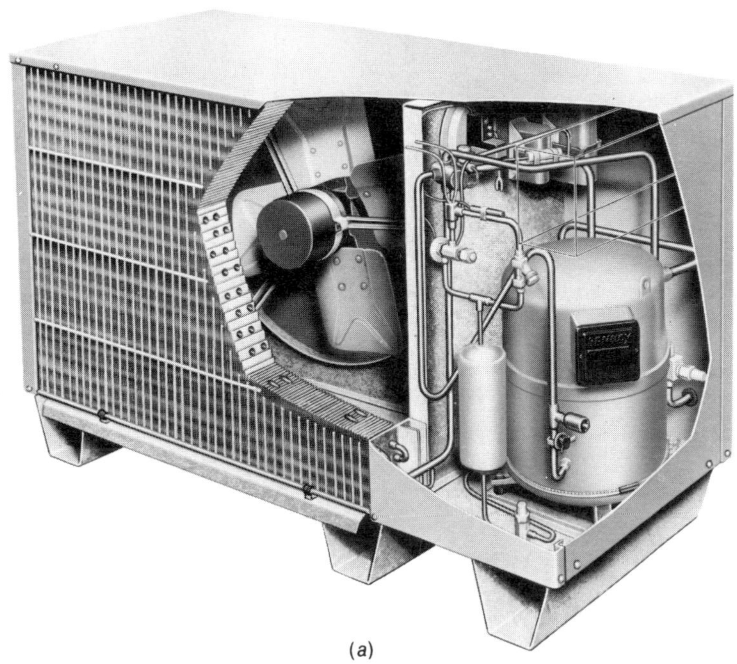

(a)

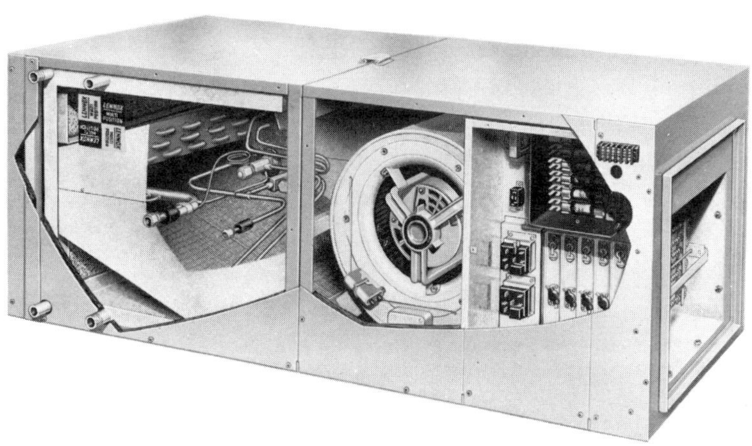

(b)

Figure 15.2 The heat pump. (*a*) Outdoor unit of the H.P. 10 series heat pump, suitable to rooftops or grade level slab installations. Note that the compressor is housed in the outdoor unit as in conventional split cooling sets. In the heat pump, however, the function of the outdoor and indoor coils reverses seasonally by means of a reversing valve. (*b*) Typical indoor blower-coils-filter unit, extremely versatile, can be installed in a closet, attic space, suspended from a ceiling, or in a crawl space. (*c*) Heat pump wiring diagram. (Courtesy of Lennox Industries, Inc.)

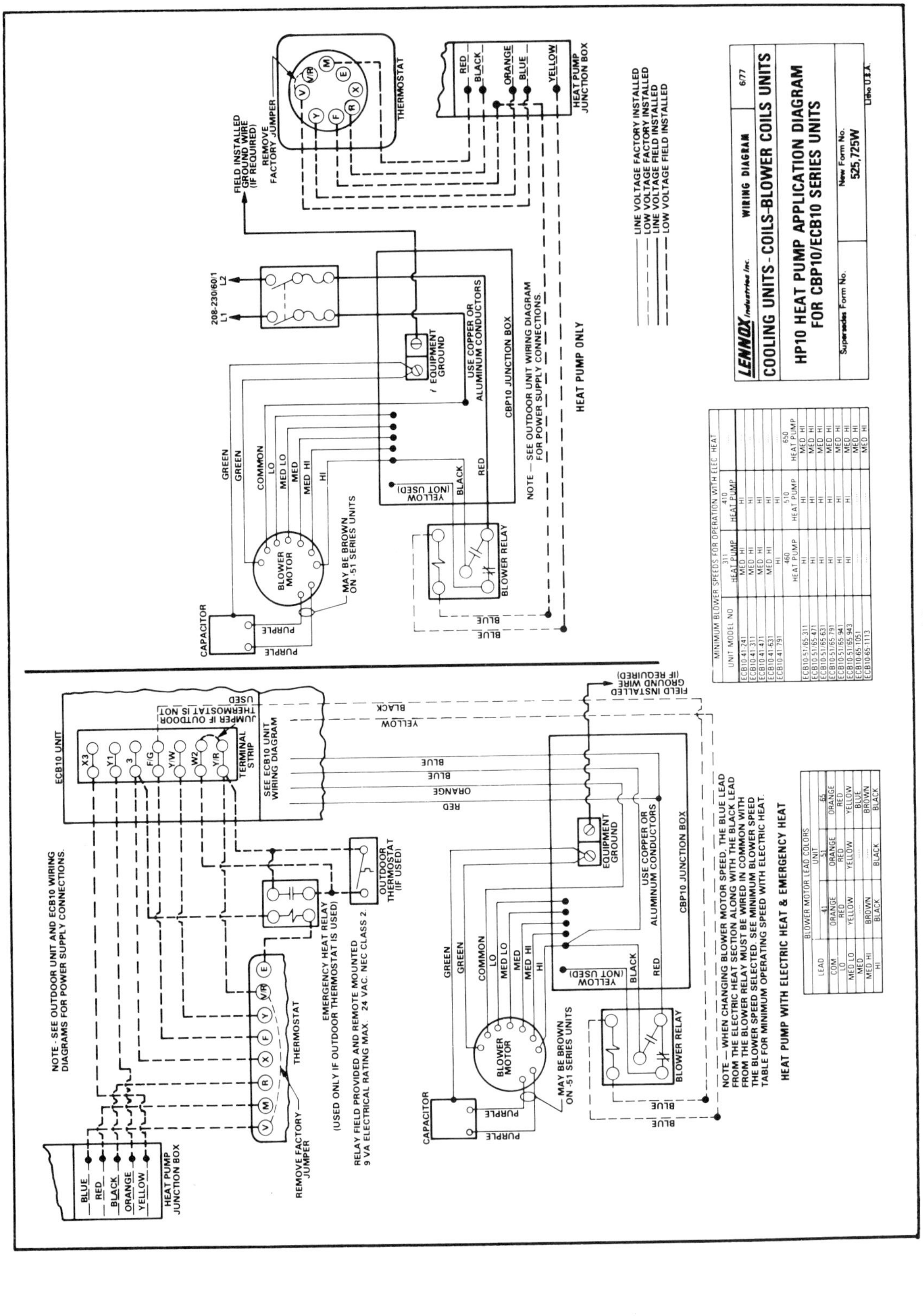

(c)

15-3 ADVANTAGES OF THE HEAT PUMP

Economy

When heating, the aim of the reverse cycle system is to get as much heat as possible from every unit of energy used. When we burn a gallon of fuel oil, we want to get as many BTU's as possible from the gallon. Any heat that goes up the chimney is loss.

The fuel used for heating with heat pumps is electricity. Whenever you convert electricity to heat with ordinary electric resistance heaters, you do this heating at a 100 percent efficiency. For every one cent of electricity you buy, you get one cent of electricity back in the form of usable heat.

The great advantage of the heat pump is that for every cent of electricity that you buy, you will get back considerably more than one cent's worth of electric heat. With a well-designed air-to-air heat pump, you will usually get back three cents' worth of electric heat for every one cent of electricity you buy to run the heat pump in 40° winter weather. The two cents of electric heat that you get for free, in this example, have been pulled out of the outdoor air by the heat pump.

The ratio between the heat output and the heat (or electric) input to the heat pump is known as the COP, which stands for "coefficient of performance."

One kilowatt hour of electricity contains 3413 BTU. If the COP of a heat pump is 3.00, then, for every kilowatt hour that you spend to run it, you will get three times this, or 10,239 BTU's of heat delivered from the machine.

As the weather gets colder, the COP of the heat pump goes down so that you get less and less heat out of the air "for free."

Service Contract

An obvious advantage to these reverse- cycle air-conditioning systems is that, since the same equipment serves both heating and cooling requirements, maintenance responsibilities for both functions can be handled by a single service contract.

Health

Another advantage is that the filtering and evaporative facilities free the interior air of excess moisture, dust, and pollen, permitting persons who normally suffer from hay fever or airborne allergies to obtain relief in the tempered atmosphere.

15-4 ALTERNATIVES

The heat pump is a possible alternative to natural gas, oil, and electrical resistance heating systems.

As stated in other chapters, the uncertain future of the availability of natural gas and heating oil has focused attention on electrically operated systems; and of the two basic systems mass-marketed today, electrical resistance heating and the heat pump, the heat pump shows more promise in terms of energy conservation. (Solar energy is not mentioned here due to the lack of mass production of the collector and because of comparable energy saving statistics.)

15-5 EFFICIENCY LOSS

The heat pump is usually sized for the cooling load of the house. As a result, the heat pump cannot fully satisfy a homeowner's needs in severe winters. When the outside temperature drops to below freezing, the heat pump starts to lose its heating efficiency. Heat pumps, at least in northern climates, require a supplemental heat source such as the standard basement gas, oil forced-air furnace, or electrical resistance heating. As mentioned in Chapter 13, supplementary electric resistance duct heaters could easily be installed in the duct work to make up for heat pump limitations during such extreme cold snaps.

15-6 CONTROL

Heating and cooling functions are controlled by a two-dial, 24-volt summer/winter thermostat in the living room, one dial for heating and one for cooling (Figure 15.3). A four-position switch on the thermostat sets its function to heat control, cooling control, fan operation, or off. Individual room temperatures may be adjusted by means of register dampers in the duct system.

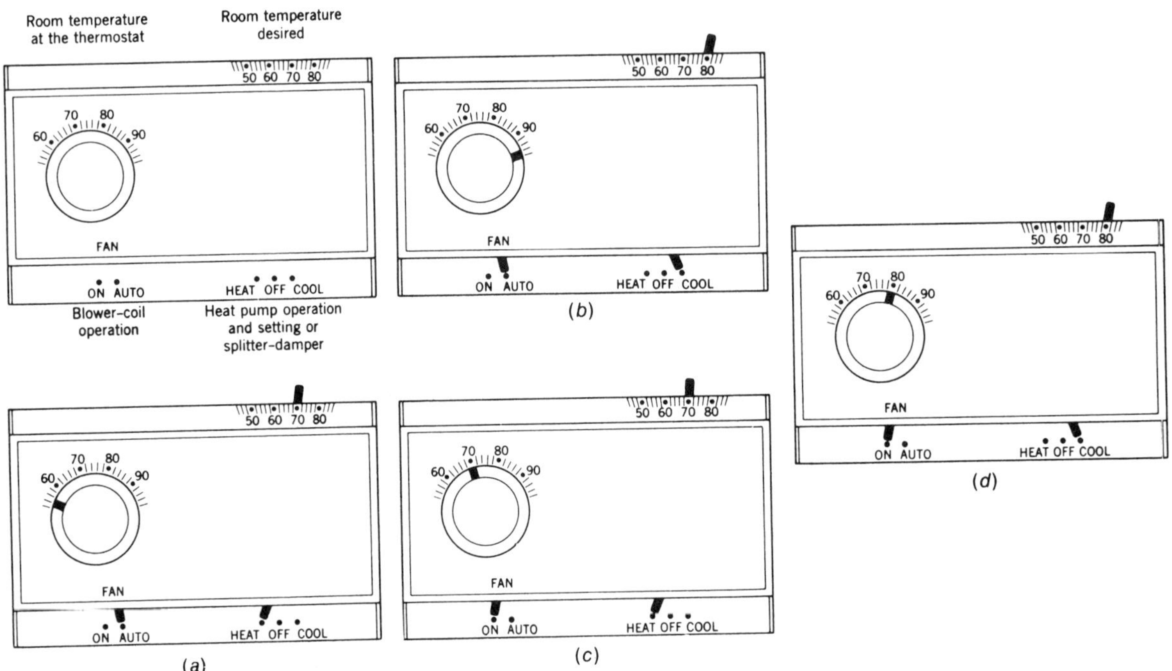

Figure 15.3 Thermostat for the control of heat pump and blower coil unit. (*a*) Turn on for heating. Room temperature at 55°F. Desired temperature set at 70°F. Heat pump set for heat. Blower set for automatic. When the heat pump starts, blower turns on automatically. When room temperature reaches 70°, both turn off. (*b*) Turn on for cooling. Room temperature 95°F, desired temperature 80°F. Blower set for automatic and heat pump for cool. Blower follows heat pump automatically. When room temperature gets down to 80°F, both turn off. (*c*) Continuous circulation—heating. (*d*) Continuous circulation–cooling. (Courtesy of Lennox Industries, Inc.)

15-7 INSTALLING THE UNIT

Slab Mounting

When installing the unit at grade level, mount it on a slab high enough above the grade to allow adequate drainage of water. The top of the slab should be located so run-off water from higher ground will not collect around the unit. The slab should have a slope tolerance away from the building of 2 degrees or 2 inches per 5 feet. This will prevent ice buildup under the unit during defrost cycle (Figure 15.4).

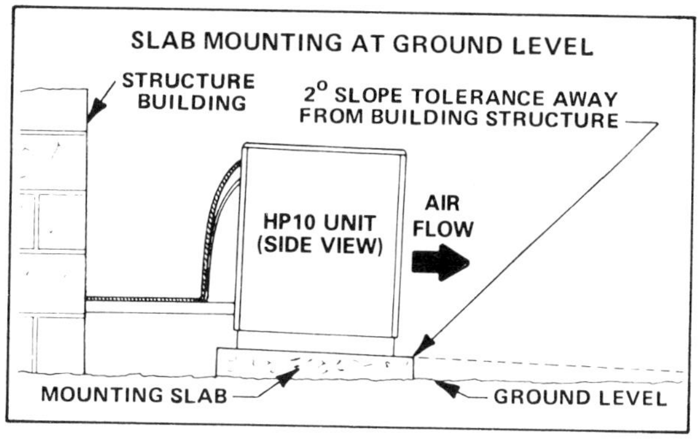

Figure 15.4 Slab mounting. When installing unit at grade level, mount on a slab high enough above the grade to allow adequate drainage of water. Top of slab should be located so runoff water from higher ground will not collect around unit. Slab should have a slop tolerance away from the building of 2 degrees or 2 inches per 5 feet; this will prevent ice buildup under the unit during defrost cycle. (Courtesy of Lennox Industries, Inc.)

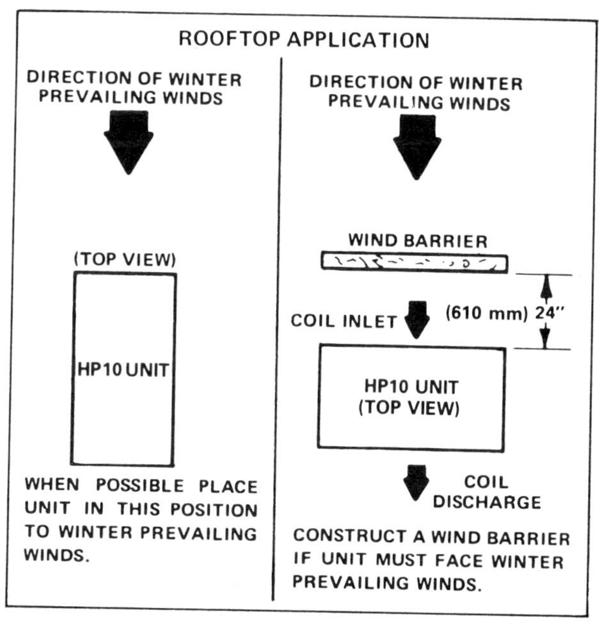

Figure 15.5 Roof mounting. When mounting unit is in installed in areas where low ambient temperatures exist, it should be placed so winter prevailing winds are not in direct line with the heat pump coil; if this is not possible, a wind barrier should be constructed. Size barrier at least the same height and width as unit and mount barrier 24 inches from the coil inlet side of the unit in the direction of the prevailing winds. (Courtesy of Lennox Industries, Inc.)

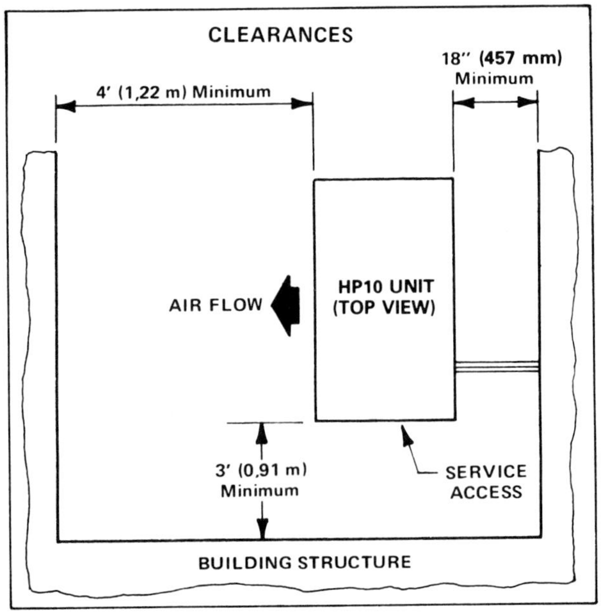

Figure 15.6 Clearances to provide service and air flow. (Courtesy of Lennox Industries, Inc.)

Roof Mounting

When the unit is installed in areas where low ambient temperatures exist, it should be placed so winter prevailing winds are not in direct line with the heat pump coil. If this is not possible, a wind barrier should be constructed. Size the barrier at least the same height and width as the unit and mount the barrier 24 inches from the coil inlet side of the unit in the direction of prevailing winds (Figure 15.5). Clearances are to provide room to service the unit and air flow (Figure 15.6).

15-8 MAINTENANCE

Because the compressor is mounted on a pad outside the building or on the roof, there should be ample room for maintenance. Maintenance requirements are mainly a matter of periodic inspections, maybe replacing filters, lubricating bearings, checking loose fan belts, keeping electrical connections tight, occasionally replacing faulty relays, and keeping contacts clean and in positive-action. See Figure 15.7 for gauge attachment.

SUMMARY

1. A heat pump is year-round comfort control from only one system.
2. Electricity is the only power/fuel source needed for the operation of the heat pump.
3. With a heat pump, there is only one system to service.
4. A heat pump may be tied in with solar energy.
5. Heating and cooling are controlled by a two-dial, summer/winter thermostat.
6. Supplemental resistance heaters may be installed for extremely cold winter days.

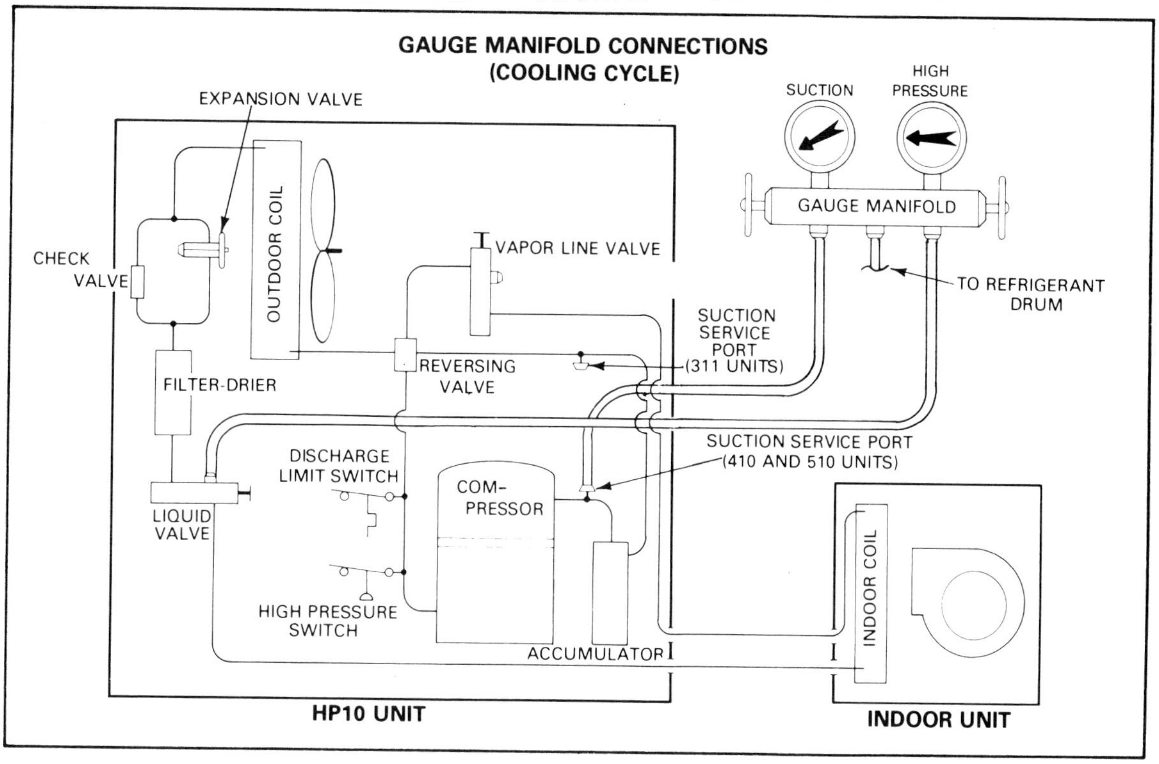

(a)

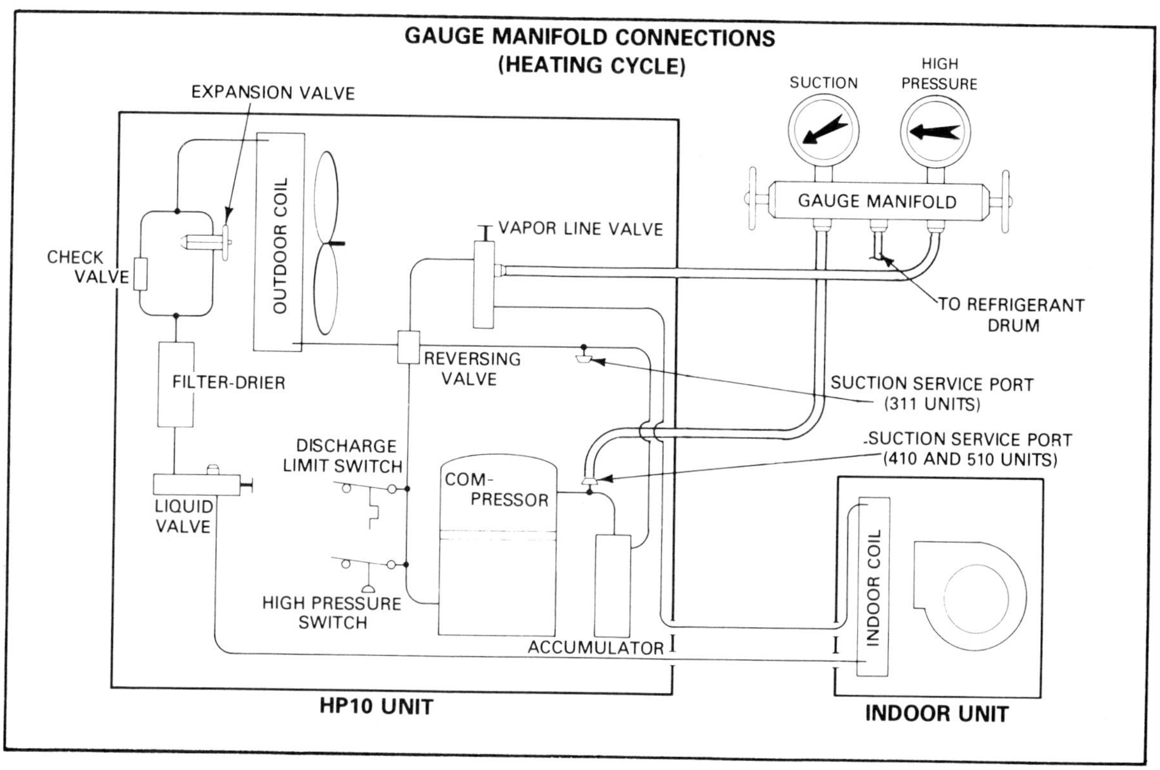

(b)

Figure 15.7 (*a*) Attaching gauge manifold, cooling cycle, checking charge or charging-connect high-pressure side of gauge manifold to gauge port on liquid line service valve. Connect suction side of gauge manifold to vapor line gauge port. (*b*) Attaching gauge manifold, heating cycle, checking charge or charging-connect high-pressure side of gauge manifold to gauge port on vapor line service valve. Connect suction side of gauge manifold to compressor suction service port. (Courtesy of Lennox Industries, Inc.)

PROBLEMS

15-1 Explain briefly how a heat pump works: (a) in winter and (b) in summer.

15-2 List three advantages of using the heat pump.

1.
2.
3.

15-3 Why is the heat pump sized for cooling load rather than for heating?

15-4 Draw a wiring diagram of a heat pump control thermostat.

15-5 List six maintenance requirements for the heat pump.

1.
2.
3.
4.
5.
6.

chapter 16

FORCED AIR HEATING

Instructional Objectives

1. To understand the operation of forced air heating.
2. To learn how warm air is circulated.
3. To learn why a perimeter type duct installation is installed for forced warm air furnaces.
4. To become familiar with the types of fuel used in today's residential heating.
5. To learn why some homeowners heat with a different type of fuel.
6. To understand the need for automatic controls.
7. To review the operation of the thermostat.
8. To understand the operation of the fan control.
9. To learn some advantages of installing an electronic air cleaner.
10. To become more familiar with the operation of the air cleaner.

Self-Evaluation Questions

Test your prior knowledge of the information in this chapter by answering the following questions. As you read the chapter, watch for the answers. When you have completed the chapter, return to this section and answer the questions again.

1. What is the purpose of the mechanical blower?
2. Why is it important to check on the fuel before installing a forced air heater?
3. Why is natural gas a popular fuel?
4. What other fuels are used besides natural gas?
5. What does the control system do?
6. How is a diaphragm valve actuated?
7. What is a stock-relay switch?

8. In an air burner installation, how is the oil ignited?
9. Where is the fan control located?
10. Give one good reason for using a room thermostat.
11. What does an electronic air filter do?

16-1 FORCED WARM AIR FURNACES

Today, most furnaces being installed circulate warm air by mechanical means and are often referred to as forced air furnaces.

Forced warm air furnaces include a fan or blower as part of the unit. The purpose of the blower is to circulate the heated air to all of the rooms. Since a mechanical blower circulates the warm air, the heat duct may be installed either in a vertical or horizontal direction.

Most forced air furnaces contain an air filter to remove impurities in the air. An electric motor, along with proper pulley and belt, drives the forced air blower. The blower forces the air across the heat exchanger up into the plenum and out through the ducts and pipes to the various rooms.

Today, the perimeter type of duct installation is used with the forced warm air furnace. This system distributes warm air to the perimeter of the house (Figure 16.1).

16-2 FUEL CONSIDERATIONS

The selection of a fuel depends on many factors. However, in most localities, we need only to choose between two, gas (which includes natural, mixed manufactured and liquified petroleum gases) and fuel oil. Some of the considerations center on economics, such as the cost of the fuel or initial cost of the heating system adapted for that particular fuel. Other considerations might be the availability of the supply, the convenience of handling the fuel, and local ordinances and zoning regulations that may prohibit the use of certain fuels.

Natural Gas

The most popular fuel being used today is natural gas, since it is used in far more new installations than any other fuel (Figure 16.2).

A vast network of gas pipelines reaches a large part of the country. These

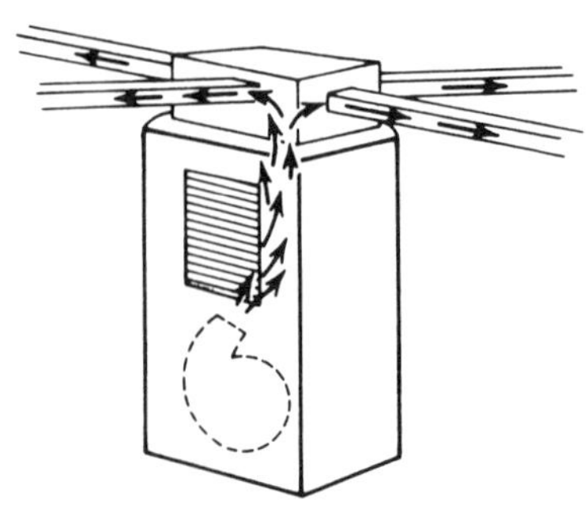

Figure 16.1 Furnace cabinet and plenum.

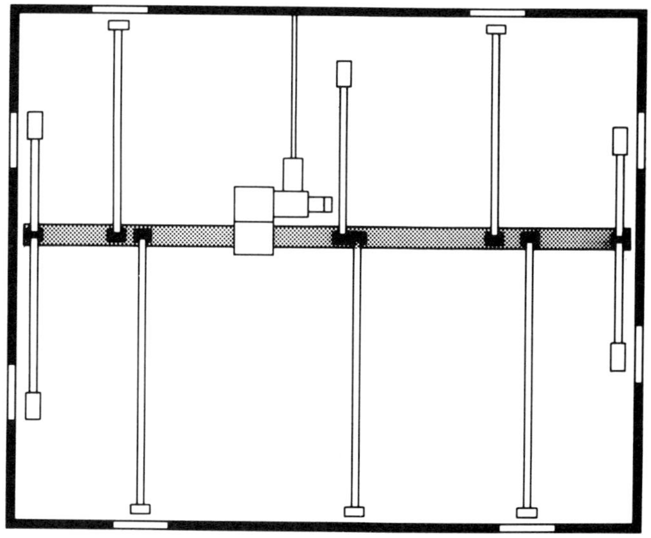

Figure 16.2 Forced-air gas furnace. GS 11 series. (Courtesy of Lennox Industries, Inc.)

pipes bring natural gas from the fields and often store it in underground locations conveniently located with relation to the gas markets. In many parts of the country, natural gas is an economical fuel to burn. Furnaces using gas are generally lower in initial cost than equipment using other fuels. The efficiency of gas-burning equipment is relatively high, and there is no requirement for fuel-storage facilities on the premises where natural gas is available.

In some parts of the country, natural gas is not available and manufactured gas is used. Manufactured gas may be made from coal or converted from fuel oil, and it is usually more expensive than natural gas. Nevertheless, it is used extensively where natural gas lines do not exist. In some places, manufactured and natural gas are mixed together to form a highly satisfactory fuel.

In many rural areas where natural gas lines do not exist, liquified petroleum gas (L.P.) or bottle gas is used. This may be either propane or butane gas. In the colder climates, propane is used because butane will not vaporize at temperatures below freezing. With the L.P. gases, it is necessary to store the fuel in tanks located on the premises but out of doors. These are pressure tanks that hold a combination of liquid and vapor.

Gas furnaces that use liquified petroleum gases are especially designed for L.P. fuel. However, these furnaces may usually be converted rather readily to use natural gas if it becomes available at a later date. Liquified petroleum gases are usually more expensive than natural gas. As a result, many consumers would prefer to have natural gas available as a fuel.

Fuel Oil

Probably the second most popular residential heating fuel is oil. However, oil equipment is generally more costly than gas, and it also requires fuel storage equipment, which adds to the installation expense (Figure 16.3).

Oil is, however, a very acceptable residential heating fuel and it is used in many parts of the country, especially where natural gas is not available or where the rate structure is such that gas is more expensive.

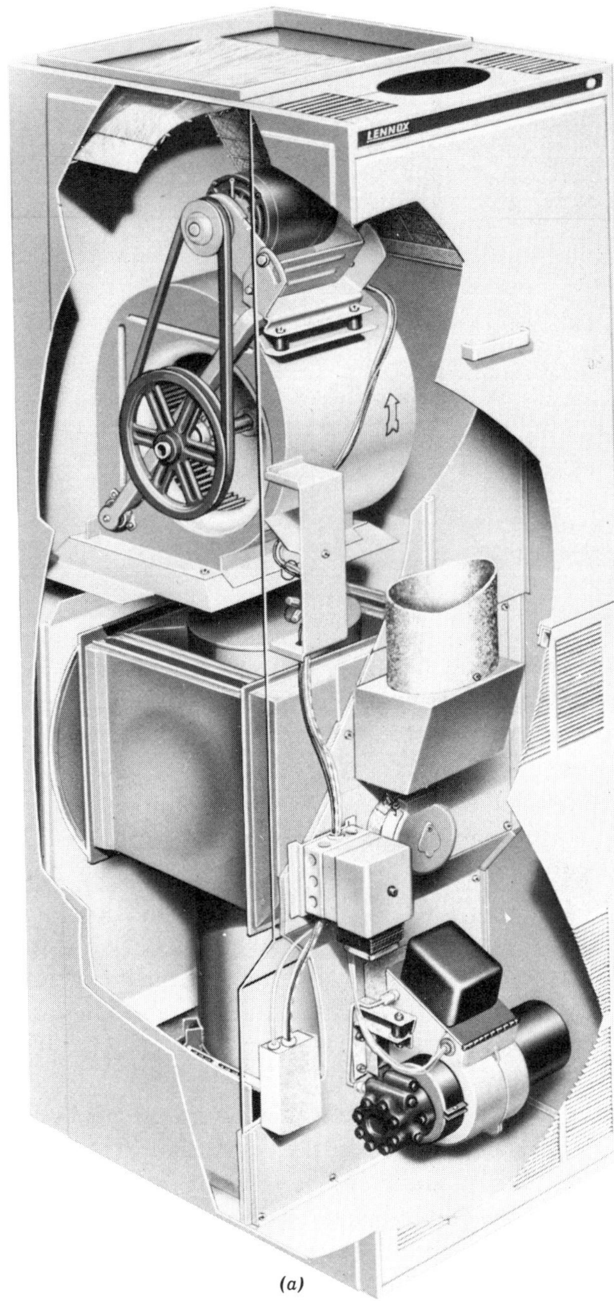

(a)

Figure 16.3 (a) Fuel-oil furnace. 0 11 R series. (b) Wiring diagram for fuel oil furnace. (Courtesy of Lennox Industries, Inc.)

Many new homes may have to burn oil because natural gas is not available to them.

16-3 AUTOMATIC CONTROLS

Like other controls, the control system performs those functions: safety, comfort, and convenience, in that order of importance.

The main factor to bear in mind when installing automatic controls for gas

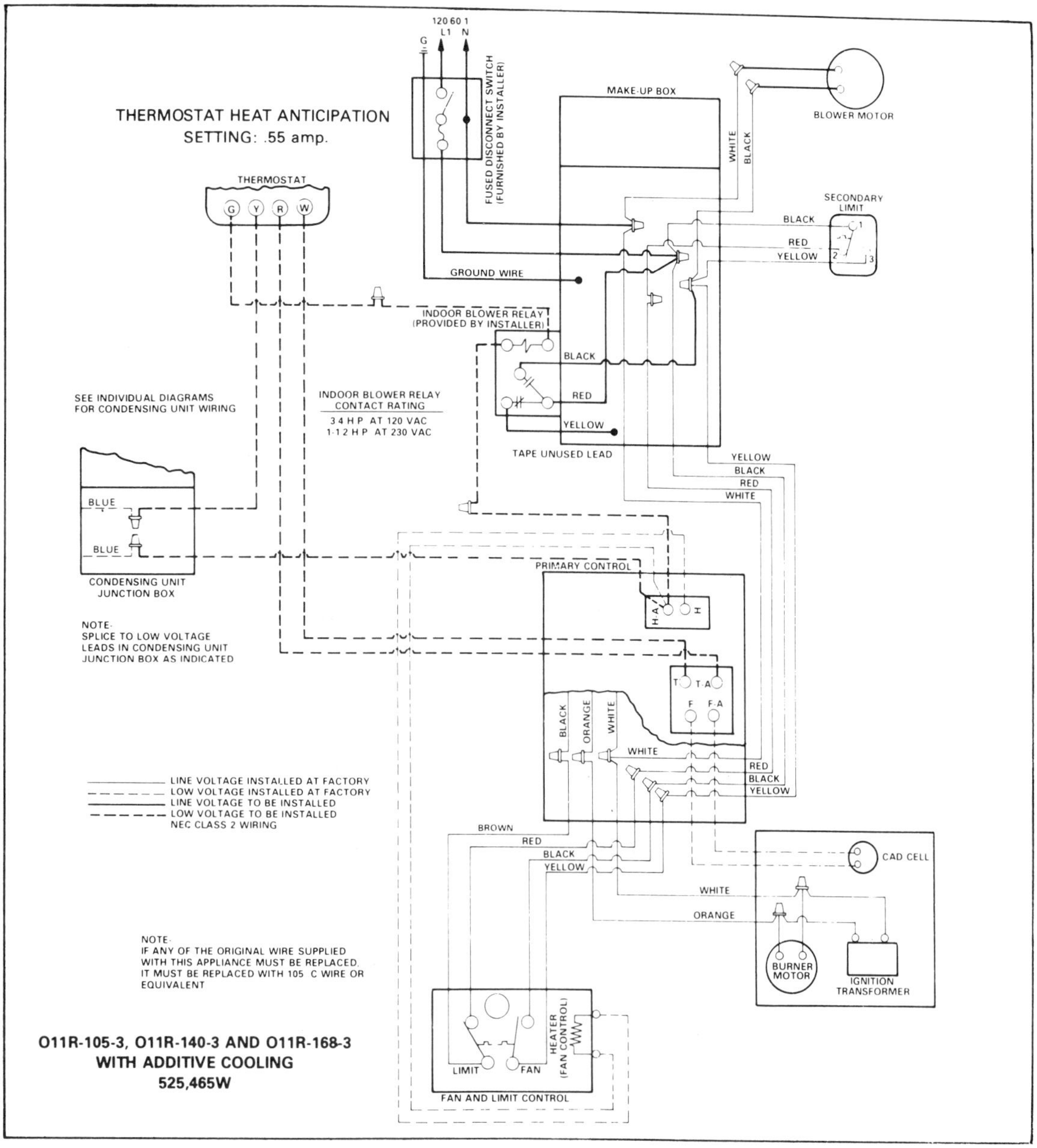

(b)

burners is that a dangerous quantity of gas must not be permitted to accumulate in the combustion chamber; there must be a safety device to stop the flow of gas in case of a pilot failure. There is a variety of safety devices for dealing with this hazard.

In general, there are two means of feeding gas to the burner when the room thermostat calls for heat: one, by means of a diaphragm valve, two, by means of a solenoid-operated valve. The diaphragm valve is pressure-actuated. A small solenoid coil electrically opens a small valve that permits gas to escape from the top of the diaphragm. When this occurs, normal pressure underneath raises the diaphragm, and gas flows to the burner. When a solenoid valve is used, completion of the operating circuit energizes the solenoid coil and pulls up a plunger, which in turn opens the valve (Figure 16.4).

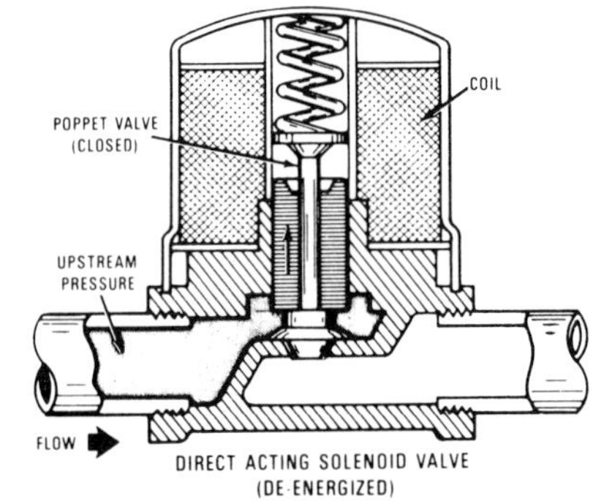

Figure 16.4 Solenoid type main gas valve is opened and closed by thermostat switching on and off magnetic coil that lifts the plunger. Spring closes valve.

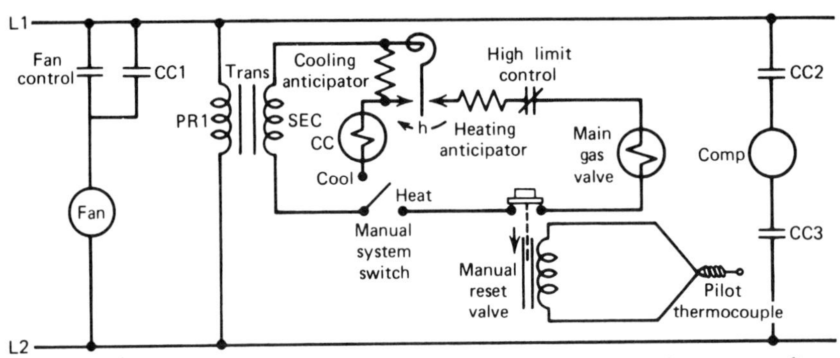

Figure 16.5 Typical control schematic for gas-fired, warm air heating system with summer cooling added.

Figure 16.5 shows a layout of the relative positions of the equipment needed for a gas-burner installation. Note that the electric safety pilot is wired to a thermocouple. A thermocouple, Figure 16.6, is made of two unlike metals that generate a small amount of electricity when heated. The current then is used to hold an electromagnetic valve open. In case of pilot failure, the metal would cool. This would stop the flow of current and close the valve.

The installation of oil-burner controls is complicated by the addition of a stack-relay switch. This switch, or protector relay, is installed in the flue pipe. It functions to shut off the burner in case the stack does not come up to a certain predetermined temperature within 45 seconds after the motor starts. The relay prevents flooding the basement with fuel oil in case the ignition system fails.

Ignition of the oil is accomplished with the aid of a spark gap that is connected to the high-voltage terminal of a transformer installed on or near the burner.

The spark may be continuous, or it can be made to cease when the oil starts to burn. The continuous spark system is generally advised when down drafts are common. With a continuous spark, it is necessary to replace the electrode more often than with the intermittent spark.

After the spark has been shut off, the oil continues to burn because the temperature of the firebrick lining is high enough to ignite the oil (Figure 16.7).

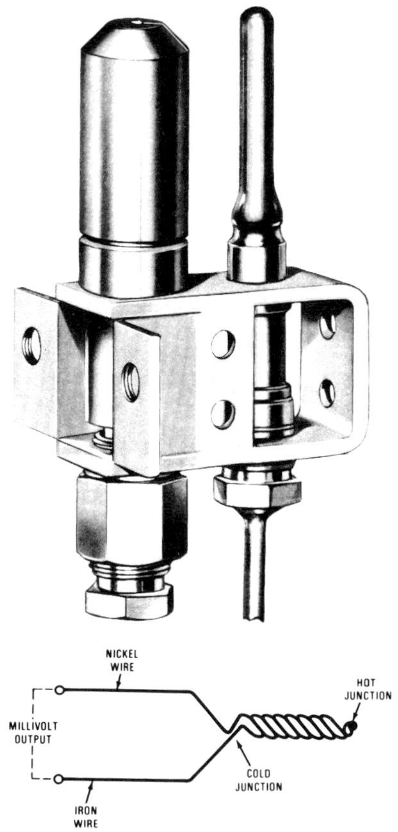

Figure 16.6 The thermocouple converts heat into electricity in sufficient quantities to power safety devices, relays, or valves. Note thermocouple mounted next to a pilot light.

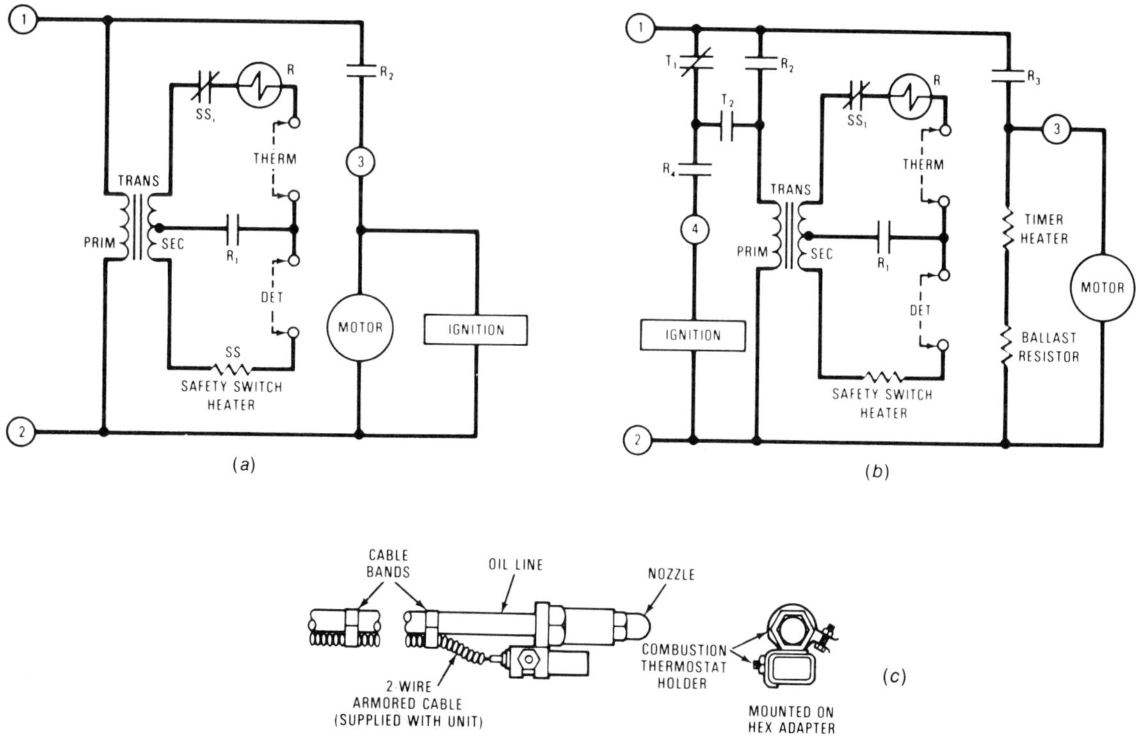

Figure 16.7 Oil-burner controls. (*a*) Constant ignition wiring hookup. With motor relay contacts closed, ignition is on throughout the call for heat. (*b*) Intermittent ignition wiring hookup. Timer heater opens contacts and stops ignition after several seconds of successful burner operation. (*c*) Combustion thermostat (switch) mounted in the burner.

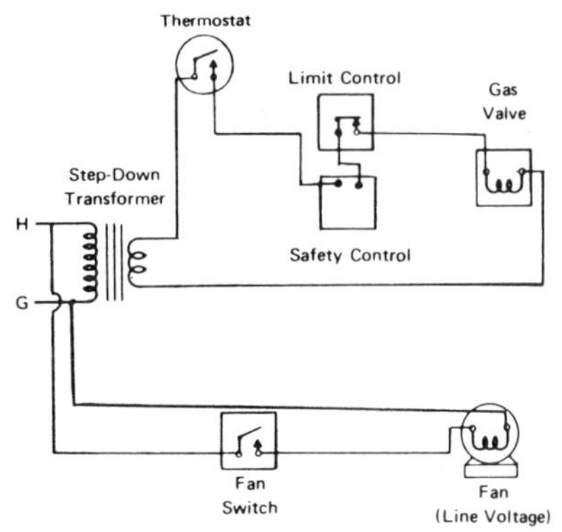

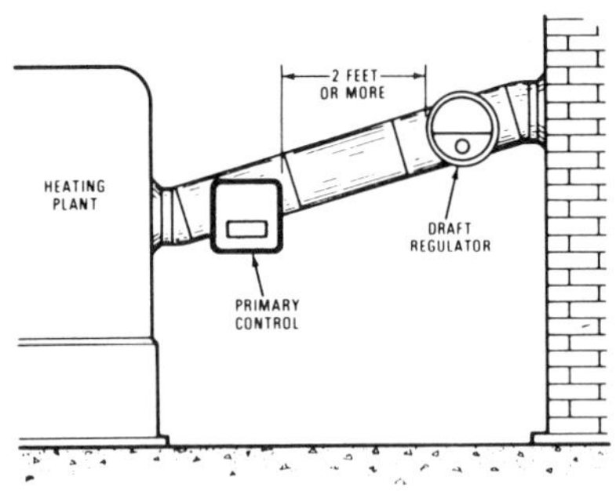

Figure 16.8 Furnace fan and limit control wiring diagram.

Fan and Limit Controls

The function of the fan control is to turn the blower on when the temperature of the air becomes warm enough to be circulated and to turn the blower off when the temperature of the air becomes too low.

The same type fan control can be employed on a heating system regardless of the fuel type used. This control responds only to the temperature of the air being circulated to the rooms. It is a comfort control, not a safety control. Obviously, if the blower starts to operate before the air has been heated to a sufficiently high temperature, little or no heat will be delivered to the conditioned space. This can result in unpleasant drafts. The same applies if circulation is continued after the air has cooled off excessively following the end of the burning "ON" cycle.

The fan control, a bimetallic element or liquid-filled tube that is exposed to the air circulating over the heat exchanger, opens and closes a line-voltage switch, turning the blower on and off. The fan control is mounted on the furnace cabinet or plenum (Figure 16.8).

Adjustable settings of the fan control are provided by means of two arms. The upper arm is set for the temperature at which the blower should turn on (about 100°F for the high side wall register and 110°F for the low side wall register). The lower arm is set for the temperature at which the blower should stop (about 15° lower than "cut in"). Such fan switches may also have a knob that permits the blower control to be changed from automatic to manual. When the blower control is on manual, the blower runs constantly.

The Limit Control

The limit control is usually combined with the fan switch for warm air furnace applications. In such a combination the limit control responds to the same temperature-sensing element as does the fan switch. When the temperature in the plenum reaches a point above the adjustable setting of the limit control, the limit switch breaks the electrical circuit, thereby causing the burner to stop (Figure 16.8). On a gas burner, the limit switch breaks an electrical circuit, causing the gas valve to close. On an oil burner, the limit switch stops the burner motor. It is common practice to set the limit control "cut out" at about 175°F on a forced warm air furnace.

The limit control never affects the blower operation even when this control is combined with the fan control of a forced air furnace. The blower will continue to operate after the burner has been turned off by the limit control until the plenum temperature has dropped to the cut-out setting of the fan control.

The Room Thermostat

The room thermostat is used only for regulating room temperature and is in command of the fire if all other controls, such as the limit switch and safety pilot control, are operating properly. For a review of the thermostat, read Chapter 13.

16-4 ELECTRONIC AIR CLEANER

An electronic air cleaner is the newest component that is being installed in cooling and heating comfort-conditioning systems (Figure 16.9).

The electronic air cleaner is capable of removing up to 95 percent of airborne dirt and irritants, odors, and tobacco smoke, and up to 99 percent of the airborne pollen present in the air that pass through it.

Basically, the electronic cleaner puts a positive electrical charge in the dust particles that pass through its filter. These particles are then attracted to negatively charged plates. Some units use a collector cell or filter, which

Figure 16.9 Electronic air cleaner. (Courtesy of Lennox Industries, Inc.)

carries small charges of static electricity that attract the dust particles. It is then necessary to use a detergent to wash the electrostatic filter to remove the deposits.

The electronic air cleaner generally has a control that monitors the unit's performance. The control automatically indicates when the filter is dirty and in need of cleaning. Some controls also have timers that permit the serviceperson or homeowner to time the drying cycle after the collector cell or filter has been washed.

SUMMARY

1. In cooler climates propane gas is often used, because butane gas will not readily vaporize at cold temperatures.
2. Natural gas is a common fuel used in residential heating.
3. All electrical controls are switches operated by either a rise or fall in temperature.
4. The limit control breaks on temperature rise, causing the burner to stop.
5. The safety pilot breaks contact on loss of heat on the thermocouple.
6. An electronic air cleaner puts a positive electrical charge in the dust particles that pass through its filter.

PROBLEMS

16-1 Name the common fuel types; list advantages and disadvantages of each.

____

____

____

____

____

____

16-2 What factors are considered when selecting a fuel for a residential heating system?

____

____

____

____

16-3 In the order of importance, list the three functions of the automatic controls.

1. ______________________________
2. ______________________________
3. ______________________________

16-4 Explain how gas is fed to the burner when a room thermostat calls for heat.

16-5 Draw a sketch showing how a safety pilot is wired to a thermocouple.

PART 5

THE ELECTRICAL SERVICE

chapter 17

SERVICE ENTRANCE EQUIPMENT—RESIDENTIAL

Instructional Objectives

1. To develop an ability to identify and define parts of an electrical service equipment.
2. To become familiar with the National Electrical Code as related to service-entrance equipment.
3. To understand the difference between an overhead service and a service lateral.
4. To become familiar with the need for grounding and bonding of the service-entrance equipment.
5. To learn why service-entrance conductors are governed by the NEC.
6. To make you aware of the utility company requirements for service laterals.
7. To learn how to ground the meter socket.
8. To understand the need for overcurrent protection against short circuits and faults or overloads.
9. To learn why the grounded neutral conductor is not switched or protected by an overcurrent device.
10. To learn why the power supplier locates the service drop and meter for an owner or builder.

Self-Evaluation Questions

Test your prior knowledge of the information in this chapter by answering the following questions. Watch for the answers as you read the chapter. Your final evaluation of whether you understand the material is measured by your ability to answer these questions. When you have completed the chapter, return to this section and answer the questions again.

1. What is the difference between the service drop and service conductors?
2. Who decides where the service drop shall be located?
3. Why must service equipment be bonded?

4. What is the purpose of a bonding jumper?
5. What is the most desirable electrode for a system ground?
6. What does the term "service" mean?
7. List the main parts of an electrical service installation.
8. Who furnishes the service drop conductors?
9. What type wire can be used for the service-entrance conductors?
10. What is the difference between an overhead service and a service lateral?

17-1 ELECTRICAL SERVICE—GENERAL (FIGURE 17.1)

The electrical service installation is the heart of the electrical system. All electrical energy supplied to power-consuming devices and appliances within the residence must pass through the electrical service-entrance equipment where it is metered, protected, and distributed through branch circuits throughout the home (Figure 17.2).

The local power supplier decides where the electrical service will enter the building and where the meter will be located. This is usually the first item to be located before any of the electrical work actually begins. However, we are considering it in this chapter because it is a part of the electrical service equipment.

The service panel should be located near rooms where the largest amount of current will be needed. This is usually in the kitchen.

The term "service" is defined by the National Electrical Code as "the conductors and equipment for delivering energy from the electricity supply system to the wiring system of the premises served."

The main parts of a complete electrical service installation may be listed as: service drop, service conductors, meter, service switch, grounding, and bonding (Figure 17.3). In succeeding paragraphs each of these parts will be defined and their function explained in detail.

17-2 SERVICE DROP

The service drop is actually not part of the wiring system in a residence. It is the connection provided by the power company from their distribution lines to the house. It is discussed in this chapter to familiarize you with the term and its importance.

In Article 100 of the NEC definitions, service drop conductors are defined as "the overhead service conductors from the last pole or other aerial support

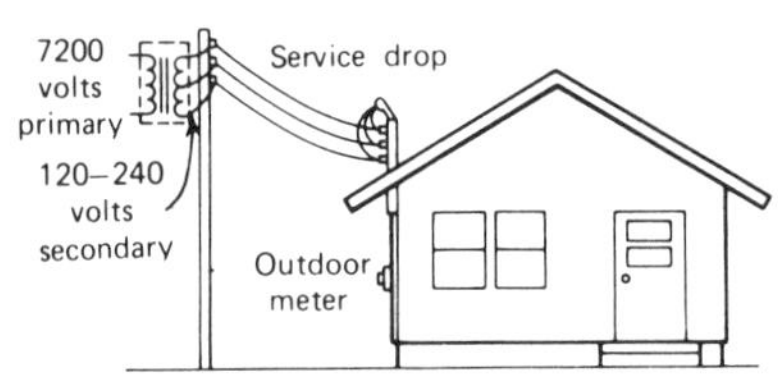

Figure 17.1 Pole mounted, center-tapped secondary, step-down transformer supplying a three-wire, 120/240 volt service to a home.

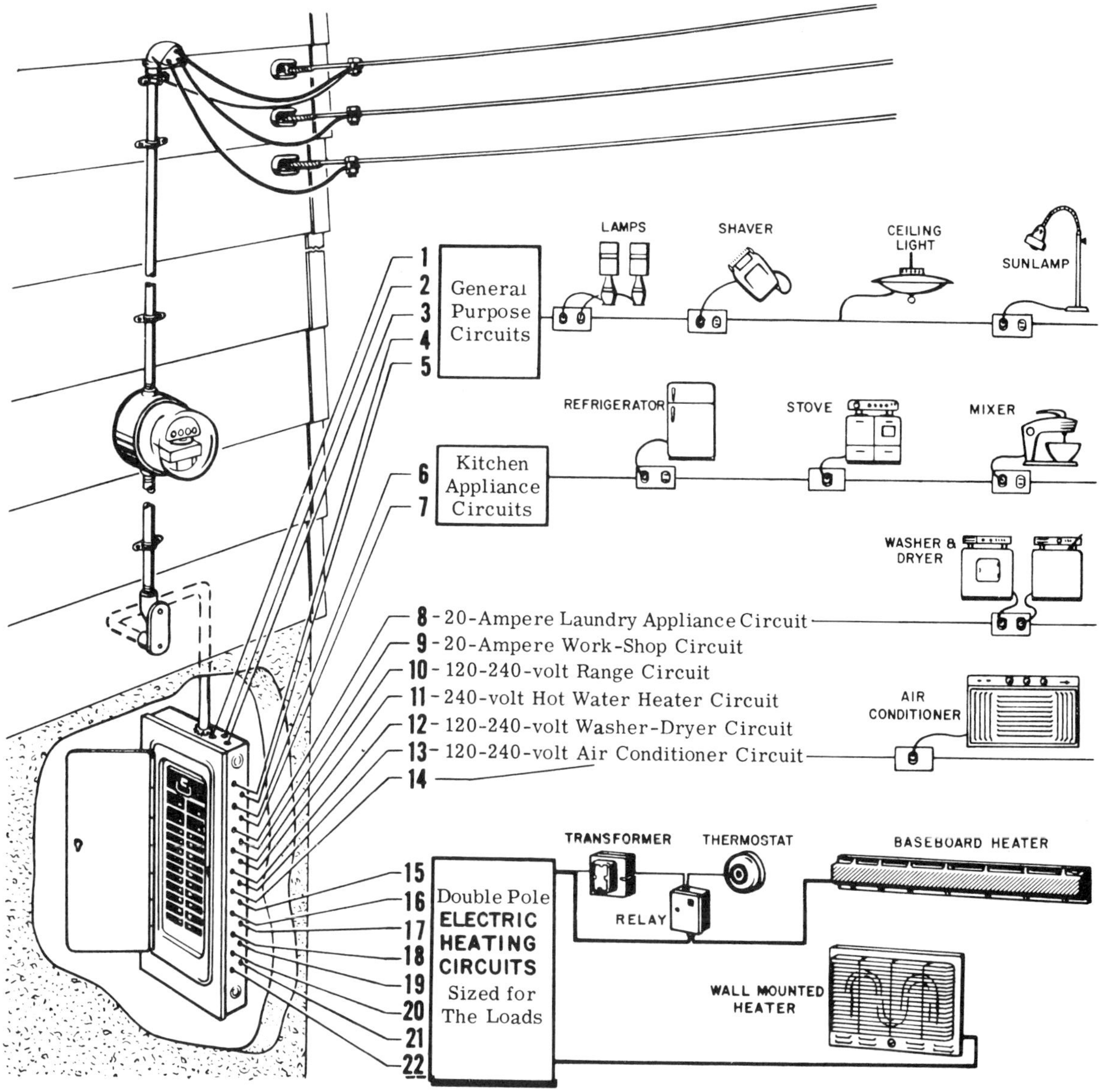

Figure 17.2 Service and distributions.

to and including the splices if any, connecting the service-entrance conductors at the building or other structure."

Service drop wires are furnished and installed by the power supplier, although the owner or electrician sometimes furnishes the insulators by which the wires are supported on the building or service mast riser. The power supplier also determines the size of the conductors, which are often smaller than the service-entrance wires.

The National Electrical Code has several rules governing the installation for the service drop.

1. NEC 230-2 In general, a building or other premises shall be supplied by only one set of service conductors.
2. NEC 230-22 Individual conductors of multiconductor cable when used as service drop shall be insulated or covered with thermoplastic, rubber or other similar material; except grounded conductor may be bare. (b) Open

1. Service drop

2. Entrance head

3. Service entrance conduit

4. Threaded hub

5. Meter base

6. Service entrance panel

7. Main disconnecting means

8. Grounding bushing

9. Equipment bonding jumper

10. Main bonding jumper

11. Grounding electrode conductor

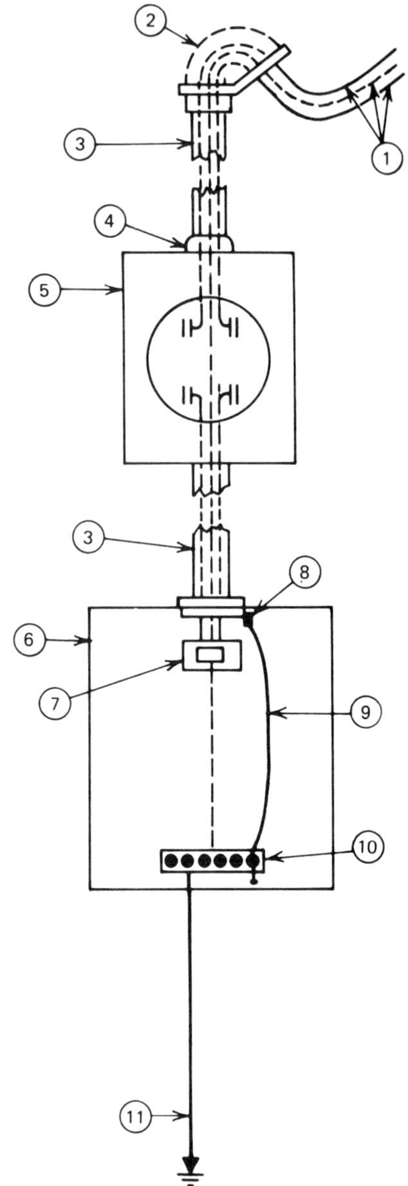

Figure 17.3 Service entrance equipment.

conductors shall be insulated or covered and have ampacity as per table 310-17 and 310-19.

3. NEC-230-24 (a.-Ex. 1) Where voltage between conductors does not exceed 300 and roof has a slope not less than 4 inches in 12 inches the clearance may not be less than 3 feet.
4. NEC 230-24 (a.-Ex. 2) Service drops 300 volts or less may pass over 4 feet of overhang if minimum of 18 inches clearance is maintained over roof and conductors are terminated in a thru-the-roof raceway or approved support.
5. NEC 230-24 and 230-26 Porcelain insulators for the service drop conductors shall be at least 10 feet above finish grade.
6. NEC 230-28 Service mast shall have adequate strength or be supported by braces or guys.
7. NEC 230-28 All raceway fittings shall be approved for the purpose.

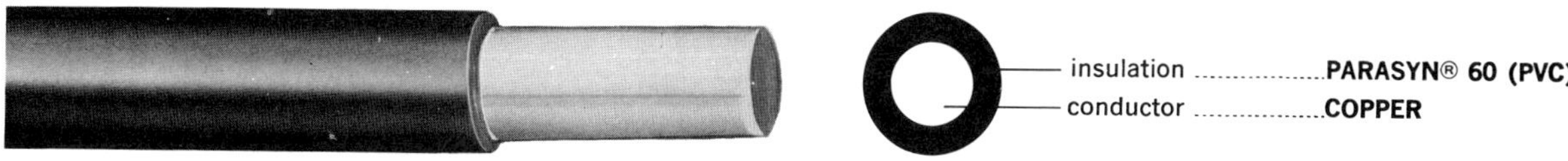

Figure 17.4 Service-entrance conductors. Type TW solid building wire. (Courtesy of ESSEX INTERNATIONAL)

17-3 SERVICE-ENTRANCE CONDUCTORS

The wires from the point where the service drops end, up to the service equipment, are called service-entrance conductors. They may be type TW, THW, RHW, or any type wire suitable for outdoor (wet) locations. They may be separate wires brought in through conduit and service head of wires made up into service-entrance cable approved for the purpose (Figure 17.4).

Service-entrance conductors are also governed by NEC.

1. NEC 230-41 (a) Service-entrance conductors shall be insulated. Exceptions 1, 2, 3: Bare copper grounded conductor acceptable if in raceway or direct burial where suitable for soil conditions,or in soil when cable has moisture and fungus-resistant covering. Exception 4: Aluminum grounded conductor without individual insulation or covering acceptable if in raceway or direct burial when part of cable assembly having a moisture and fungus-resistant outer covering and approved for the purpose.
2. NEC 230-42 (a) Service-entrance conductors shall have ampacity to carry loads as per Article 220.
3. NEC 230-42 (b) Ungrounded conductors. 230-42 (b-1), (b-2) In single family dwellings with six or more 2-wire branch circuits or with initial computed load of 10 kW or more, conductors shall have ampacity of not less than 100 amperes 3-wires.
4. NEC 230-42 (b-3) Not smaller than 60 amperes for other loads. Exception 1: Not smaller than No. 8 copper or No. 6 aluminum for more than two 2-wire branch circuits. Exception 2: Not smaller than No. 8 copper or No. 6 aluminum if by special permission for loads limited by demand or by source of supply. Exception 3: Not smaller than No. 12 copper or No. 10 aluminum for limited loads of a single-branch circuit but never smaller than the branch-circuit conductors.
5. NEC 230-54 Raintight service head required.
6. NEC 230-54 (c) Service heads located above service drop conductors. Exception: where impracticable to locate service head above drops, it may be located not more than 24 inches to one side.
7. NEC 230-54 (e) Service conductors brought out through separately bushed holes in service heads.
8. NEC 230-43 Service-entrance conductors may be installed as (1) open wiring on insulators; (2) rigid metal conduit; (3) electrical metallic tubing; (4) service-entrance cables; (5) wireways; (6) busways; (7) auxiliary gutters; (8) rigid nonmetallic conduit; (9) cablebus; or (10) mineral-insulated metal-sheathed cable.
9. NEC 230-26 In no case shall the point of attachment to the service drop be less than 10 feet above the finished grade.
10.

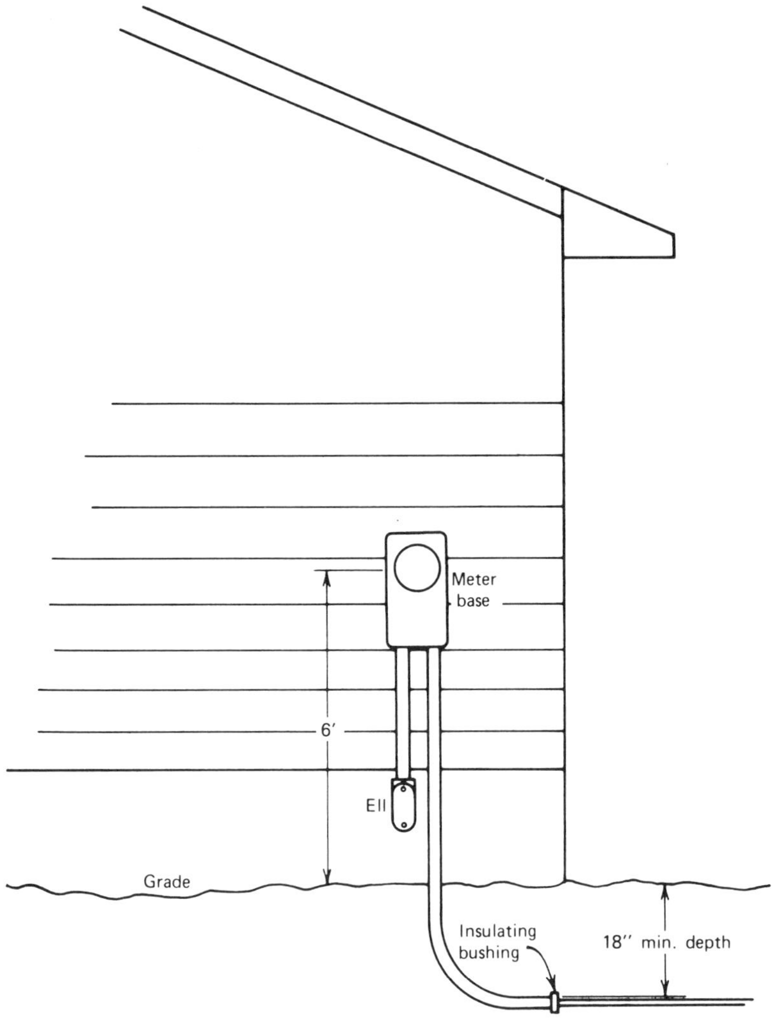

Figure 17.5 Underground service lateral.

17-4 SERVICE LATERALS (FIGURE 17.5)

In most new residential developments and subdivisions, the choice is made to bring the service into the house underground. When the power supplier's conductors to a building are installed underground, they constitute a service lateral.

The term "service lateral" includes the underground cable installed from the point of connection to the utility company's system. The conductors usually terminate at the base of a pad-mounted transformer or in some instances a pedestal that is placed at or near the rear of the owner's property line, or it could be located in some other inconspicuous location in the development.

The service lateral conductors are usually furnished and installed at the owner's expense.

The underground service installation is covered by the National Electrical Code requirements.

1. NEC 230-30 Service lateral conductors shall be insulated. Exceptions 1, 2, 3: Bare copper grounded conductors acceptable if in raceway, or direct burial

where suitable for soil conditioning, or in any soil when cable has moisture and fungus-resistant covering. Exception 4: Bare aluminum grounding conductor acceptable if in raceway, or direct burial if in a cable with moisture and fungus-resistant covering.

2. NEC 230-31 Conductors shall have sufficient ampacity to carry the load. Service laterals shall not be smaller than No. 8 copper or No. 6 aluminum, except for limited loads. Not smaller than No. 12 copper or No. 10 aluminum or limited loads.

When the service lateral conductors are installed by the electrician, the installation must conform to NEC 230-48 and 230-49, which refer to the protection of the conductors against damage and the sealing of underground conduits where they enter the house. NEC 230-49 refers to 300-5, which covers all situations relative to underground wiring.

17-5 THE METER SOCKET

The utility company decides on the location of the meter. In most cases it will be the outside type, with its socket exposed to the weather. This location is helpful to the meter reader, since access to the inside of the house is not needed.

The meter socket is sometimes furnished by the utility company and sometimes by the owner. In any event it is installed by the electrician. The electrician must fasten the meter socket to the building, install the service conduit and wires, and then cut in conductors and attach to the clips in the meter socket. The supply conductors are connected to the upper terminals and load conductors to the bottom.

The neutral conductor may be grounded in the meter socket and in the service panel. However, local utility company requirements prevail.

In some localities, a combination of meter socket, service disconnect switch, and circuit breakers in the same enclosure is favored (Figure 17.6).

The recommended height of a meter socket is between 5 feet and 6 feet, at eye level.

17-6 SERVICE DISTRIBUTION PANEL

All of the circuits you have studied so far are known as branch circuits. Chapter 4 instructed us, "a branch circuit is the circuit conductors between the final overcurrent device protecting the circuit and the outlets."

Types of Circuits

1. A general-purpose branch circuit is a circuit that supplies a number of outlets for lighting and small appliances.
2. An appliance branch circuit is a circuit that supplies energy to one or more outlets to which kitchen appliances are connected; no lighting fixtures are permitted on this circuit.
3. An individual branch circuit is a circuit that supplies power to only one utilization equipment, such as a kitchen electric range.

(*a*)

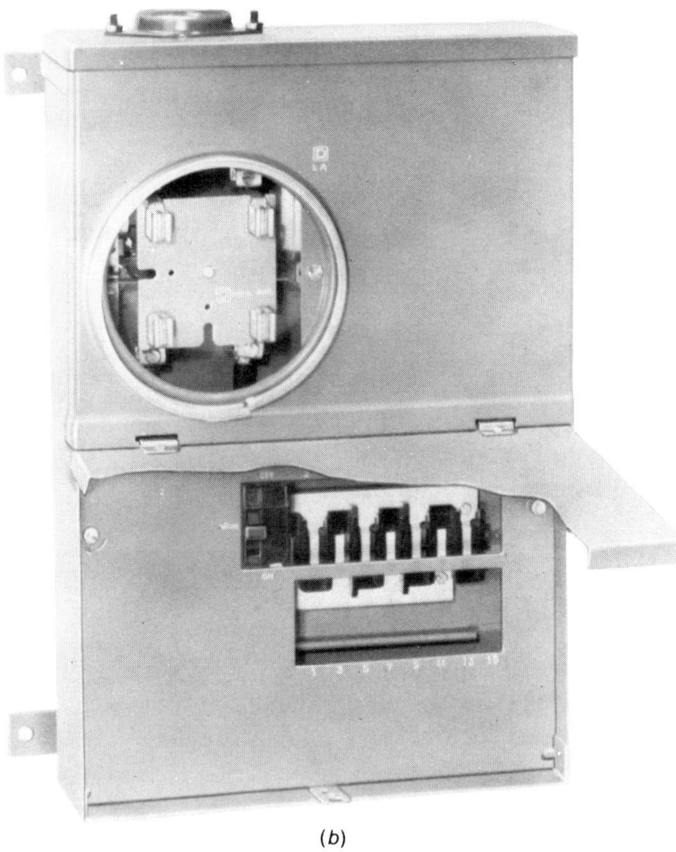

(*b*)

Figure 17.6 (*a*) Combination rainproof underground service, surface-mounted meter base, service disconnect switch-circuit breakers. (*b*) Combination rainproof service entrance with branch circuit breakers. (*c*) Combination rainproof surface-mounted main circuit breaker and service entrance, with threaded conduit hubs on top, knockouts available for bottom line and load. (Courtesy of Square D Company)

(c)

In Chapter 8, review the drawings of the branch circuit cable runs. The diagrams indicated a cable with an arrow point and number. This represented the number of the circuit and cable home run to the main power distribution panel.

Types of Service Panels

There are many types of service distribution panels used in residential wiring; however, there are only two types of overcurrent protection: the fuse and the circuit breaker.

The circuit breaker panel is usually installed with a main disconnect circuit breaker and facilities to mount plug-in circuit breakers (Figure 17.7). The circuit breaker is a device that combines the function of an overcurrent circuit protective device with that of a switch. Basically it is a switch equipped with a tripping mechanism that is activated by excessive current (Figure 17.8). The important fact to remember is that circuit breakers, like fuses, react to excess current. However, the breaker is not self-destructive. After the breaker opens to clear the fault, it can simply be reclosed and reset. This means that replacement problems are eliminated.

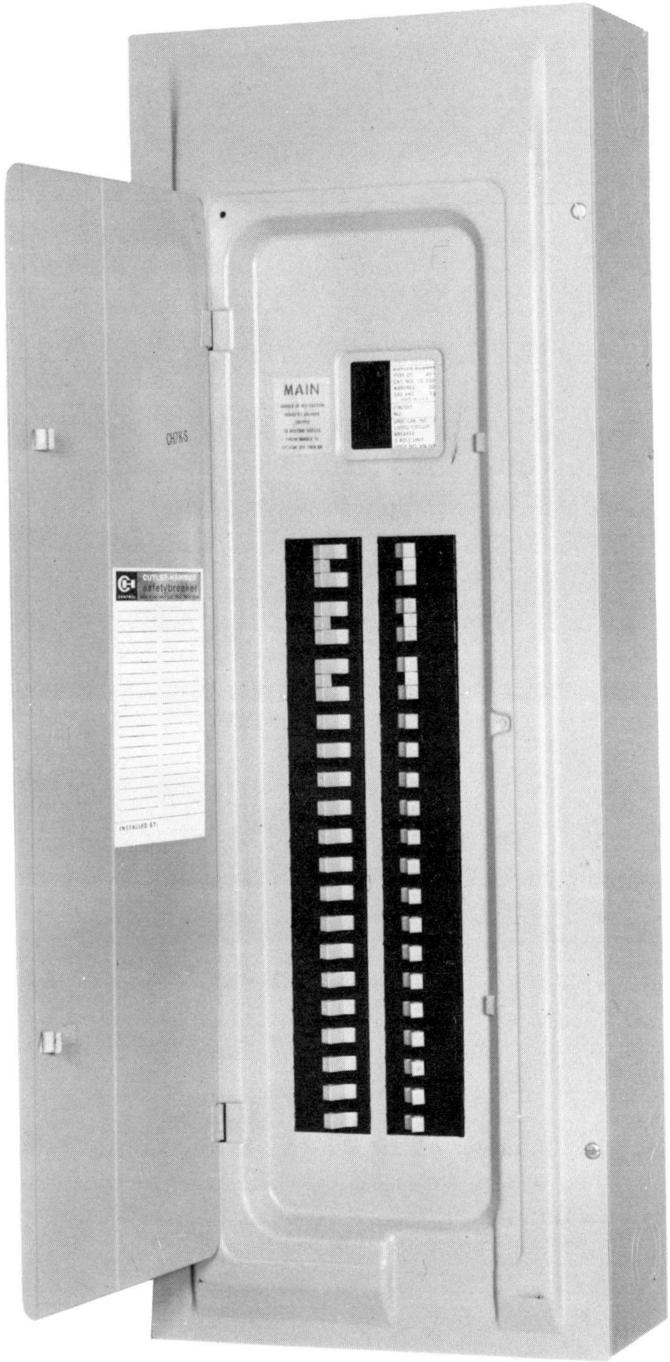

(a)

Figure 17.7 Main disconnect panels. (*a*) Service disconnect panel showing main disconnect switch, cover, and fill of safety-breakers. (Courtesy of Cutler-Hammer, Incorporated) (*b*) Flush type load center with main circuit breaker and neutral terminals. Complete panel before installation of trim. (Courtesy of General Electric Company–Circuit Protective Devices Department) (*c*) Internal wiring diagrams for breaker type service equipment.

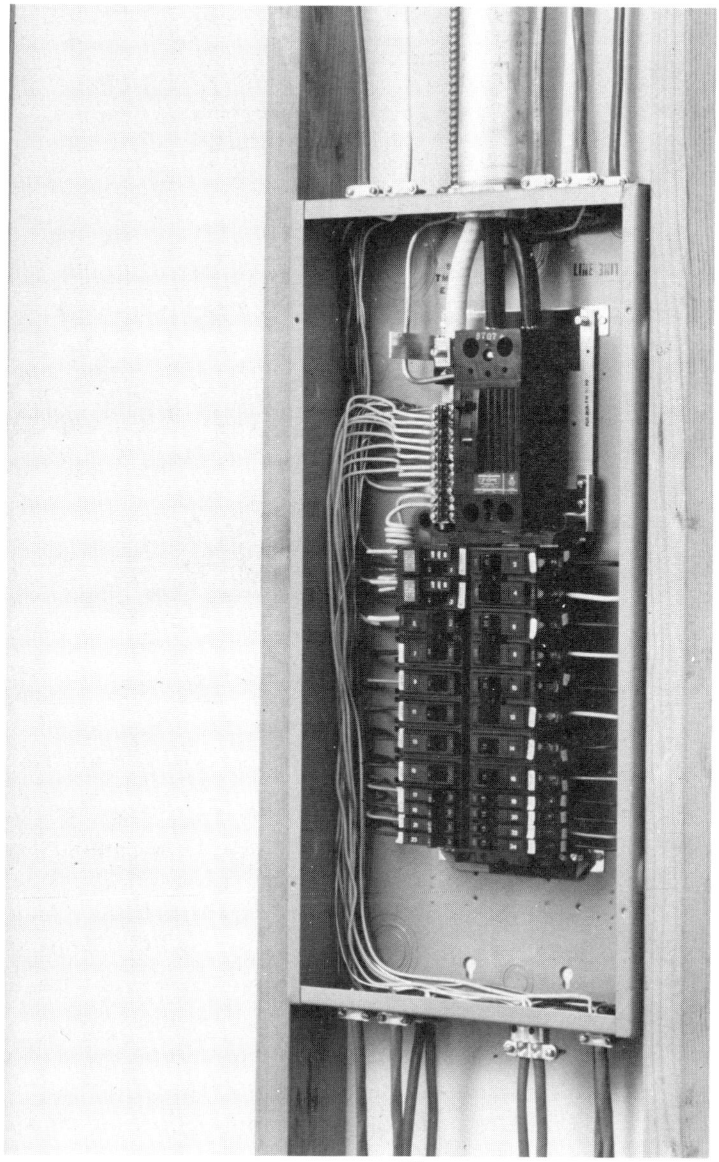

(b)

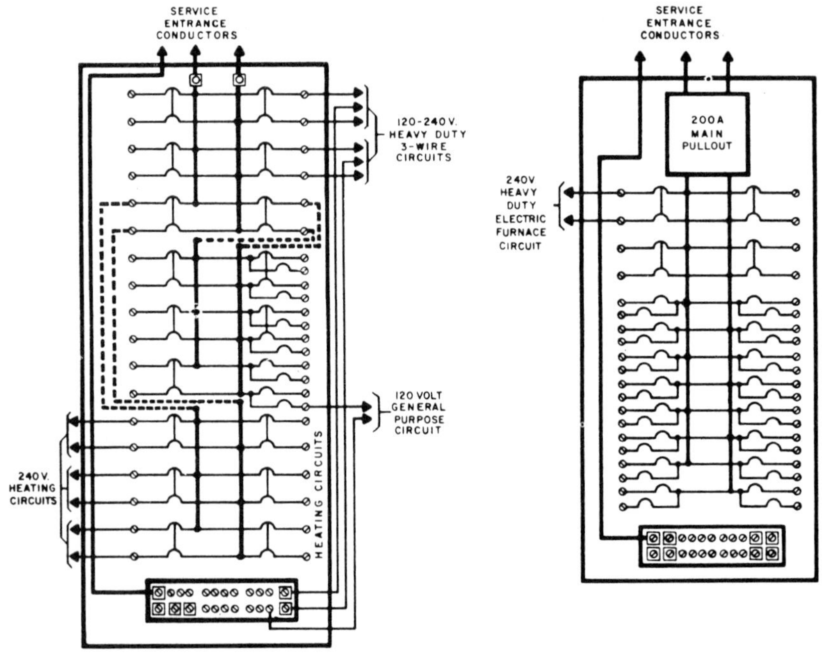

(c)

To install a circuit in a panel, the slot at the rear of the breaker is engaged over the protruding metal stud in the panel. The breaker jaws are firmly pushed onto the bus-bar tab. Breakers must be clearly marked to show whether they are in the open (OFF) or closed (ON) position. Trip the breaker to the OFF position and you are ready to connect the branch circuit wire. Use a 15-ampere breaker for No. 14 AWG wire, 20-ampere breaker for No. 12 wire, 30-ampere for No. 10 wire, and so on.

Circuit breakers are made in single-pole and double-pole with voltage and current ratings to match the circuit conductors.

In modern installations, the devices to control and protect individual branch circuits are always included in the same distribution panel cabinet and must bear the UL label and must be marked as suitable for service equipment.

Fuse panels for main service distribution panels usually consist of a main pullout fuse, range pullout fuse, and other branch circuit plug fuse holders (Figure 17.9). When installing a new fuse panel, the NEC states that only "Type S fuses must be used." Type S fuses have a different base size for identifying the current range of the fuse; 15-ampere, 20-ampere, and 30-ampere. Type S fuses are so designed that they cannot be used in any fuse box other than a type S fuse holder or a fuse holder with a type S adapter inserted (Figure 17.10). NEC 240-53.

It is the responsibility of the electrician to determine the ampere rating of the circuit and select the proper size type S adapter. The adapter is then inserted into the Edison-base fuse holder. The adapter makes the fuse holder nontamperable and noninterchangeable. The proper size fuse can now be inserted into the adapter, and the wire for that circuit may be connected to the fuse holder terminal.

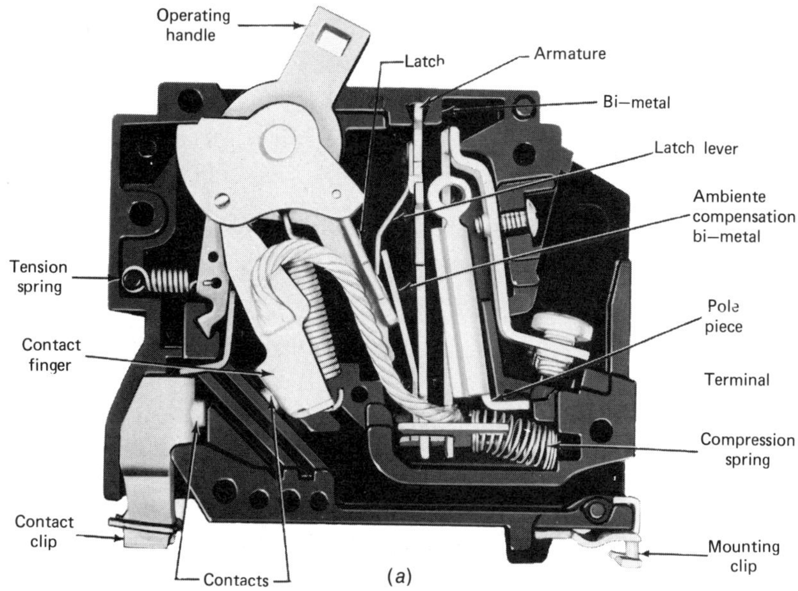

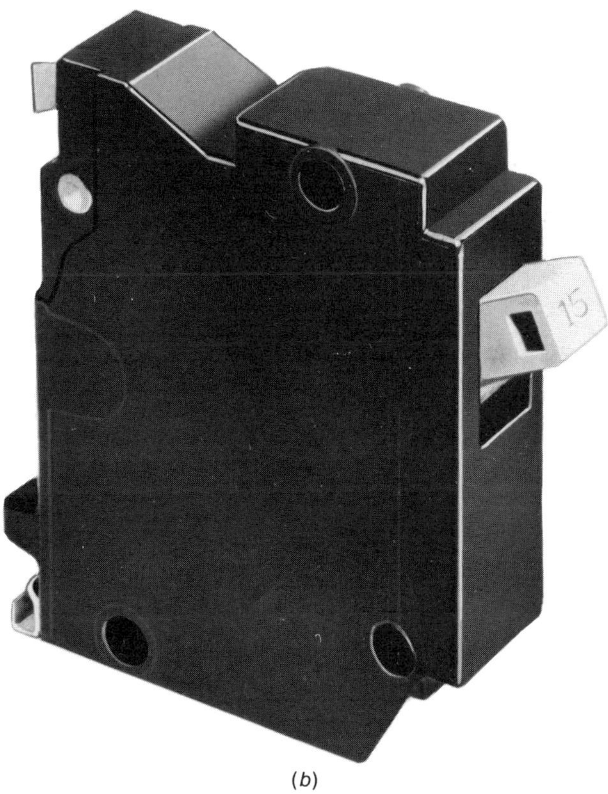

(b)

Figure 17.8 (*a*) Typical circuit breaker (open view). (*b*) Single pole breaker. (*c*) Double pole 15-ampere breaker. (Courtesy of Cutler-Hammer Incorporated)

(c)

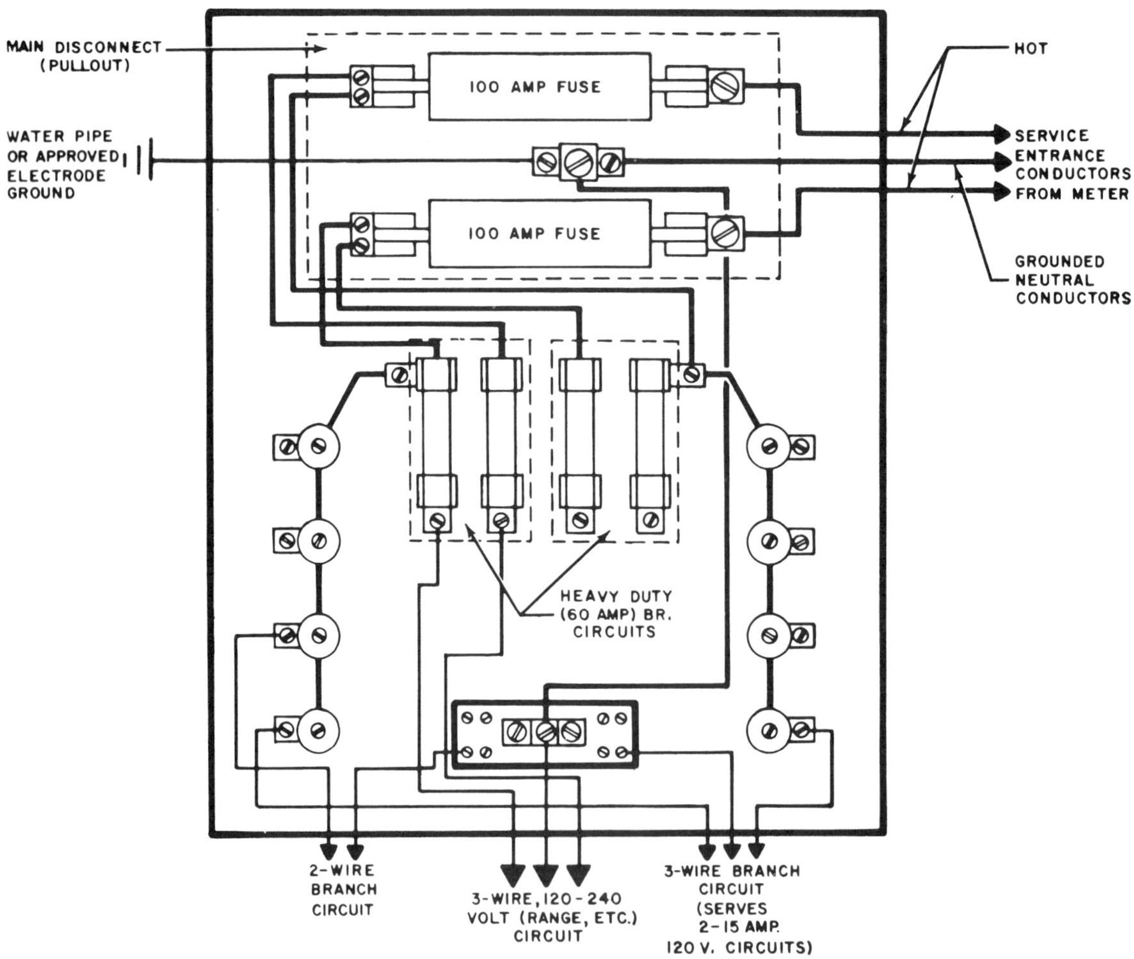

Figure 17.9 Schematic of a typical 100-ampere fuse main disconnect and load center.

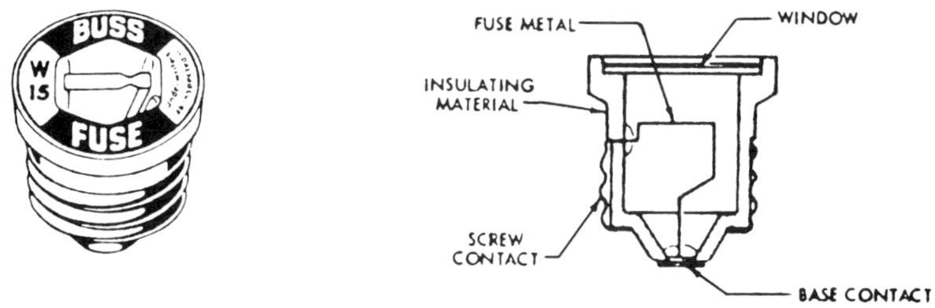

(a)

Left: Plug fuse designed to pass safely 15 amps. Note fuse link in hexagonal shaped window. Right: Cross section of plug fuse. (Bussman Mfg. Co.)

Adapter to accept special fuse sizes. Barb on lower front prevents easy removal of adapter. (Bussman Mfg. Co.)

(b)

Type "S" fustat for use with special adapter and certain amperage ratings. (Bussman Mfg. Co.)

(c)

Figure 17.10 Fuses. (*a*) Plug fuse. (*b*) Type "S" fustat. (*c*) Adapter to accept special fuse size. (Courtesy of Bussman Manufacturing Company)

Some residential equipment such as counter-mounted cooking tops, ovens, ranges, clothes dryers, water heaters, and electric space heaters operate on 240 volts. This requires a different type of fuse: the ferrule contact cartridge type. A ferrule cartridge fuse is made in two different physical sizes. The smaller of the two is produced in current ratings up to 30 amperes. The larger is available in 35- to 60-ampere ratings and is made in 5-ampere units (35, 40, 45, 50, 60) (Figure 17.11).

When currents in excess of 60 amperes are flowing in a circuit (usually the main fuses), a more rugged knife-blade cartridge type fuse is used to protect the entrance conductors.

The ground-fault circuit interrupter receptacle is covered in Chapter 8, page 128. However, in this section the circuit breaker that protects both the equipment and people will be introduced for discussion. It is a circuit breaker with a built-in ground-fault circuit interrupter (GFCI). It provides the same branch circuit overcurrent protection as a standard circuit breaker; however, it provides added protection from ground-faults. It is designed and manufactured to automatically open a circuit when a fault current is 5 mA or more

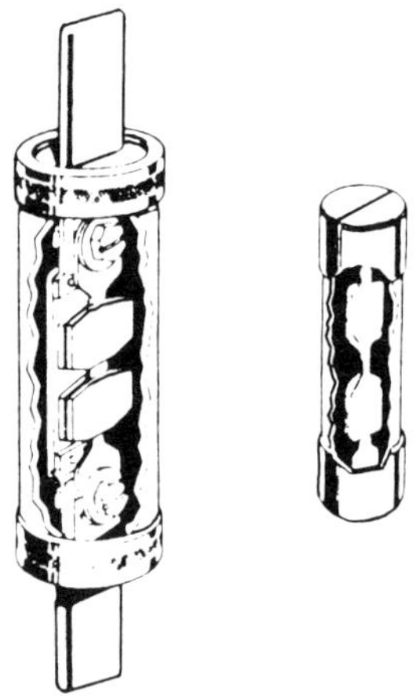

Cutaway of fuse types available in renewable and non-renewable links. Left: Blade-contact fuse available in sizes larger than 60 amps. Right: Furrule-contact cartridge fuse available in sizes to 60 amps. Both fuses shown are renewable type. Fuse links may be replaced. (Bussman Mfg. Co.)

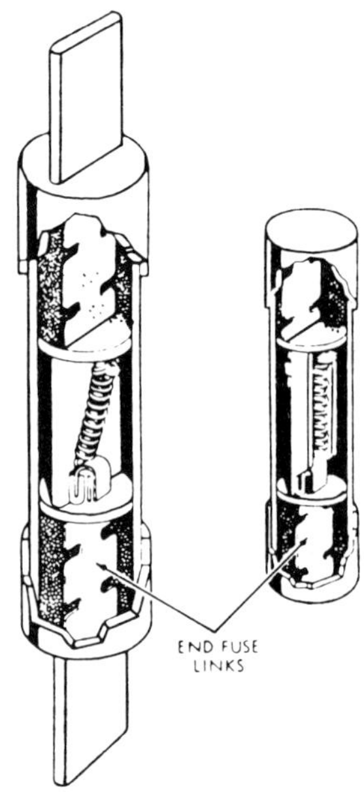

Cutaway view of dual element fuses (non-renewable). Both sizes are of the time-delay type. (Bussman Mfg. Co.)

Figure 17.11 Cartridge fuses. (Courtesy of Bussman Manufacturing Company)

The GFCI breaker is a self-contained unit that may fit directly into the distribution panel (Figure 17.12). It operates on the principle that the current leaving a circuit is equal to the current entering that circuit. Both supply conductors of the circuit pass through a highly developed transformer. When there is no leakage current, the magnetic fields around the supply conductors cancel one another. No voltage is produced in the transformer.

Should a leak or fault develop, more current is entering the circuit on one supply conductor than is leaving on the other. This magnetic imbalance causes a voltage to be induced into the transformer coils. An amplifier increases the strength of the voltage and uses it to trip the circuit breaker. A fault current as low as 0.002 amperes will trip the GFCI breaker.

GFCI circuit breakers or a GFCI receptacle are required in the following areas:

NEC 210-8 (a) In residential occupancies, on all 120-volt single phase, 15-ampere, 20-ampere receptacle outlets outdoors, in garages, and in bathrooms.

NEC 305-4 (a) On the construction site (usually the temporary power pole).

NEC 210-8 (b) Hotel and motel bathrooms in guest rooms.

NEC 680-6 Swimming pools: lighting fixtures and lighting outlets.

NEC 680-20 Underwater lighting fixtures (wet-niche).

NEC 680-31 Storable swimming pools.

NEC 680-41 (a) Supplying fountain equipment.

NEC 680-5 Swimming pool transformers.

NEC 555-3 Marinas and boatyards.

NEC 517-14 Health care facilities.

NEC 215-9 Feeders.

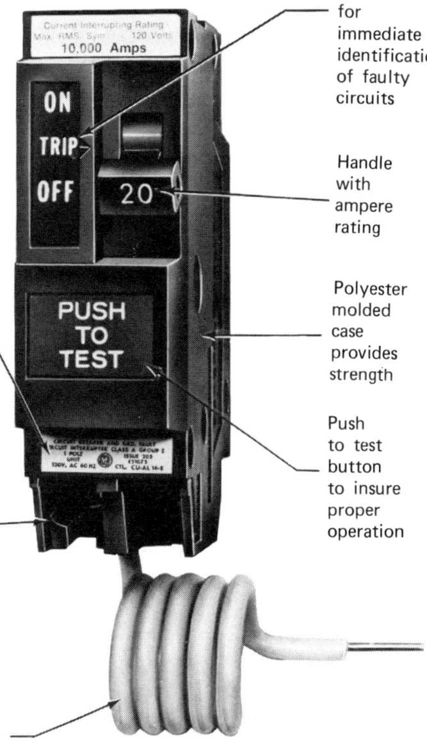

Figure 17.12 Ground-fault circuit interruption feature is built into a single-pole molded case circuit breaker of the type usually used to protect normal receptacle circuits. These circuit breaker units will fit into any ordinary load center, replacing a conventional circuit breaker. (Courtesy of General Electric Company)

17-7 GROUNDING

The purpose of grounding is safety. According to the NEC, a wiring system must afford protection to life and property against faults caused by electrical disturbances, lightning, failure of electrical equipment that is a part of the wiring system, or failure of equipment and appliances that are connected to the wiring system.

For this reason, all metal enclosures of the wiring system, as well as the noncurrent-carrying or neutral conductors, should be tied together and reduced to a common earth potential.

Grounding falls into two categories: (1) system grounding and (2) equipment grounding.

1. *System Grounding.* System grounding means the connection of the neutral conductor of the wiring system to the earth. Its purpose is to drain off any excessive high voltage that may accidentally enter the system as a result of an electrical disturbance, lightning, an insulation breakdown in the supply transformer, or an accidental contact between the service conductors and high tension wires. An accidental grounding of one of the current-carrying conductors will result in a short circuit causing a fuse or circuit breaker to open, thereby disconnecting the line conductors.

2. *Equipment Grounding*. Equipment grounding, or grounding noncurrent-carrying parts of the installation, means the steel raceway itself, service equipment panel, and metal enclosures of equipment, like the frames of ranges or motors.

   The purpose of this grounding is to prevent a voltage higher than earth potential on the enclosures or equipment, reducing the danger of shock in case a life-conductor comes in contact with these conductive parts.

   The grounding is accomplished by running a wire from the neutral connection in the main service switch or, in some cases, the meter socket, to the water piping system on the street side of the water meter. In addition the water pipe shall be supplemented by an additional electrode, which may be bonded to a grounding electrode conductor.

The NEC again rules here:

NEC 250- 81 Where available on the premises, a metal underground water pipe shall always be the grounding electrode.

NEC 250-92 (a) Metal grounding electrode conductor enclosures shall be electrically continuous from cabinet to grounding electrode.

NEC 250-92 (a) If metal enclosure is not electrically continuous it must be bonded at each end to grounding conductor.

NEC 250-115 Ground clamp approved for the purpose shall be used to connect grounding electrode conductor to grounding electrode.

NEC 250-112 Connection should be on street side of water meter or bonding required around valves, meter, unions, etc.

Size of grounding wire is found in NEC Table 250-94.

17-8 BONDING

All service equipment must be properly bonded. The purpose of installing bonding bushings and bonding jumpers on service-entrance equipment is to assure a low impedance path to ground should a fault occur on any of the service-entrance conductors.

1. NEC—Definitions: Main bonding jumper is the connection between grounded circuit conductor and equipment grounding conductor at service.
2. NEC 250-79 (a) (b) Equipment bonding jumper—main and equipment bonding jumper shall be (a) copper or other corrosion-resistant metal: (b) attached as per NEC 250-113 and 250-115.
3. NEC 250-79 (c) Equipment bonding jumper on supply side of service and main bonding jumper. Sized as per Table 250-94 but not less than 12½ percent of largest phase conductor.

SUMMARY

1. All service equipment wiring must adhere to the National Electrical Code.
2. The local utility company is responsible for locating the meter and service drop.

3. The meter socket is grounded for protection.
4. Grounded neutral conductors are not switched or protected by an overcurrent device.
5. Circuit breakers and fuses are safety devices to protect wire against short circuits and overloads.
6. The meter is located outside of the house for convenience.
7. Service laterals are protected by the NEC.
8. Circuit breaker panels are widely used in residential wiring.
9. To reset the circuit breaker, move handle to reset position.
10. The purpose of grounding a service is for safety.

PROBLEMS

17-1 When a circuit breaker trips due to a short circuit or overload, how is power restored? Explain.

____

____

____

____

17-2 A branch circuit wired with No. 12 wire would be protected with a fuse rated at (circle one):

a. 30 amperes b. 15 amperes c. 20 amperes

17-3 A branch circuit is the circuit conductor between the final overcurrent device protecting the circuit and the outlets. It is known as an appliance branch circuit when it supplies (circle one):

a. a single receptacle outlet for one large appliance only.

b. a number of outlets for lighting and all appliances.

c. a number of outlets for appliances only.

17-4 List six parts of the electrical service.

1. ______________ 4. ______________
2. ______________ 5. ______________
3. ______________ 6. ______________

17-5 Explain the function of a ground-fault circuit interrupter breaker.

____

____

____

____

chapter 18

THE SERVICE AND EQUIPMENT—COMMERCIAL

Instructional Objectives

1. To become familiar with the National Electrical Code as related to service-entrance equipment.
2. To understand the need for overcurrent protection.
3. To identify the need for overcurrent protection.
4. To learn the advantages of a three-phase system.
5. To become familiar with the basic methods of metering.
6. To identify the parts of a circuit breaker.

Self-Evaluation Questions

Test your prior knowledge of the information in this chapter by answering the following questions. Watch for the answers as you read the chapter. Your final evaluation of whether you understand the material is measured by your ability to answer the questions. When you have completed the chapter, return to this section and answer the questions again.

1. How many service-drops or laterals may be supplied to a building?
2. Does the NEC require ground-fault protection in commercial buildings?
3. What are the advantages of three-phase voltage systems in commercial wiring?
4. Why are current transformers used in high-voltage metering?
5. What is a fuse?
6. What is the difference between a ferrule-contact and a knife-blade contact type fuse?
7. What is a circuit breaker?
8. How do you reset a circuit breaker?
9. How are panelboards identified?

10. Does the NEC have restrictions to the use of panelboards used for lighting and appliances?

18-1 THE ELECTRIC SERVICE

Commercial buildings are sometimes equipped with 115/230-volt service as used in residential installations. The larger capacity 115/230-volt services (400 amperes) have more than enough circuits for lighting, receptacles, and motors these small, commercial facilities need.

Service to the Building

It would be well at the start of this section to review the NEC requirements for services.

NEC Article 230-2 states, "a building shall be supplied by only one set of service-drop or service lateral conductors." See Figure 18.1 for service drop and Figure 18.2 for service lateral.

This is one rule where we want to take a good look at the exceptions. There are seven exceptions that would allow additional drops or laterals to a building.

Exceptions No. 1 and 2 allow a separate drop or lateral for a fire pump, or emergency electric service equipment, as required for some types of commercial buildings.

Exception No. 3 (Multiple-occupancy buildings). By special permission from the authority having jurisdiction, more than one set of service drops or laterals shall be permitted in multiple-occupancy buildings (such as a multi-

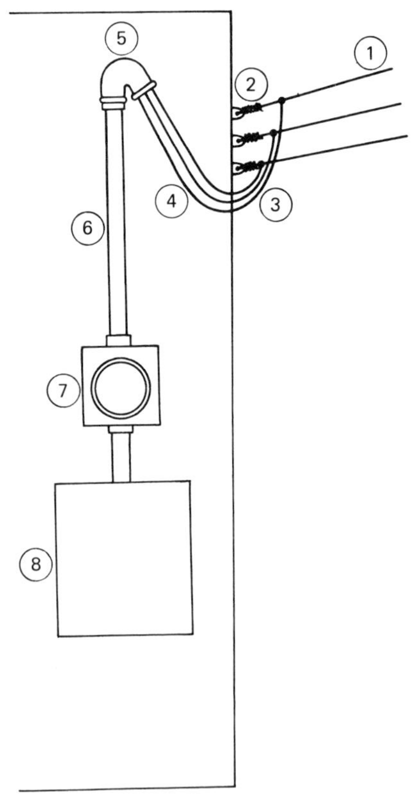

Figure 18.1 Overhead service. (1) Service drop conductors. (2) Point of attachment. (3) Drip loop. (4) Service-entrance conductors. (5) Service head. (6) Service-entrance conduit. (7) Meter and metering equipment. (8) Service equipment.

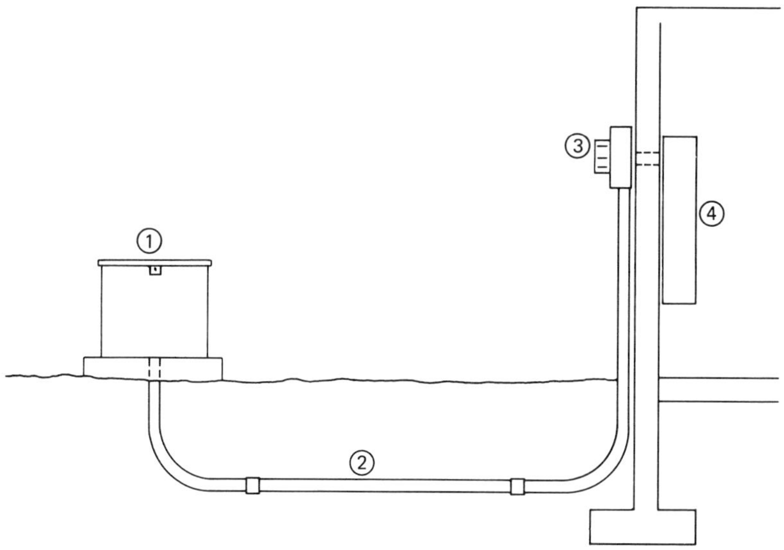

Figure 18.2 Service lateral. (1) Fed from pad-mounted transformer. (2) Service lateral conduit. (3) Meter base and metering equipment. (4) Service equipment.

ple-dwelling or shopping center) where there is no available space for service equipment accessible to all the occupants. Two or more separate sets of service-entrance conductors may be permitted to be tapped from one service drop or lateral.

Exception No. 4 (Capacity requirements). Two or more service drops or service laterals shall be permitted where the capacities are in excess of 3000 amperes, and at a supply voltage of 600 volts or less. Two or more services shall be permitted where the load requirements of a single-phase installation are greater than the serving agency normally supplied through one service. Again, two or more services shall be permitted when special permission is granted by the authority having jurisdiction.

Exception No. 5. By special permission from the authority, two or more services may be installed for a single building or other structure that is sufficiently large to make two or more services necessary.

Exception No. 6. Services having different characteristics, such as for different voltages, phases and frequencies. Different uses, such as being on different rate schedules.

Exception No. 7. Underground sets of conductors, sizes 1/0 and larger running to the same location, and connected together at their supply end, but not connected together at their load end. This shall be considered to be one service lateral.

18-2 SPECIAL REQUIREMENTS FOR HIGHER VOLTAGE

Fuses and circuit breakers in general provide better protection against short-circuit and ground-fault conditions on services of 230 volts or less than on 480-volt services. However, in some cases, a 480-volt service has been destroyed when a really severe fault occurred.

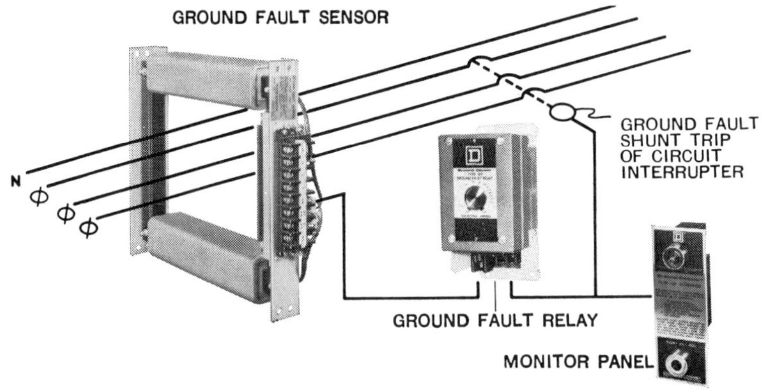

Figure 18.3 Ground-fault protection equipment. Shows type GP, used on power distribution systems that include a grounded conductor (neutral or grounded phase). When circuit conditions are normal, current from all phase conductors is neutral (if used) and adds up to zero; and the sensor produces no output signal. If a ground fault occurs, the current adds up to equal the ground-fault current, and the sensor provides an output proportional to the ground current. This output is measured by the ground-fault relay which activates the ground-fault shunt trip mechanism of the circuit interrupting devices. (Courtesy of Square D Company)

The 1984 NEC now requires additional ground-fault protection.

NEC Article 230-95 Ground-fault protection of equipment service disconnecting means rated at 1000 amperes or more, solidly grounded wye electrical service with a voltage more than 150 volts to ground, and where voltage phase to phase does not exceed 600 volts. Ground-fault protection shall be provided for each such service disconnecting means.

The maximum setting of the ground-fault protection shall be 1200 amperes, the maximum time delay shall be one second for ground-fault currents equal to or greater than 3000 amperes. The protection shall operate to cause the disconnecting means to open all ungrounded conductors of the faulted circuit (Figure 18.3).

18-3 HIGHER VOLTAGES

Large commercial buildings usually have several installations requiring motor-driven equipment. These motors often need far more current than lighting circuits. When being started, a motor will draw three to five times its normal operating current, which places an even greater load on the service equipment. A motor designed to operate on a higher voltage will produce the same horsepower but with a decrease in line current, usually about half the amperes.

Motors and equipment designed to operate on even higher voltages need still less current. When the current is reduced, the wire size in the motor winding can also be reduced. This allows for a higher voltage motor to be much smaller in size than a lower voltage motor, using the same horsepower.

Special voltage systems are available for commercial use. The most commonly used system for modern commercial buildings is the three-phase, four-wire wye. This system has the advantage of being able to provide three-phase power, and permitting lighting to be connected between any of the three phases and the neutral. Voltages available with this type of system are 120/208, 277/480, and 347/600 volts. See Figure 18.4.

Another advantage of the three-phase system is that motors can be

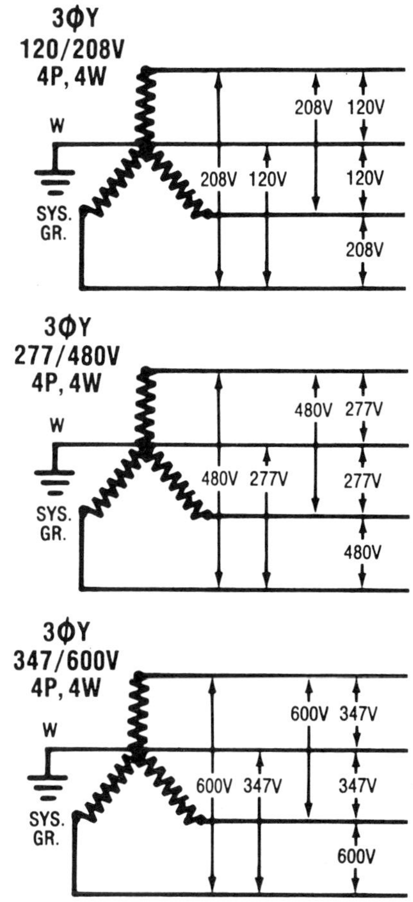

Figure 18.4 Special voltage system for commercial buildings. Illustration shows a three-phase, four-wire wye. (Courtesy of Crouse-Hinds Company, Arrow-Hart Div.)

reversed easily. A three-phase motor will operate in the opposite direction if any two of the three motor leads are interchanged.

A word of caution. When working on a three-phase service, take care not to interchange any of the hot wires. If the hot wires are interchanged, every motor in the building will have its direction of rotation reversed.

18-4 METERING

The electrician working on commercial buildings seldom makes metering connections. This is a job for the utility company supplying the power. However, the electrician should be familiar with the basic methods of metering.

The three-phase system needs a special kilowatt-hour meter containing two voltages and two current coils for recording power. A potential transformer and two current transformers are installed on the high-voltage lines and leads are brought to meter. Review Chapter 21, Section 21-7 on current transformers.

For loads of 200 amperes or less, the service conductors are connected to the service drop and run directly to the meter base. For multiple dwellings, the meters are usually installed as a part of the service-entrance equipment. See Figure 18.5

Figure 18.5 Grouped metering is ideally suited for multimetering applications where a main device is used for incoming single-phase, three-wire, 120/240 volts AC, or three-phase, 4-wire, 120/240 volts AC service. (Courtesy of Square D Company)

18-5 SWITCHBOARDS

Article 100 of the NEC defines a switchboard as:

> a large single, frame, or assembly of panels on which are mounted on the face or back or both, switches, overcurrent and other protective devices, buses, and usual instruments. Switchboards are generally accessible from the rear as well as from the front and are not intended to be installed in cabinets.

Safety enclosed boards are used for most new installations. Most of these boards are of the unit or sectional type. They consist of a combination of the desired number and type of standardized unit sections. Each section is a factory assembled combination of a formed steel panel and apparatus mounted on a steel framework. See Figure 18.6.

On switchboards supplied from a four-wire delta connected system, where the midpoint of one phase is grounded, that phase bus bar or conductor having the higher voltage to ground shall be marked. See Figure 18.7.

The phase arrangement on three-phase buses shall be A, B, C, from front to back, top to bottom, or left to right as reviewed from the front of the switchboard or panelboard. The B phase shall be that phase having the higher voltage to ground. See Figure 18.8.

Location of Switchboards

Switchboards of the safety enclosed type may be placed in any nonhazardous location without any restrictions. However, there should be sufficient access and working space to maintain all electric equipment.

Grounding

Bonding of a service raceway is illustrated here, by the way of an example.

When a ground develops beyond point A, Figure 18.9, which represents the service overcurrent device, fault current is readily interrupted before

Figure 18.6 Switchboard. Multisection switchboards are available with ratings of 1200, 1600, or 2000 amperes at 600 volts or less. The main section can be selected as either a main molded case circuit breaker or a main Bolt-Loc fusible switch and is furnished with a ground-fault protection system when required. The distribution section contains an I-Line branch circuit breaker panel of double row construction. The enclosures are 91.5 inches high, 24 inches deep and 30 through 48 inches wide. All sections are completely accessible for easy installation and maintenance. (Courtesy Square D Company)

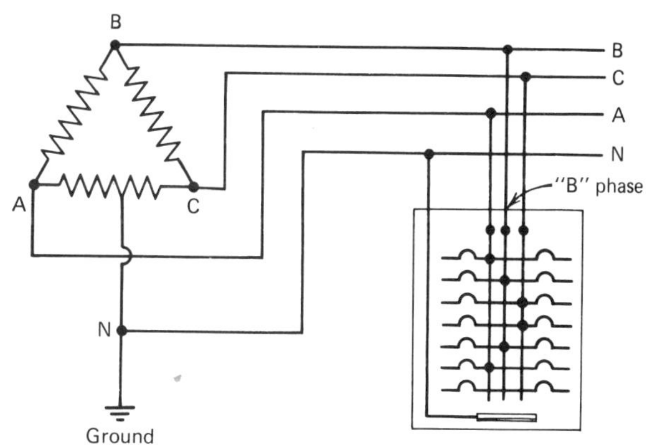

Figure 18.7 Drawing shows a four-wire delta system, where the midpoint of one phase is grounded. The phase having the higher voltage to ground shall be marked as the "B" phase.

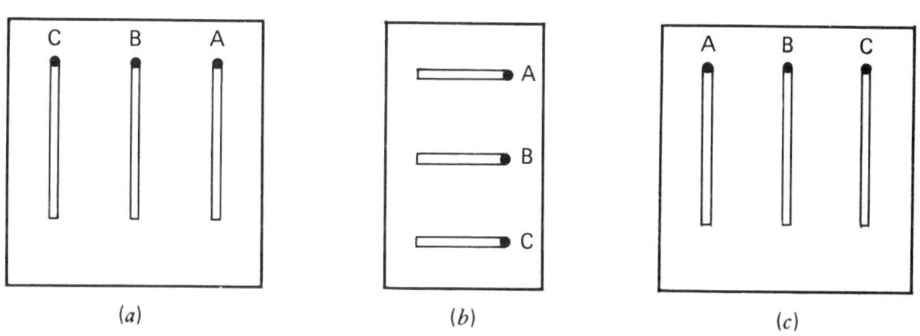

Figure 18.8 Illustration shows phase arrangement on a three-phase bus. (*a*) ABC from front to back. (*b*) ABC from top to bottom. (*c*) ABC left to right as reviewed from the front of the switchboard.

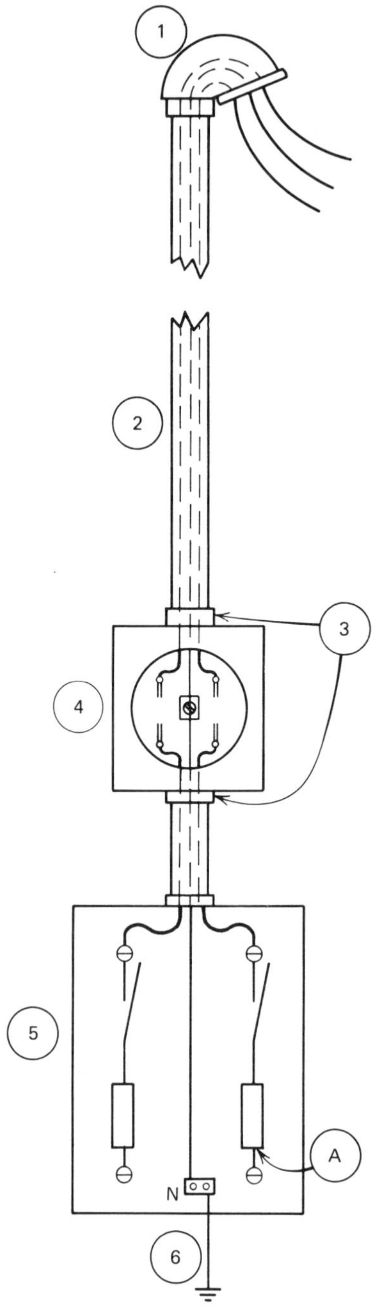

Figure 18.9 Grounding service raceway. (1) Service head. (2) Service conduit. (3) Threaded hubs. (4) Meter base. (5) Service switch showing point "A" at fuses. (6) Ground conductor.

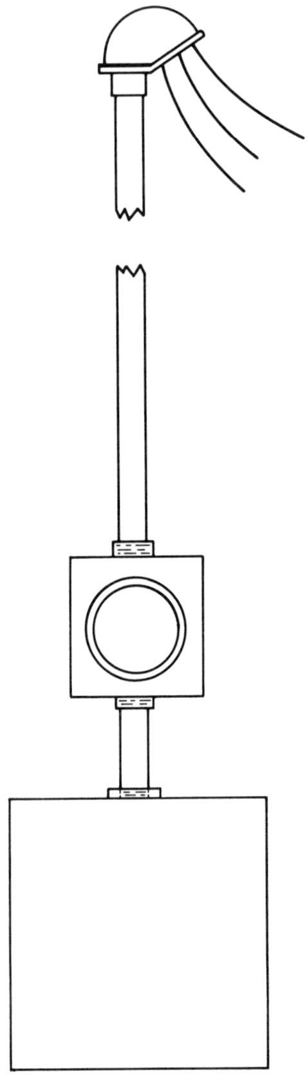

Figure 18.10 Shows threaded hubs used with meter base and service equipment.

serious damage results. Should the ground occur in the service run ahead of A, however, there is nothing to limit flow of current except the utility company's primary overcurrent unit, which may be quite remote from the service location.

It is for this reason that either a threaded hub (Figure 18.10) or bonding must be relied on here. Resistance of the ground circuit must be made as low as possible, the object being to increase the rate of flow, if possible, so that the utility company's remote protective device will operate to disconnect the service-drop conductors. Connection between the service equipment and the ground electrode should be as direct as possible, for a like reason.

Figure 18.11 Light panel. (Courtesy of Square D Company)

Figure 18.12 Power panel. (Courtesy of Square D Company)

18-6 LIGHTING AND POWER BRANCH CIRCUIT PANELBOARDS

On commercial projects, the electrical supply feeds into several panelboards, oftentimes broken up into lighting panels to serve light fixtures and general-purpose receptacles (Figure 18.11), and power panels to supply electricity to all appliances and equipment on the project (Figure 18.12). Quite often there are several of each type of panel, depending on the size of the project.

Figure 18.13 Shows a 480-volt main disconnect section. (Courtesy of Square D Company)

Figure 18.14 Lighting and appliance panel. (Courtesy of Cutler-Hammer, Incorporated)

In such a design, the circuits are labeled with a circuit number and panel number. For example C-5, LP-2 would indicate circuit 5 from light panel 2.

It is necessary to have a feeder from the service entrance to the main disconnect panel (Figure 18.13) which is in turn connected to the lighting and power panels by feeders. Each feeder must be sized with the NEC.

Lighting and appliance panelboards (Figure 18.14) are usually the kind you will encounter most often in smaller commercial buildings. However, there are special restrictions to their use. See Figure 18.15.

1. NEC 384-14 defines a lighting and appliance panelboard as "one having more than 10 percent of its overcurrent devices rated at 30 amperes or less, for which neutral connections are provided."
2. NEC 384-15 states, "no more than 42 overcurrent devices (in addition to the main overcurrent devices) may be installed in any one lighting and appliance branch-circuit panel."

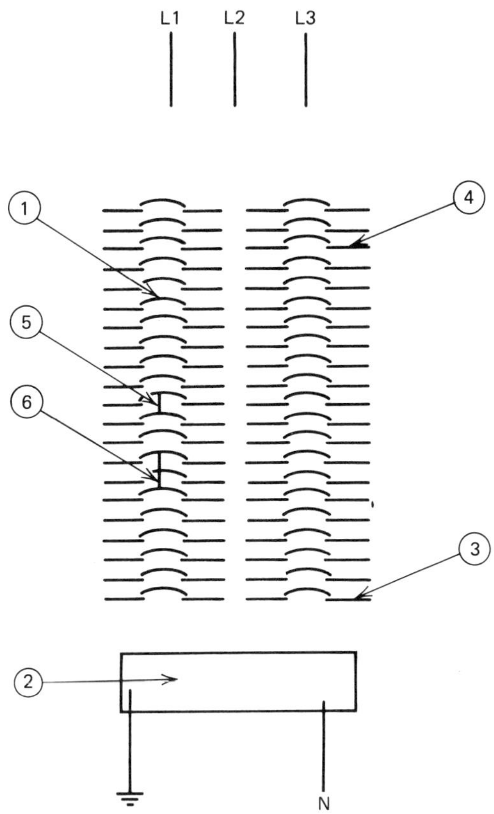

Figure 18.15 Light and appliance panelboard, illustrating six restriction points.

3. A lighting and appliance branch-circuit panelboard shall be provided with physical means to prevent the installations of more overcurrent devices than the number for which the panel was designed, rated, and approved. In other words if "standard" width breakers can be replaced by the half-size or "thin-wafer" breakers, this is taken into account in the total number of spaces allowed in the panelboard and its ampere rating.
4. For the purpose of Article 384-15, a two-pole circuit breaker shall be considered two overcurrent devices.
5. A three-pole circuit breaker shall be considered three overcurrent devices.
6. NEC 384-16 states that each panelboard shall be individually protected on the supply side by not more than two main circuit breakers, or two sets of fuses (Figure 18.16).

   The overcurrent protection shall have a combined rating not greater than that of the panelboard.

   EXAMPLE

   For a panelboard rated at 400 amperes, use:

   1. Two 200-ampere overcurrent devices, or
   2. one 300-ampere overcurrent device and one 100-ampere overcurrent device, or
   3. any combination not exceeding the 400 amperes.

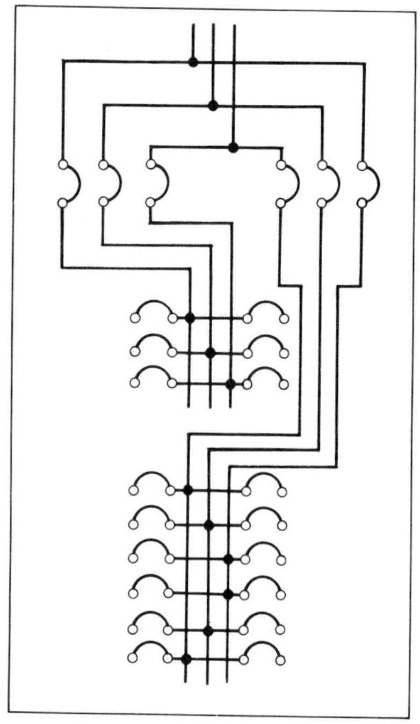

Figure 18.16 Lighting and appliance branch-circuit panelboard. Shows the panel protected on the supply side with two main circuit breakers.

18-7 STEADY-BURNING LOADS (CONTINUOUS LOADS)

NEC Article 384-16 requires that overcurrent protective devices ahead of panelboards be installed in commercial buildings where loads continue for long periods of time (3 hours or more). The total load on any overcurrent device in such a panelboard shall not exceed 80 percent of its rating.

There is an exception to the rule that states, "where the assembly including the overcurrent device is approved for continuous duty at 100 percent of its rating, this type of panelboard shall not be required to be derated to 80 percent."

Since these provisions are sometimes found confusing look at an example:

Assume an office building has a floor area of 9000 sq ft. Table 220-2(b) NEC specifies a unit load of at least 3½ watts per sq ft (for office space), or a total of 31.500 watts

$$9000 \times 3.5 \text{ watts} = 31.500 \text{ watts}$$

At 230 volts, the current equals

$$31.500 \div 230 \text{ volts} \quad \text{or} \quad 137 \text{ amperes}$$

If there are twelve 20-ampere circuits on a three-wire, 11/230-volt panelboard according to NEC 210-23(b), they cannot be loaded above 80 percent of 20 amperes, or 16 amperes. The total current that they can handle is equal to 6 × 16 amperes, or 96 amperes on either side of the three-wire feeder. These lighting circuits will furnish the area with lighting equal to 3.5 watts per sq ft. However, they will only be loaded to 80 percent of their rating. The feeder, meanwhile, will be supplying current which loads the conductors to only 80 percent of their rating.

If the connected load consists partly of lighting units having fluorescent lamps, ballasts, and incandescent lamps, it must be remembered, when branch circuits supply lighting loads having ballasts or transformers, the load calculations must be based on the total volt ampere rating and not on the total wattage ratings of the lamps.

EXAMPLE

Four 40-watt fluorescent lamps = 160 watts. However, each has an ampere rating of 1.6 amperes at 115 volts = 184 volt-amperes. This could be rounded off to 200 volt-amperes. Add this value to the wattage values of the incandescent lamps to determine the total load.

18-8 THREE-PHASE PANELBOARDS

In some commercial installations there is a need for three phases. For such an application a panelboard is used with what is called a "delta circuit breaker." This three-pole breaker (Figure 18.17) is constructed so that two of its line poles are plugged in or bolted onto the ungrounded bus bars of a conventional 120/240-volt, three-wire panelboard. However, the board is supplied by a 120/240-volt, four-wire three-phase delta supply, which provides 120/240 volts with a grounded neutral for three-wire or two-wire

(a)

Figure 18.17 Three-phase panelboard. (*a*) Panel complete with circuit breakers. (*b*) 400-ampere line and neutral lugs. (2) 1/0 300 MCM or (1) No. 4 600 MCM AL/CU. (*c*) 42 pole spaces. (Courtesy of Cutler-Hammer, Incorporated)

(b)

single phase loads. The third or "high leg" of the delta supply is connected directly to a line terminal of the three-pole delta breaker. This provides a three-phase supply for a single branch circuit or feeder.

Location and Installation

All panelboards should be located as near the loads that they supply and control as possible. Mounting heights should be such that the distance from the floor or working level to the top of the uppermost overcurrent device is not more than 7 feet, and the distance from the lowest overcurrent device to the floor is not less than 6 inches.

18-9 CIRCUIT PROTECTIVE DEVICES

It is impossible for an electronic current to flow through a wire without heating the wire. As the number of amperes increases, the temperature of the wire also increases. As the temperature of the wire increases, its insulation may become damaged by the heat, thus causing a breakdown of the insulation.

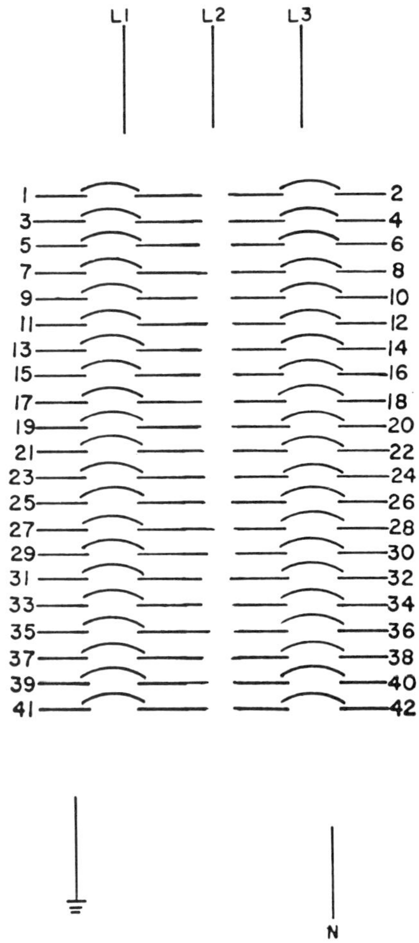

(c)

It is necessary to limit the current through a wire to a maximum value, one that is safe for a given size and type wire. The maximum number of amperes that a wire can safely carry for a continuous load is called the "ampacity" of the wire. Review Section 4.4. Article 100 of the NEC defines ampacity as "current-carrying capacity of electric conductors expressed in amperes."

Any device that opens the circuit when the current in a wire reaches a predetermined number of amperes is called an "overcurrent device."

The fuse and the circuit breaker are the circuit protective devices most frequently involved in the job of making sure a circuit operates within its designed limits.

Fuses

The NEC requires all power wiring systems to be properly fused. This is necessary to protect the wiring and equipment on the circuits, as well as to ensure personal safety.

Fuses in electrical circuits are similar in purpose to the safety valve on the steam boiler. When steam pressure in a boiler rises so high that it is unsafe, the safety valve opens to relieve the pressure. In electrical circuits, whenever the current becomes more than the wires can stand without overheating and burning the insulation, the fuse burns out or blows; this opens the circuit to prevent damage. If the fuse did not open the circuit, such excess current would burn out the wiring or cause a fire or injury and result in costly repairs.

A fuse is basically a short length of metal ribbon made of an alloy with a low melting point, and of a size that will carry a specified current indefinitely, but will melt when a large current flows. When the ribbon inside the fuse melts, the fuse is said to "blow."

Fuses are made in many different styles and sizes for different voltages and current loads, but they all operate on the same general principle, that is opening the circuit by melting the piece of soft metal which becomes overheated when excessive current flows through it. Fuses are safety valves for electric circuits.

When the circuit becomes overloaded, the fuse opens. If there is a short circuit, the fuse opens. When the fuse opens, there is an open circuit, electric current cannot flow and there is zero current. With zero current, no equipment on the line can operate.

Define: Overload and Short Circuit

Overload An overload is a current, larger than normal, but flowing within the normal current path. Overloads are usually in the range of one to six times normal. Rapid removal of overloads is not generally necessary except for certain delicate components such as semiconductors. The chief damage expected under overload conditions is thermal and most often appears as insulation deterioration and failure.

Short Circuits A short-circuit current is distinguished from an overload by virtue of the current flowing outside of the normal path. Usually this current is much larger than overload values, ranging from tens of hundreds of times normal. Great damage can be expected quickly under short-circuit conditions and takes the form of thermal magnetic and arcing damage. Time is critical in the removal of short-circuits from the system.

A short-circuit occurs when the electric current does not go through its intended path, that is, through the load. A good example would be a lamp cord with broken insulation between the wires. This would cause the electric current to bypass the lamp and return to the source through the bare wire. The bare wire offers a lower resistance path than the lamp.

Ohm's law shows what happens in a short circuit:

$$I = \frac{E}{R}$$

If E is equal to 100 volts divided by R of 100 ohms, then I is 1 ampere. This would be a normal situation for a 100-watt lamp.

$$\frac{100 \text{ volts}}{100 \text{ ohms}} = 1 \text{ ampere}$$

Now what happens when two bare wires touch and have a resistance of only 0.01 ohm?

$$\frac{100 \text{ volts}}{0.01 \text{ ohm}} = 10{,}000 \text{ amperes}$$

This is the type of current encountered in some short circuits, and 10,000 amperes is a lot of current.

Fuses Have Four Things to Do in a Circuit

1. The fuse must sense faults.
2. The fuse must open quickly and clear itself when a short circuit occurs.
3. The fuse must also sense a normal and harmless overload, but must open if the overload becomes excessive or prolonged.
4. The fuse must not change or alter the characteristic of the circuit during normal operation.

Voltage Rating

The voltage rating of a fuse must be equal to or greater than the voltage of the circuit in which the fuse is applied.

The voltage rating is the ability of the fuse to extinguish quickly the arc after the fuse element has melted and to prevent the system open-circuit voltage from restricting across the open fuse element.

The fuse voltage rating is not a measure of its ability to withstand a specified voltage while carrying current.

For power systems of 600 volts or less, fuses of a higher voltage rating can be applied on circuits of a lower system voltage; for instance, a 600-volt fuse can be used on a 220-volt system.

Standard fuse voltage ratings are 600 volts, 300 volts, 250 volts, and 125 volts.

Ampere Rating

Every fuse is designated by an ampere rating which has been determined under specific test circuit conditions.

In selecting a fuse ampere rating consideration should be given to the type of load and code requirements for the specific application. To provide reliable overload and short-circuit protection the ampere rating of the fuse should normally not exceed current carrying capacity of the circuit.

There are some specific circumstances where the fuse ampere rating is permitted to be greater than the current carrying capacity of the circuit, for example, a motor circuit where dual-element fuses are generally permitted to be sized at 175 percent and nontime delay fuses sized at 300 percent of the motor full-load amperes.

Generally, the fuse ampere rating and switch combination should be selected at 125 percent of the load current.

Types of Fuses

Cartridge Fuses There are two basic types of fuses.

1. *Renewable Fuse Link.* When the link burns out the end caps are easily removed, the burned link is taken out, and a new link of the same current capacity is installed.
2. *One Time Fuse (Nonrenewable).* When this type of fuse blows, the cartridge is removed, discarded, and replaced with another fuse of the same voltage and current rating (Figure 18.18).

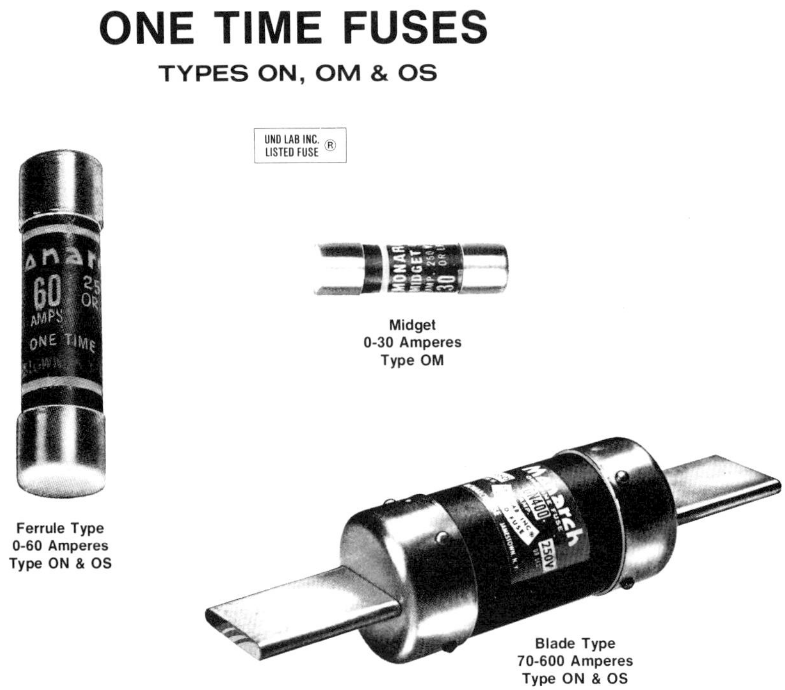

Figure 18.18 Nonrenewable fuses.

Cartridge Fuse Class, Size, and Ratings

Present NEMA and UL standards indicate that standard National Electrical Code type generally one-time or renewable cartridge fuses are designated as class H. Such fuses are classified at interrupting ratings up to 10,000 amperes (Figure 18.19).

Class J, L, G, or K These classes have an interrupting capacity rating above 10,000 amperes. These fuses are "high-interrupting capacity" fuses, and indicate a fuse interrupting rating at some value above 10,000 to about 200,000 amperes.

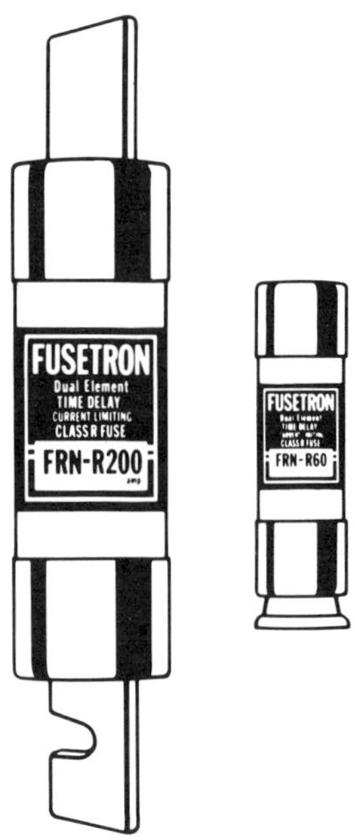

Figure 18.19 Cartridge fuses.

Both J and L class fuses are current limiting, high-interrupting capacity types. The interrupting ratings are 100,000 or 200,000 amperes. The designated rating is marked on the label.

Class J fuse dimensions are different from the standard class H cartridge fuses of the same voltage and current ratings. They require special fuse holders that will not accept noncurrent limiting fuses. See Figure 18.19. This requirement complies with NEC Article 240-60(b): "Fuse holders for current-limiting fuses shall not permit insertion of fuses which are not current-limiting."

What is meant by current-limiting and noncurrent-limiting? A current-limiting fuse is a fuse that when interrupting current in its current-limiting range, reduces the short-circuit current flowing in the faulted circuit to a magnitude substantially less than that which would flow if the current limiting fuse were not in the circuit.

UL states the "current-limiting" indicates that a fuse, short-circuit current at rated voltage, will start to melt within 90 electrical degrees and will clear the circuit within 180 electrical degrees (½ cycle).

Class J fuses of 60 amperes or less are ferrule type. From 61 to 600 amperes, they have slots in the fuse knife blades to permit bolted or knife-blade connection to fuse holders.

Class L fuses are divided into several different current (amperage) classifications with various fuse blade mounting hole dimensions. The dimensions vary according to fuse size, permitting bolted connection to fuse holder.

Class K fuses have interrupting ratings from 50,000 to 200,000 amperes at various peak let-through currents and maximum let-through energy conditions. These fuses are divided into three groups, K1, K5, and K9.

All UL-listed class K fuses have the same dimensions as conventional class H 250-volt or 600-volt, 0- to 600-ampere fuses. Because of this interchangeable feature, class K fuses are not labled "current-limiting," even though qualifications for K1 fuses closely match the qualifications for class J fuses.

Circuit Breakers

A circuit breaker is a device which after an overcurrent condition causes the breaker to open the circuit, can be reset by placing the handle in the off position and then to the on. See cutaway view of a circuit beaker (Figure 18.20).

One of the greatest advantages of a circuit breaker is when it trips, it can be reset by hand and put in operation again without replacing any parts. Many times breakers are used to control lighting as well as to provide overcurrent protection. When circuit breakers are used to control fluorescent fixtures, the breaker shall be marked SWD and approved for switching (NEC 240-83). Figure 18.21 shows a typical circuit breaker.

The important fact to remember is that circuit breakers, like fuses, react to excess current. However, the breaker is not self-destructive. When short-circuit conditions occur, the electrical resistance in a circuit is reduced to a very low value, and the current flow in the circuit rises to a very high level in a very short time (in microseconds; that is millionths of a second). As the alternating current rises to a maximum value in its cycle the circuit breaker senses this drastic increase and opens the circuit.

Commercial circuit breakers are made in single-pole, double-pole, and

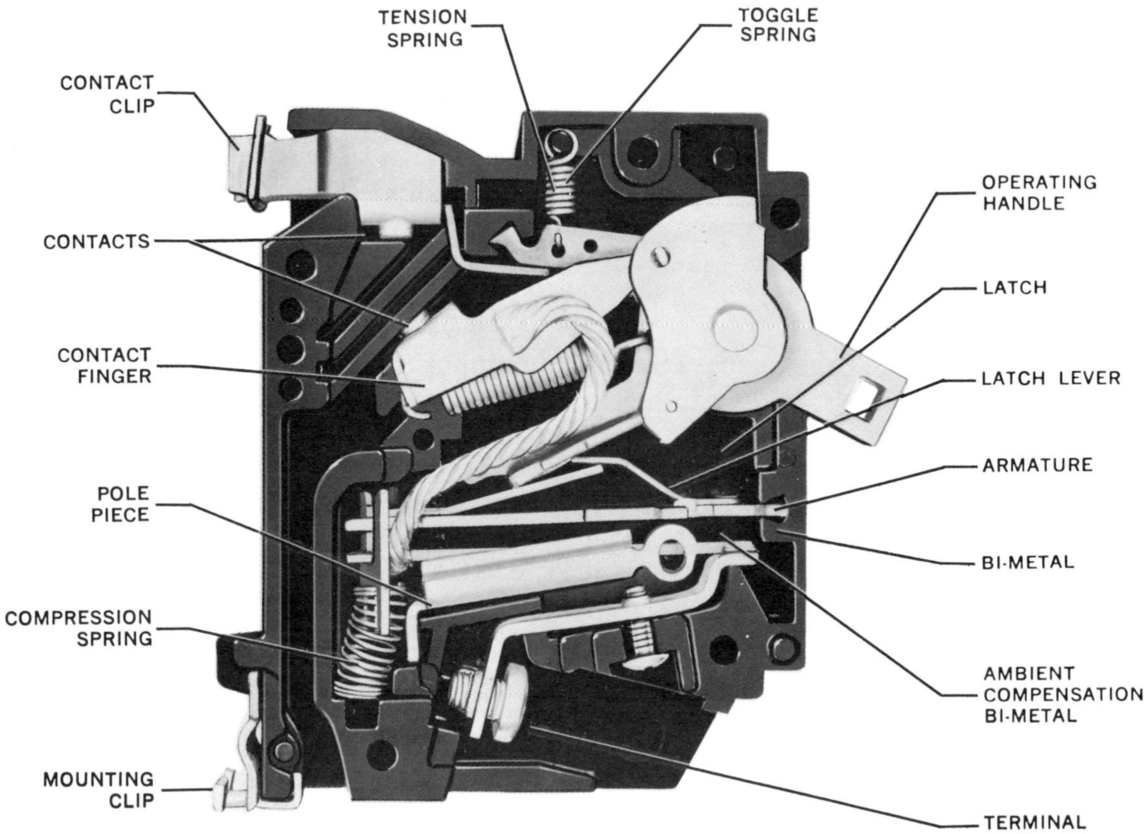

Figure 18.20 Open view circuit breaker. Under severe short-circuit or overload conditions, a safety breaker trips as fast as 8/1000 of a second. This tripping standard is achieved by the magnetic armature assembly which instantly disengages from the latch lever notch to allow the powerful tripping spring to snap the contacts open. (Courtesy of Cutler-Hammer, Incorporated)

Figure 18.21 Single-pole circuit breaker (one pole space). (Courtesy of Cutler-Hammer, Incorporated)

three-pole units, with voltage and current rating to match the circuit conductors. Figure 18.22 shows a two-pole breaker with a common trip, requiring two spaces in the panelboard. Figure 18.23 is a three-pole, common trip breaker which can control three lines, requires three places in the panel.

To install a circuit breaker in a panel, the slot at the rear of the breaker is engaged over the protruding metal stud in the panel. The breaker jams are

Figure 18.22 Double-pole circuit breaker (two pole spaces). (Courtesy of Cutler-Hammer, Incorporated)

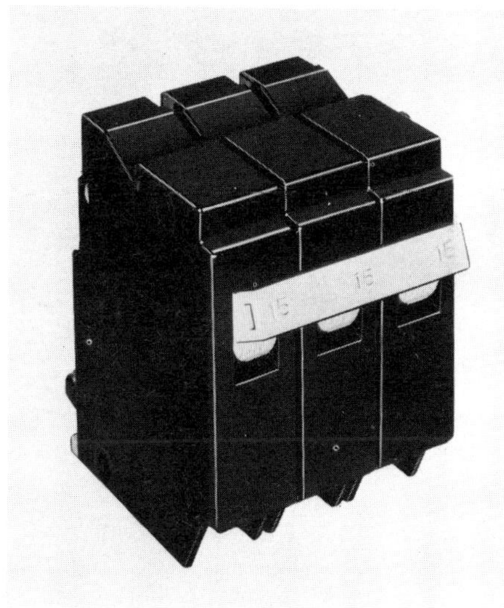

Figure 18.23 Three-pole circuit breaker (three pole spaces). (Courtesy of Cutler-Hammer, Incorporated)

firmly pushed onto the bus-bar tab. Breakers must be clearly marked to show whether they are in the open (off) or closed (on) position. Trip the breaker to the off position and you are ready to connect the branch circuit wire. Use a 15-ampere breaker for No. 14 AWG wire, 20-ampere breaker for No. 12 wire, and a 30-ampere breaker for No. 10 wire, and so on.

In modern installations, the devices to control and protect individual branch circuits are always included in the same distribution panelboard and must bear the UL label and must be marked as "suitable for service equipment."

SUMMARY

1. Electricians need to be familiar with the methods of metering.
2. The service-entrance equipment for commercial buildings consists of the service-conductor, metering equipment, main disconnect switch, and secondary distribution switches.
3. Power and light panelboards are labeled with circuit numbers and panel numbers.
4. The NEC has special restrictions for lighting and appliance panelboards.

5. A fuse is designed to open the circuit when too much current is flowing or when there is a sudden rush of current of dangerous proportion.
6. One of the advantages of a circuit breaker is when it trips, it can be reset by hand and put in operation again without replacing any parts.
7. The devices to control and protect individual branch circuits are always included in the same distribution panelboards and have UL labels "suitable for service equipment."

PROBLEMS

18-1 Sketch an overhead service. Identify the eight important parts of the service equipment.

1. ______
2. ______
3. ______
4. ______
5. ______
6. ______
7. ______
8. ______

18-2 Sketch a service lateral, and identify four main parts of the service equipment.

1. ______
2. ______
3. ______
4. ______

18-3 The most commonly used electrical system for modern commercial buildings is the three-phase, four-wire wye. Draw this system, show the Y, A, B, C phases, ground, 208-volts, and 120-volt connections.

18-4 You are looking at the front of a switchboard. Draw a sketch showing the three-phase bus bar. Identify A, B, C phases.

18-5 On switchboards supplied from a four-wire delta connected system, the midpoint of one phase is grounded. The phase bus bar or conductor having the higher voltage to ground shall be marked. Draw a delta system, show A, B, C phases, ground, and identify phase having voltage to ground.

18-6 According to the NEC, how many service drops are allowed to a building (use rule, not exception)?

____

____

____

____

18-7 Do motor loads require more or less current than lighting loads (circle one)?

18-8 Special voltage systems are available for commercial buildings. The most commonly used system for modern commercial buildings is the ______________________________ .

18-9 In some commercial buildings there is a need for three-phase power. For such an installation a panelboard is used. What are the breakers called? __________ __________ __________.

18-10 What would happen if a fuse did not blow and open the circuit?

____

____

18-11 Define an "overload." ______________________________

____

____

18-12 Fuses have four things to do in a circuit. List the four things.

____

____

____

____

chapter 19

SERVICE LOAD CALCULATIONS

Instructional Objectives

1. To learn how to calculate the size of the service-entrance conductor.
2. To learn how to use the National Electrical Code in calculating service-load calculations.
3. To determine by the National Electrical Code when a neutral conductor may be reduced in size.
4. To provide the basic information to calculate a dwelling.
5. To become more familiar with the NEC.

Self-Evaluation Questions

Test your prior knowledge of the information of this chapter by answering the following questions. Watch for the answers as you study the chapter. Your final evaluation of whether you understand the material is measured by your ability to answer these questions. When you have finished the study of this chapter, return to this section and answer the questions again.

1. What is the unit load per square foot of floor space for the general lighting load of a dwelling?
2. What demand factors are allowed when computing a general lighting load in a dwelling?
3. Is the above demand factor used when computing the lighting load in residential occupancies?
4. Why is it important to calculate the service load?
5. Is the size of the neutral service conductor ever reduced?
6. What will result if the service conductors are too small?

19-1 IMPORTANCE OF SERVICE LOAD CALCULATIONS

Adequate service capacity is an absolute necessity. When calculating and installing electric space heating, the loads are sized to operate at full capacity, during design temperature conditions, for relatively long periods of time (NEC 220-15). If an undersized electrical service is installed, full capacity operation will reduce the power available to the household appliances such as the electronic range and oven or counter-mounted cooking top, dishwasher, trash compactor, hot water heater or the television set. If overloading the service becomes severe enough it can cause conductor heating and contribute to premature insulation breakdown and thus cause a fire hazard. To avoid overloading the electrical service, especially where electrical space heating equipment is to be installed, service load calculations are mandatory. Most requests for electrical inspection permits require them. Review Figure 2.4.

19-2 STANDARDS FOR SERVICE LOAD CALCULATIONS

According to the National Electrical Code:

> In determining the lighting load on the watts per square foot basis, the floor area shall be computed from the outside dimensions of the dwelling. Not included in this area are: open porches, garages, and unfinished or unused spaces (unless these spaces are adaptable for future use; then they shall be computed in the floor dimensions). [See Chapter 8, Section 8-1.]

NEC Table 220-2 (b) shows 3 watts per sq ft.

NEC Section 220-16 (a) rates the two small appliance circuits at 3000 watts (1500 watts each circuit).

NEC Table 220-19, Column C, shows the load for two cooking units to be 65 percent of the sum of the nameplate rating of the two units.

NEC Section 220-16 (b) rates the laundry circuit at 1500 watts.

19-3 CALCULATE THE MINIMUM SERVICE LOAD

With the above code sections and tables we are able to calculate the minimum service load for a dwelling 40 feet by 25 feet, or 1000 square feet. (Figure 19.1.)

STEP 1

Calculate the lighting load requirements.
1000 sq ft × 3 watts per foot = 3000 watts
(two circuits for lighting load)

STEP 2

Calculate the small appliance circuits code states.
We use two circuits = 3000 watts.
So far in our calculations we have four circuits and 6000 watts.

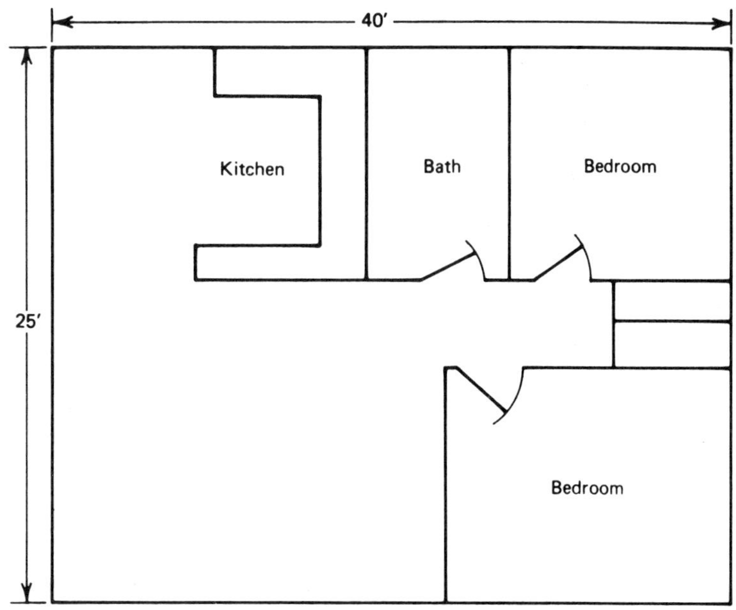

Figure 19.1 Typical single-family dwelling floor plan.

STEP 3

It is reasonable to believe that not all lights and outlets for the appliance circuits will be in use at the same time, so a demand factor is used to calculate the feeder demand factor.

NEC Table 220-11 shows the first 3000 watts of lighting and appliance circuits shall be computed at 100 percent. The remainder of the 3000 watts are to be computed at 35 percent.

EXAMPLE 1

In step 1 we showed a total of 6000 watts.

First 3000 watts computed at 100 percent	=	3000 watts
Remainder, 3000 watts computed at 35 percent	=	1050 watts
Our 6000 watts are now reduced to		4050 watts

STEP 4

Feeder load for the oven and countertop.

Oven is rated at	5000 watts
Countertop rated at	6500 watts
	11500 watts

NEC Table 220-19 Column C shows that we can take a demand factor of 65 percent of the sum of the nameplate rating of the two units.

EXAMPLE 2

Again we use the demand factor—65 percent
65 percent of 11,500 watts = 7475 watts.
Our 11,500 watts now reduced to 7475 watts.

STEP 5

Compute the feeder load.

1. Lighting and appliance circuits	4050 watts
2. Countertop and oven	7475 watts
3. Dishwasher (cannot be reduced)	1500 watts
4. Garbage disposer unit	750 watts
5. Exhaust fan	300 watts
	14,075 watts

STEP 6

Compute the feeder size.
The utility company has supplied a 115/230 volt overhead service to the house. The total computed feeder load is 14,075 watts.

EXAMPLE 3

14,075 watts ÷ 230 volts = 61.19 amperes.

Table 310-16 of the NEC shows that 61.19 amperes requires a No. 6 AWG type THW copper conductor good for 65 amperes.

Table 3A, Chapter 9 of the NEC shows that a 1-inch conduit is required for three No. 6 AWG , Type THW, copper conductors.

19-4 CALCULATE SERVICE LOAD FOR 1500 SQ FT DWELLING (FIGURE 19.2)

General lighting load 1500 sq ft × 3 =	4500 watts
4500 ÷ 115 volts = 39.1 amperes (3–15 ampere circuits)	
Allowance for two appliance circuits (2 circuits—20 amperes)	3000 watts
Laundry circuit NEC 220-3(a) (1 circuit—20 amperes)	1500 watts
NET TOTAL	9000 watts
Feeder demand factors	
First 3000 watts at 100 percent	= 3000 watts
Remainder 6000 watts at 35 percent	= 2100 watts
NET TOTAL	5100 watts
12-kW range load (Table 220-19)	8000 watts
NET LOAD WITH RANGE	13,100 watts

Utility installed 115/230-volt three-wire service.

EXAMPLE 4

13,100 watts ÷ 230 volts = 57 amperes.

NEC 230-42(b) Net computed load exceeds 10 kva, so service conductors shall be at 100 amperes.

19-5 REDUCING THE FEEDER SERVICE NEUTRAL

Many electricians prefer to reduce the service neutral conductor. However, for most installations and especially for a dwelling equipped with an electric range, electric clothes dryer, and electric hot water heater, a neutral conductor

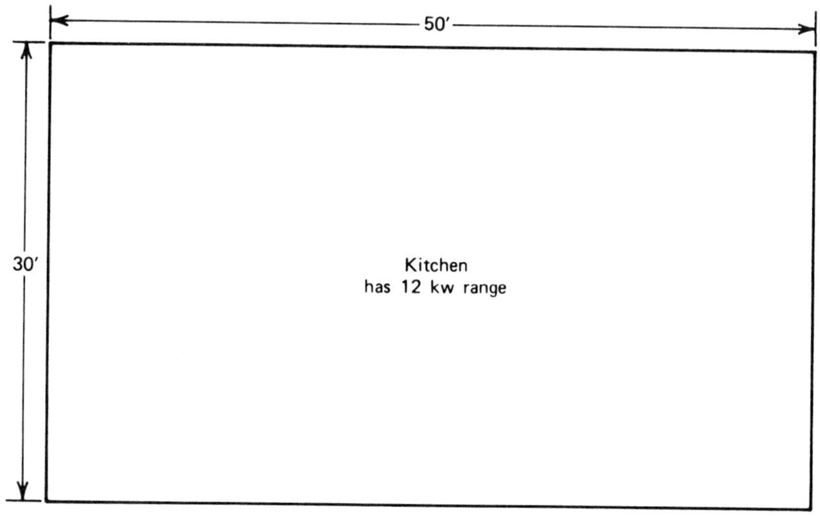

Figure 19.2 Floor plan for calculating service load.

one size smaller than the ungrounded conductors is usually more than adequate. It is recommended that the correct neutral size first be determined using the procedure outlined in this chapter.

NEC 220-22 will allow us to reduce the service neutral.

EXAMPLE 5

Using the calculated service load for 1500 sq ft dwelling in 19.4.
Lighting and appliance load 5100 watts
Range load reduced 70 percent, 8000 × 70 percent = 5600 watts
10,700 watts
10,700 watts ÷ 230 volts = 46.5 amperes.

SUMMARY

1. When calculating electric space heating equipment, loads are figured to operate at full capacity.
2. Severely overloading the service could cause conductor heating and insulation breakdown.
3. Floor area is computed from the outside dimensions of a dwelling.
4. Small appliance circuits are rated at 3000 watts.
5. Demand factors are used when calculating service feeders.
6. The National Electrical Code has many requirements relative to service load calculations.

PROBLEMS

19-1 Calculate the lighting load for a 1575 sq ft dwelling. Show calculations and NEC sections.

19-2 Complete the lighting, two-appliance, and laundry circuits, for a new 1625 sq ft single-family dwelling. Show calculations and Code sections.

19-3 What is the load for an electric range rated at 16 kW?

19-4 When the service-entrance calculation results in a computation of 10 kva or more, what is the minimum size service permitted by the NEC? Give Code section.

19-5 Calculate the service load for a 1550 sq ft dwelling. The kitchen has a 750-watt garbage disposer, 1200-watt dishwasher, and 350-watt exhaust fan. Show calculations in space below.

19-6 Calculate THW copper conductor and conduit size for problem 19-5. Do not reduce neutral.

19-7 What load rating may be used for an electric range rated at not over 12 kW?

19-8 How many circuits will be needed for problem 19-5?

____

19-9 When wiring a house, adequate service capacity is an absolute

__________.

19-10 If an undersized service is installed, what happens when it is loaded to full capacity?

____

____

____

____

19-11 Is it possible to reduce the size of the service neutral conductor?

__________.

PART 6

WIRING COMMERCIAL BUILDINGS

chapter 20

MULTIPLE-FAMILY DWELLING

Instructional Objectives

1. To become familiar with the National Electrical Code when applied to multiple dwellings.
2. To learn why demand factors are used when figuring load calculations.
3. To learn how to calculate the service for an apartment house.
4. To become more familiar with the demand factor for countertops and ovens in multiple dwellings.
5. To learn how to install modular metering equipment.

Self-Evaluation Questions

Test your prior knowledge of the information in this chapter by answering the following questions. Watch for the answers as you read the chapter. Your final evaluation of whether you understand the material is measured by your ability to answer the questions. When you have completed the chapter, return to this section and answer the questions again.

1. Does the NEC require GFCI receptacles to be installed in multiple-family dwellings?
2. How do you figure the lighting load of an individual apartment unit?
3. What type cable may be used in a multiple-family dwelling?
4. According to the NEC, what is the rule for the feeder demand factor when figuring lighting and appliance loads?
5. How do you figure the load for fixed appliances such as garbage disposers, dishwashers, and so forth?
6. How do you determine the number of branch circuits for a single apartment unit?
7. How do you derate the neutral feeder?

20-1 A MULTIPLE-FAMILY DWELLING

A multiple-family dwelling is defined as, "a building containing three or more dwelling units."

> NEC Article 230-72(c). In a multiple-occupancy building, each occupant shall have access to their own disconnecting means. A multiple-occupancy building having individual occupancy above the second floor shall have service equipment grouped in a common accessible location. The disconnecting means shall consist of not more than six switches or circuit breakers.
>
> Multiple-occupancy buildings that do not have individual occupancy in accordance with Article 230-2 Exception No. 3 and each such service may have not more than six switches or circuit breakers.

The words "individual occupancy above the second floor" may be a little confusing to the newcomer. However, in a three-story and higher multiple-family dwelling, multiple occupancy units shall have access to the disconnecting means located in their unit of the apartment building. In the example in this chapter of a 12-unit, three-story dwelling, all of the disconnecting means are located in the modular metering equipment located in the basement utility room. See Figure 20.1. Notice the service equipment is controlled by a separate main circuit breaker ahead of the metering equipment.

In practically all multiple-family dwelling construction there is a single service drop or lateral to the entire building. In most of the apartment units, each tenant pays for the power consumed, because there is a separate meter and disconnect for each unit regardless of the number of units plus a meter and disconnect for the "house loads." A subfeeder feeds the load center in each individual apartment. The load center is the panelboard that controls the branch circuits as found in residential construction.

To determine the number of branch circuits for any single apartment, proceed as you would for a single-family dwelling. Review Electrical Wiring—Residential (Section 12.4). For lighting, allow 3 watts per sq ft. To this

Figure 20.1 Typical modular metering equipment. The cabinet at left contains the main service disconnect, in this case a circuit breaker. Adjacent, and fed from this, are two sections of meter pan, one 8 and one 4. Below each meter socket is a circuit breaker that is the main protection for the apartment involved. More sections can be added as needed. (Courtesy of Square D Company)

is added the two 20-ampere appliance circuits for the kitchen and dining areas. A separate circuit must be provided for each special appliance, such as dishwasher or cooking top. The number of duplex receptacles is determined by the same procedure as in a one-family dwelling. The 1984 NEC requires ground-fault circuit protection (GFCI) for the bathroom and for the receptacle installed outside where there is a direct grade level access to the dwelling unit; however, GFCI is not required as an outside receptacle where the apartment is separated from the grade level by a balcony.

NEC Article 210-8 (a)1–2. Ground-fault circuit protection. (a) Dwelling units. (1) All 125-volt, single-phase 15- and 20-ampere receptacles installed in bathrooms shall have ground-fault circuit interrupter protection.

(2) All 125-volt, single-phase, 15–20 ampere receptacles installed in garages shall have ground-fault circuit interrupter protection. Exception No. 1: Receptacles which are not readily accessible. Example: garage door opener, or receptacle behind a deep freeze.

(3) All 125-volt, single-phase, 15- and 20-ampere receptacles installed outdoors where there is direct grade level access to the dwelling unit and to the receptacles shall have ground-fault circuit interrupter protection.

20-2 PLANNING AN INDIVIDUAL APARTMENT

Our example is a 12-unit, three-story apartment house. All apartments are identical, having two bedrooms, living room, bath, kitchen, dining area, hall, and closet.

The kitchen is equipped with a wall-mounted oven rated at 4800 watts, a 6700-watt counter-mounted cooking top, and a 350-watt hood exhaust fan over the cooking top unit (Figure 20.2). A garbage-disposer rated at 600

Figure 20.2 Delux Hood exhaust fan. (Courtesy of Nutone Division, Scovill Manufacturing Company)

(a)

Figure 20.3 (*a*) Safety breaker load center. (*b*) Main lug load center. (*c*) 24 pole spaces. (Courtesy of Cutler-Hammer, Incorporated)

watts and a dishwasher rated at 1500 watts, located under the cabinet drain board, are also included. The load center, Figure 20.3 is a flush type, located in the hall.

The house lights are connected to a time clock located in the utility room. The house receptacle circuit supplying the duplex receptacles in the halls is on a separate circuit and terminates at the house load center located next to the service modular metering equipment (Figure 20.4). Also located in the utility room is the laundry circuit, suitable for an automatic washer and gas clothes dryer.

Type NM (nonmetallic-sheathed cable) may be used in a multiple-family dwelling or other structure not exceeding three floors above grade.

> NEC Article 336-3. Remember to support NM cables at intervals not exceeding 4½ feet and within 12 inches from every cabinet, box, or fitting.

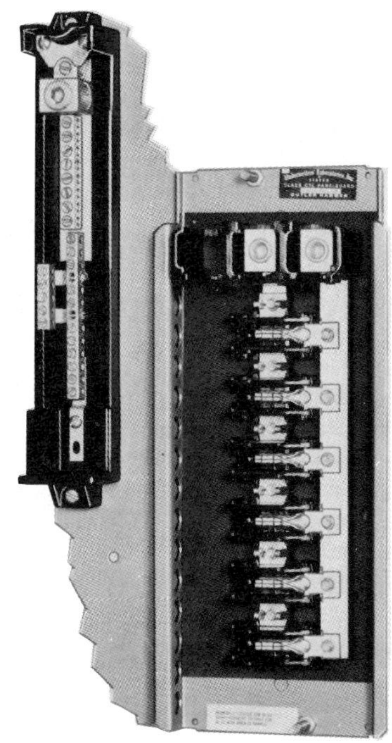

(b)

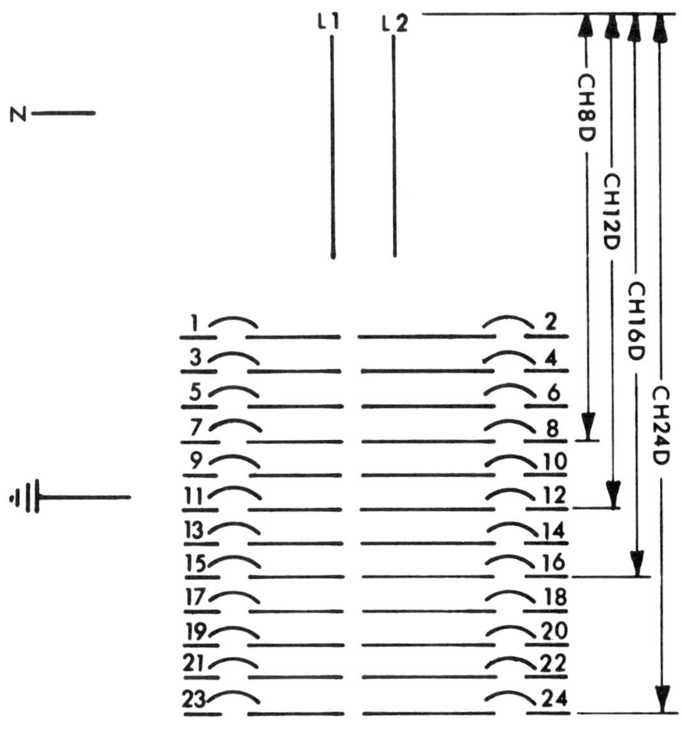

L1
L2
N
CH8D
CH12D
CH16D
CH24D
1
2
3
4
5
6
7
8
9
10
11
12
13
14
15
16
17
18
19
20
21
22
23
24

(c)

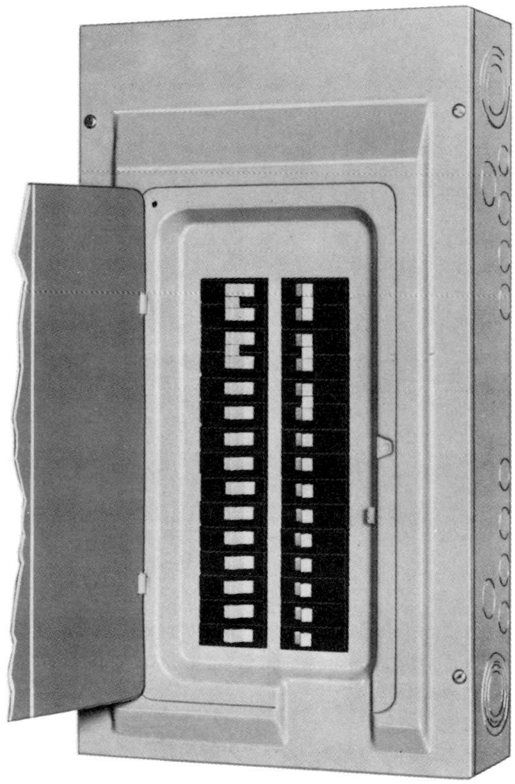

(a)

Figure 20.4 House load center. (*a*) Panel complete with single-and double-pole circuit breakers. (*b*) Main lug center load. (*c*) Diagram of main lug center load, 30 pole spaces. (Courtesy of Cutler-Hammer, Incorporated)

20-3 LOAD CALCULATIONS FOR INDIVIDUAL APARTMENT

The unit load specified for dwelling occupancies in the NEC Table 220-2(b) is 3 watts per sq ft. The NEC 220-16(a) allowance for two small appliance circuits is 3000 watts. Load required for two cooking units NEC Table 220-19, Column C, is 65 percent of the sum of the nameplate ratings. On the basis of the NEC sections, the load for each of the 12 apartments is determined as follows:

Area of each apartment 20 × 40 feet = 800 sq ft

Lighting requirements equal

800 × 3 watts =	2400 watts	2 circuits
Allowance for appliance circuits	3000 watts	2 circuits
	5400 watts	4 circuits

Feeder demand factor (NEC Table 220-11)

First 3000 watts at 100 percent	3000 watts
Remainder, 2400 watts at 35 percent	840 watts
Total	3840 watts

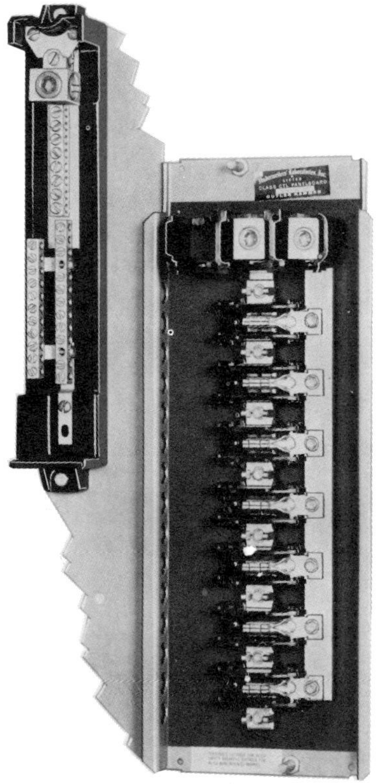

(b)

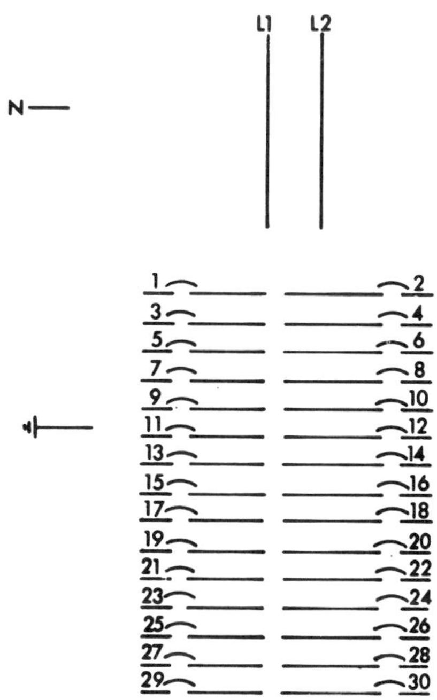

L1
L2
N
1
2
3
4
5
6
7
8
9
10
11
12
13
14
15
16
17
18
19
20
21
22
23
24
25
26
27
28
29
30

(c)

Feeder load for oven and countertop

Oven rated at	4800 watts
Cooktop rated at	6700 watts
Total	11500 watts

Feeder capacity 0.65 × 11500 watts = 7475 watts

Feeder load:

Lighting and utility	3840 watts
Oven and cooktop	7475 watts
Dishwasher	1500 watts
Garbage disposer	600 watts
Exhaust fan	350 watts
	13,765 watts

Service feeder size to each apartment:

Feeder size (115/230 volt service)

13,765 watts ÷ 230 volts = 59.8 or 60 amperes

NEC Table 310-6 states that 60 amperes requires a No. 6 type RHW, THW, THWN, or XHXW, copper.

Table 3A, Chapter 9 shows that three No. 6 THW conductors require a 1-inch conduit.

Three No. 6 type THW copper conductors shall be installed from the meter panel to each panel located in each of the apartments.

20-4 SERVICE CALCULATIONS

To calculate the service for a 12-unit apartment and house load requires the rules of the National Electrical Code. NEC Article 220-2(c) (4) states, "that a load of 180 volt-amperes per outlet shall be allowed for outlets other than for general illumination."

The provisions of the above article shall not be applicable to receptacle outlets provided for the connection of cord-and-plug connected equipment as provided for in Article 400-7 of the NEC.

Usually there are plans and specifications for such occupancies and these will indicate the various outlets required. Quite often the wattage of the outlet will be indicated. When this is done, the load calculations shall be based on the wattage value shown by the plans. If the wattages are not shown, then the load calculations are based on the value of Article 220-2(c) above.

Many plans are created without any indication on them as to the wattage values for each outlet. The minimum values of 180 volt-amperes and 600 volt-amperes is usually assumed. The installation is far from satisfactory both from the standpoint of wiring design and from the lighting system that is obtained from this minimum load calculation.

On the basis of the nominal voltage of 120 volts, the minimum ampacity for each value shown (180 and 600 volt-amperes) would be:

180 volt-amperes ÷ 120 volts = 1.5 amperes
600 volt-amperes ÷ 120 volts = 5.0 amperes

Receptacle outlets located in the halls and work areas would be much better on a separate circuit than on general illumination.

> NEC Article 220-17. Four or more fixed appliances other than electric ranges, clothes dryers, space heating equipment, or air-conditioning units are connected to the same feeder or service in a multi-family dwelling. A demand factor of 75 percent may be applied to the load. [The garbage disposers, dishwashers, and exhaust fans qualify here.]

NEC Article 230-42 governs the service conductor size. "Conductors shall be of sufficient size to carry the load." This article refers to Article 220, so that service conductors for the 12 units are calculated as follows:

Area to be lighted approximately 9600 sq ft
Building 80 × 40 feet = 3200 sq ft × 3 floors =
9600 sq ft × 3 watts per sq ft

Lighting requirements equals	
9600 × 3 watts	28,800 watts
Appliance circuits:	
12 × 3000 watts	36,000 watts
Laundry circuit:	1,500 watts
House night-lights:	
12 × 100 watts	1,200 watts
6 receptacles in halls:	1,500 watts
Total	69,000 watts

Load for lighting and appliance circuits	
First 3000 watts at 100 percent	3,000 watts
Remainder, 66,000 watts at 35 percent	23,100 watts
Total	26,100 watts

Loads for countertops and ovens	
12 × 4800 watts =	57,600 watts
12 × 6700 watts =	80,400 watts
Total	138,000 watts

NEC Table 220-19, Column C, permits a demand factor for the 24 ranges at 26 percent.

0.26 × 138,000 = 35,880 watts

Loads for garbage disposers, exhaust fans and dishwashers:

Garbage disposers	12 × 600 =	7200 watts
Exhaust fan	12 × 350 =	4200 watts
Dishwashers	12 × 1500 =	18000 watts
		29,400 watts

NEC Article 220-17 allows for the 75 percent demand factor:

0.75 × 29,400 = 22,050 watts

Service load:	
Lighting and appliance	26,100 watts
Cooking units	35,880 watts
Appliances	22,050 watts
Total	84,030 watts

Service-entrance conductors:

84,030 watts ÷ 230 volts = 365.35 amperes
Table 310-16 shows 500-MCM type THW
Copper conductors to be suitable for 380 amperes.

In some cases, the neutral feeder may be smaller than the other service conductors.

NEC Article 220-22 allows a 70 percent demand factor on that portion of neutral feeder load which supplies an electric range or equivalent cooking device.

Neutral conductor:

Lighting and appliance	26,100 watts
Cooking units	
0.7 × 35,880 watts	25,116 watts
Fixed appliances	22,050 watts
Total	73,266 watts

73,266 watts ÷ 230 volts = 318 amperes
Table 310-16 shows that 400-MCM type THW copper conductor is acceptable for the neutral.

Service size:

Two 500-MCM and one 400-MCM type THW copper conductors, all in a 3-inch conduit (Table 3A).

SUMMARY

1. In multiple-occupancy buildings, each occupant shall have access to their own disconnecting means.
2. To determine the number of branch circuits for any single apartment, proceed as you would for a single-family dwelling.
3. Receptacle outlets located in the halls and work areas would be much better if installed on a separate circuit and not on general illumination circuit.
4. A separate circuit must be provided for each special appliance such as a dishwasher or cooking top.
5. The NEC requires ground-fault circuit protection for the receptacles in the bathrooms and outside the apartment if on grade level.
6. Ground-fault protection is not required outside where the apartment is separated from the grade level by a balcony or porch.
7. Nonmetallic-sheathed cable (NM) may be used in a multiple-family dwelling or other structure not exceeding three floors above grade.
8. The load limit specified for dwelling occupancies in the NEC is 3 watts per sq ft.
9. The NEC governs the service conductor size. "Conductors shall be of sufficient size to carry the load."
10. When calculating the service, there are conditions when the neutral feeder may be smaller than the other service conductors.

PROBLEMS

20-1 A developer developed a six-unit apartment. Each apartment is 710 sq ft with the following appliances: electric oven rated at 4800 watts, cooktop rated at 6700 watts, dishwasher at 1500 watts, garbage disposer at 600 watts, and exhaust fan at 350 watts. What is the total load calculation for 1 unit?

Demand load

710 sq ft at ________ watts = ________ watts

________ ________ ________ watts

TOTAL TOTAL

____

________ ________

________ ________ ________ watts

TOTAL

____

________ ________ watts

________ ________ watts

TOTAL watts

________ ________ watts

____

________ ________

________ ________

________ ________

________ ________

TOTAL watts

Feeder size (115/230-V service)

________ amperes (1 unit)

20-2 Calculate Problem 20-1 for six units.

Area ________ sq ft at ________ watts = ________ watts

6 appliance circuits × ________ watts = ________ watts

________ = ________ watts

Night lights, house meter

6 × 100 watts = ________ watts

6 receptacle outlets in halls = ________ watts

________ TOTAL

Lighting and appliance NEC ________

________ at ________ = ______ watts

________ at ________ = ______ watts

TOTAL

Load for ________ and ________

______ × ______ watts = ________

______ × ______ watts = ________

TOTAL watts

Table ______, Column ______, permits demand factor of

______ percent for ______ ______ = ______ ×

______ =

Load for:

Dishwashers ________ = ________

Disposals ________ = ________

Exhaust fans ________ = ________

Water heater 1 × 5000 watts = ________

TOTAL

NEC ______ 4 or more appliances

. ______ × ______ watts = ________

Service load:

Lighting and appliances ______ watts

Cooking units ______ watts

Fixed appliances ______ watts

TOTAL watts

Service-entrance conductors:

______ watts ÷ 230 volts = ______ amperes

Table 310-16 shows ______ MCM type ______

20-3 Calculate the six units without cooktops and ovens. What are the total calculations for the six units? Show your work.

20-4 Define a multiple-family dwelling. ______________________________

____

____

20-5 In a multiple-occupancy building, each occupant shall have access to their own ____ .

20-6 You are wiring an apartment complex. How do you determine the number of branch circuits for a single apartment? ______________

____

____

____

20-7 What type wiring would be used in the installation of an apartment unit, in a building not exceeding three floors above grade? ________ .

20-8 According to NEC, what is the unit load for dwelling occupancies?

____ .

20-9 Receptacle outlets located in the halls and work area of a multi-family dwelling would be much better on a ______________ circuit.

20-10 How are the house lights of an apartment building controlled?

____ .

chapter 21

FARM WIRING

Instructional Objectives

1. To develop an ability to calculate a farm service.
2. To learn how to compute farm loads other than dwelling units.
3. To become familiar with percent demand factor used when computing total farm loads.
4. To understand the need for system grounds.
5. To learn some advantages of an emergency electric service.
6. To understand the purpose of a current transformer.
7. To understand the need for a transfer switch when installed in an emergency electric farm service.

Self-Evaluation Questions

Test your prior knowledge of the information in this chapter by answering the following questions. Watch for the answers as you read the chapter. Your final evaluation of whether you understand the material is measured by your ability to answer the questions. When you have completed the chapter, return to this section and answer the questions again.

1. Who decides the location of the meter?
2. Does the NEC permit a demand factor for farm loads?
3. Is a service disconnecting means required for individual buildings?
4. What is a current transformer?
5. Where are current transformers usually installed?
6. Is it necessary for system grounds and equipment grounds in various buildings to be bonded to the underground water piping system?
7. What is the purpose of a current transformer?
8. Why is an emergency electric service important to the farm?
9. Describe a transfer switch.
10. Where is a transfer switch used?

21-1 FARM RESIDENCE

Designing a farm electrical system involves the same basic procedures used for commercial projects. The NEC includes specific minimum requirements for the number and spacing of outlets, certain demand factors, and the location and number of certain types of circuits, for dwelling type occupancies. All other design considerations remain applicable, such as convenience, flexibility, safety, adequate voltage drop, and allowances for spare circuit capacity.

As in other types of installations farm wiring should provide for expansion of the electrical system.

A farm dwelling electrical system must serve a variety of loads from water pump to space heating. While not all farm homes are equipped for electrical space heating, many are. Air conditioning equipment has become more and more common, and appliances, particularly in the kitchen and laundry areas, may be major loads that the electrical system must be specially designed to service safely and efficiently. Otherwise the farm home electrical system is designed and installed under the same conditions and codes as the city residence.

21-2 FARM WIRING

In a typical farm installation, wires from the power supplier's transformer end at a pole in the yard. The trend now is for the utility company to bury its wires underground. The buried wires constitute a service lateral. The lateral ends where the wires enter the location of the meter base (Figure 21.1).

Figure 21.1 (*a*) 200-ampere single-phase mater socket for both overhead. (Courtesy of RONK Electrical Industries, Inc.) (*b*) Standby generator connections to conventional pole meter installation. (*c*) Self-contained pole meter installation—no disconnect. (*d*) 200-ampere single-phase meter-rite pole top with cover hanging. (Courtesy of RONK Electrical Industries, Inc.) (*e*) Standby generator connections to current transformer-type pole meter installation with pole-top double-throw switch.

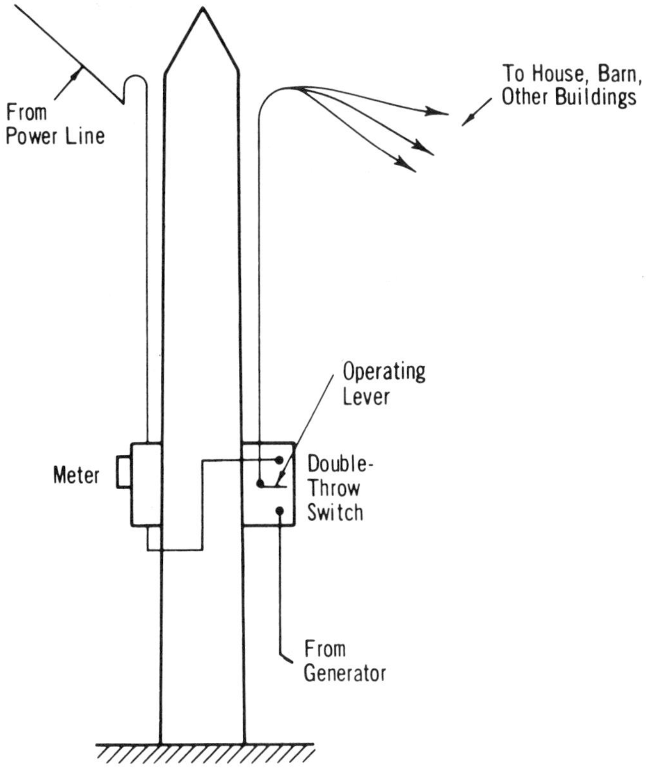
From
Power Line
To House, Barn,
Other Buildings
Operating
Lever
Meter
Double-
Throw
Switch
From
Generator

(b)

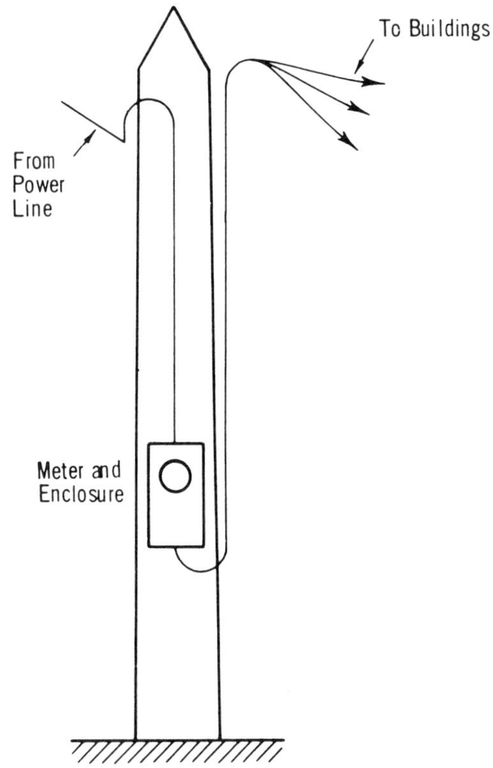
To Buildings
From
Power
Line
Meter and
Enclosure

(c)

(d)

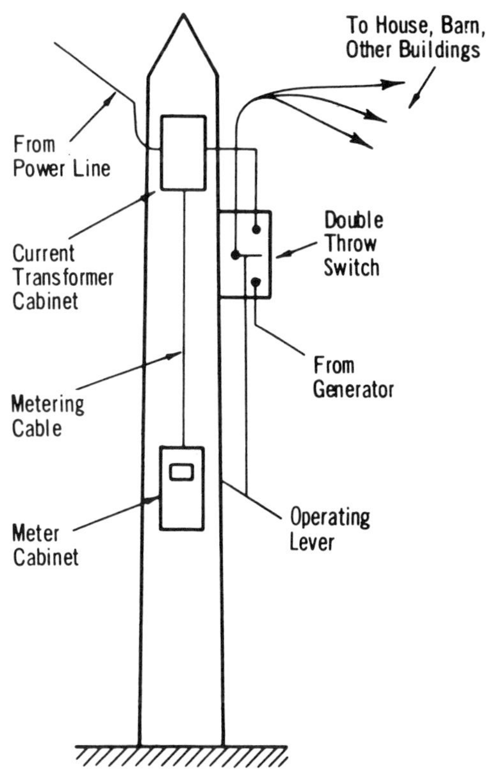
To House, Barn,
Other Buildings
From
Power Line
Current
Transformer
Cabinet
Double
Throw
Switch
From
Generator
Metering
Cable
Meter
Cabinet
Operating
Lever

(e)

The power supplier decides the location of the meter; if the meter is located on a pole in the yard the service lateral or service drop will be installed by them.

The wires from the point where the service drop ends, up to the service equipment, are called "service-entrance wires." They may be type TW, RHW, THW, or any other type suitable for outdoor locations. Review Chapter 4.

Because of the widely scattered pattern of farm buildings, an exterior distribution system is an inherent part of every farm wiring layout. The branch feeders must originate at some central distribution point and either go overhead or underground to the individual building electric service locations.

Special provisions for branch circuits and feeders for farm wiring systems are given in NEC Articles 220-40 and 41, "Methods of computing farm loads." Demand factors for farm loads other than dwellings are listed in NEC, Table 220-40 (Table 21.1). Total design farm loads are based on demand factors for the largest and for successively smaller loads as shown in NEC Table 220-41 (Table 21.2).

TABLE 21.1 Methods for Computing Farm Loads Other Than Dwelling Units (NEC Table 220-40)

Ampere Load at 230 Volts	Percent Demand Factor
Loads expected to operate without diversity, but not less than 125 percent full-load current of the largest motor and not less than the first 60 amperes of load	100
Next 60 amperes of all loads	50
Remainder of other loads	25

TABLE 21.2 Methods of Computing Total Farm Loads (NEC Table 220-41)

Individual Loads Computed in Accordance with Table 220-40	Demand Factor Percent
Largest load	100
Second largest load	75
Third largest load	65
Remaining loads	50

21-3 TOTAL LOAD CALCULATIONS

The total load calculations determine the size of the conductors from the service to the separate buildings.

Proceed as follows using the load in amperes at 240 volts, for each building.

1. Highest of all loads in amperes for an individual building ________ amperes at 100 percent. ______ amperes
2. Second highest ampere load ________ amperes at 75 percent. ______ amperes
3. Third highest ampere load ________ amperes at 65 percent. ______ amperes

4. Total of all other buildings (except home) ______ amperes at 50 percent. ______ amperes
5. Total of preceding. ______ amperes
6. House computed at 100 percent. ______ amperes
7. Grand total including house. ______ amperes

Note. If two or more buildings have the same function, consider them as one building.

EXAMPLE 1

If there are two hog houses requiring 50 and 60 amperes, respectively, consider them as one building requiring 110 amperes and enter 110 amperes to line No. 1 above.

Remember, the dwelling load must be added to the farm loads to obtain the total service load.

Where branch circuit capacity must be held to a minimum, the building service should be large enough to permit circuit additions or extensions in the future. It is easier and less expensive to add circuits than to change services.

Provisions may be desired for disconnecting the feeders to individual buildings at the main distribution site. A service disconnecting means is required at the building; however, a disconnect at the outer end would permit the conductor to be de-energized in case the feeders should break during a storm, short circuit, or need repair.

It is important that all system grounds and equipment grounds in the various farm buildings be bonded and connected to the well casing or underground water piping system (NEC Article 250-81). This interconnection assures that all conducting surfaces will be at the same potential. In the case of a fault, the fault current will be sufficient to open the protective device rapidly if the service equipment enclosure is bonded properly to the grounded neutral conductor by the main bonding jumper.

21-4 FARM BUILDINGS

Dairy Structures

Stanchion Barn

The most common application of electricity is for lighting, but a well-planned system must include convenient outlets for general use of clippers, tools, special lamps, spot heaters, portable fans, livestock waterers, water warmers, and similar devices. Electrically powered equipment such as feed handling units should have individual circuit service equipped with lockout switch devices for safe servicing.

Lighting Outlets For face-out stanchion arrangements (Figure 21.2*a*) place lighting outlets along the centerline of the litter alley, one outlet directly behind every other stall divider.

As an alternate face-out arrangement lighting system, install a continuous row of two-lamp, standard white fluorescents in semidirect fixtures. Mount along alley centerline for even lighting of cow udders.

For face-in arrangement, place lighting outlets about 12 inches to the rear of gutter and directly behind every other stall divider as shown in Figure 21.2*b*.

To light feed alley, place the lighting outlets along centerline about 12 feet apart (approximately 1.5 times lamp mounting height).

Provide a separate wall switch for each group of lights.

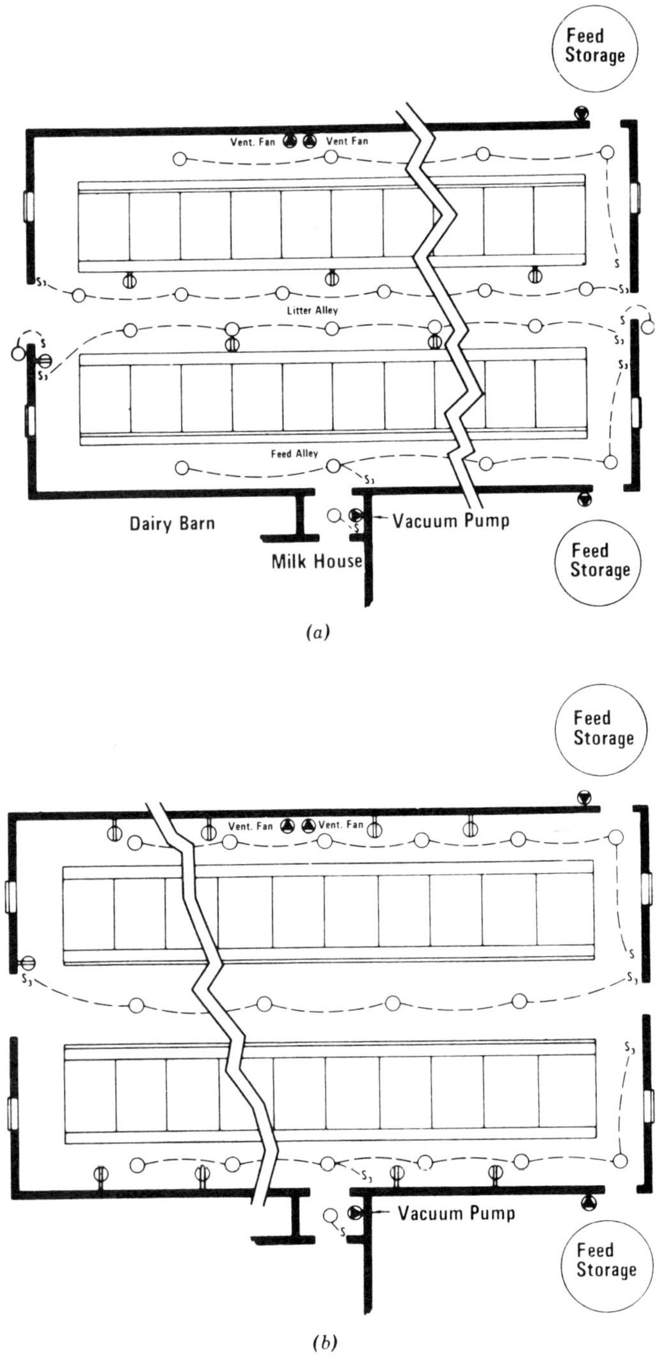

Figure 21.2 (*a*) Lighting for stanchion barn, face-out arrangement. (*b*) Lighting for stanchion barn, face-in arrangement. (*c*) Lighting for milking room (single row of cows). (*d*) Lighting for milking room (two rows of cows). (*e*) Milkhouse lighting.

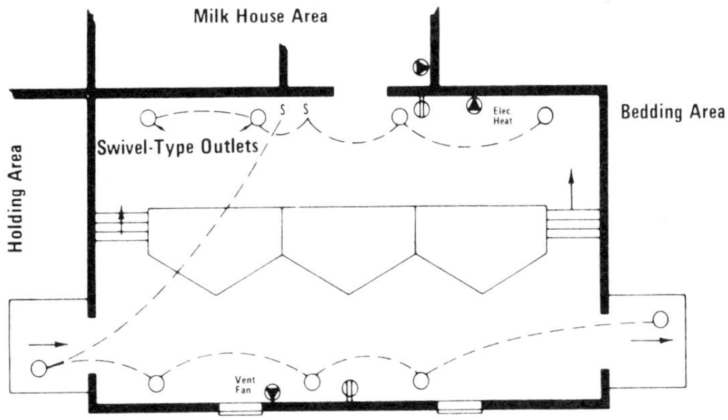
Milk House Area
Elec. Heat
Bedding Area
Swivel-Type Outlets
Holding Area
Vent Fan

(c)

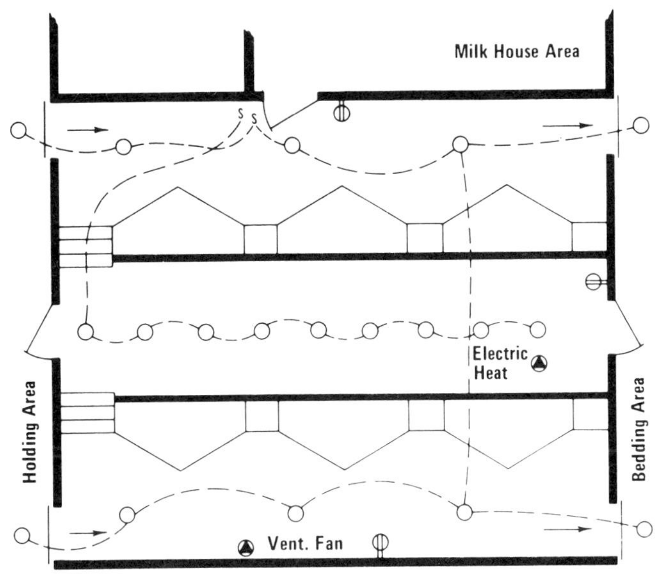
Milk House Area
Electric Heat
Holding Area
Bedding Area
Vent. Fan

(d)

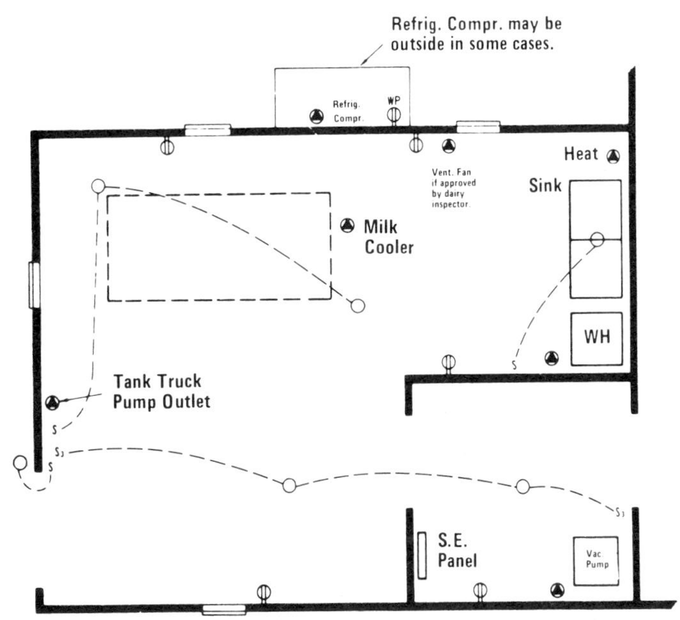
Refrig. Compr. may be outside in some cases.
Refrig. Compr.
WP
Heat
Sink
Vent. Fan if approved by dairy inspector
Milk Cooler
WH
Tank Truck Pump Outlet
S.E. Panel
Vac Pump

(e)

Outdoor and haymow lights should be equipped with pilot lights to indicate when lights are "ON."

It is a good idea to check with the local power supplier, dairy inspector, and insurance company representative about regional, state, and local regulations.

Convenient Outlets Provide 20-ampere duplex receptacles about 20 feet on center so that an outlet can be reached from any location within the litter alley, with a 25-foot cord.

They should be about 6 feet above the floor to minimize damage by animals. These outlets can be on outside walls where cows face in. Where cows face out, they may be more convenient on structural posts, if such are available. Add a few outlets in the feed alley for supplemental lighting, as required.

Special Purpose Outlets Provide an outlet at each point where planned equipment is to be served on an individual branch circuit. Examples are water heaters, refrigeration compressors, feed handling equipment, ventilating fans, gutter cleaners, and so forth. Outlets for extra summer ventilation fans may also be needed. Requirements for these outlets are determined by the load to be served; some are 120 volts, some 240 volts; and amperage will vary. Circuit wire size depends upon current, voltage, and distance. Select from Table 310 in the 1984 NEC; however, use nothing smaller than No. 12 AWG copper or equivalent.

The vacuum pump circuit is usually included in the plans for this area, but it could be included with the milk house circuits if desired. See Figure 21.2*e*.

Milking Parlor

Major wiring requirements relate to proper heating and lighting. Weatherproof convenience outlets will be needed for general-purpose use because of damp interior environment. Individual circuits are needed for space heating and ventilation fans.

Lighting Because of a wide variety of stall arrangements in milking parlors, a specific recommendation for lighting system is best made after review of specific plans. However, a well-lighted space is mandatory.

A good lighting plan generally results when one outlet is provided for each 36 sq ft of working area. This means about 2.5 outlets per stall, depending on lamp type. If fluorescent lights are selected, lamps should extend over 80 percent of the room length. Allow one row of lamps for the operator and one row for each row of stalls. Locate lamps so the operator does not work in shadows. Plan one lighting outlet at each animal entrance to the room. Provide weatherproof (WP) wall-switch control for all lighting. Where incandescent lights are used behind cows (in other than "herringbone" parlors), place lighting outlets about 1 foot from outer wall so as to light cow udders. See Figure 21.2*c* and *d*.

Convenient Outlets Place one 20-ampere duplex, weatherproof receptacle at each end of the operator's work area. A weatherproof outlet may also be required for a waste pump located under the pit floor.

Special Purpose Outlets While ventilating fans often require the approval of a local dairy inspector, at least one individual circuit and outlet should be provided for air moving equipment. An individual circuit, with controls, is also needed for the vacuum pump. In each case, select system components as prescribed for a motor circuit.

Heat for the operator's work area is normally needed. Heating cable in the pit floor is often desirable. Many operators prefer quick heat provided with overhead infrared units (heat lamps, metal sheath heaters, or quartz-tube lamps). Provide one or more heating circuits with appropriate controls.

One heating system might include a 3500-watt quartz-tube lamp located along the centerline of the operator area. One such lamp for each 10 linear feet of work space provides a satisfactory heat source. When using two such lamps, serve through a controller of suitable ampere rating, with the holding coil controlled by a SPST wall switch (a three-wire 120/240-volt circuit should be considered for two or more heat units).

When using infrared heat lamps, plan for 50-watts per sq ft of pit floor area. Heat lamps should be positioned about 7 to 7½ feet above the floor, and 26 to 32 inches in front of curb, mounted as to warm head, shoulders, and back of the operator when he or she is attaching milker to cow.

Heating cables embedded in concrete ramps and unprotected pen floor may be used for ice and snow melting. Install 40 to 60 watts per square foot for ice melting. One or more 120-volt, 20-ampere circuits should be provided for motors on automatic feeding systems. Feed auger motors should have built-in manual reset overload protection and individual fuse protection of time delay type.

Milk House or Milk Room

Local regulations may require that the vacuum pump be placed in a vestibule or dry closet area. Adequate illumination, hot water for sanitizing, and power for refrigeration are major considerations for milk room wiring needs. Allow for wiring expansion to take advantage of alternate energy sources, energy conserving heat exchangers, and so forth.

A ceiling-suspended or wall-mounted heater should be thermostatically controlled and directed toward the wash-up area.

Lighting Fluorescent lamps or incandescent bulbs with suitable diffusers to minimize glare, are suggested for inside areas. Vapor-tight fixtures should be used over steam-laden, wash-up areas. Weatherproof (WP) receptacles are needed for damp and wet areas. Provide three WP switches, one for the outside light, one for wash and storage area outlets, and a third for tank wash-up lights.

Two 150-watt PAR lamps in swivel-base fixtures should be mounted to illuminate the tank interior. Lamps must not be located directly over the tank. (See Figure 21-2*e*.)

Convenient Outlets Provide 20-ampere duplex outlets with WP covers for each work area. Outlets for clock, radio and desk lamp are suggested for the dry record-keeping room. Locate main switch panel in this dry area. In all other areas, place outlets high to avoid adverse effects from splashing.

Special Purpose Outlets Plan an individual circuit with a 240-volt outlet for each milk cooler, and an individual circuit with control for pipeline milking system units. Also provide a 240-volt outlet for the water heater. Provide an outlet for electric pumps on the milk pickup tank-truck. The type of outlet required will depend upon the configuration of plug cap used on truck-pump motor cord. Provide a manual switch adjacent to and for control of this outlet. An outlet for the wash-up pump, and an outlet for the water pressure booster pump may be required.

Where toilet facilities are required, electrical outlets should include at least one:

Lighting outlet over washbasin, with wall-switch control.

Duplex convenient outlet, grounded type.

Special outlet for space heater.

Farm Shop and Machine Shed

Farm Shop

Place lighting outlets about 10 to 12 feet on center and 4 to 6 feet from walls. This will allow one outlet for every 100–150 sq ft. Provide separate wall-switch control for each row of lights. Use three-way switches to avoid backtracking.

Local Lighting Place two outlets over each 10 feet of work bench and provide wall-switch control. Two 150-watt reflector flood bulbs about 50 inches above the bench will provide excellent lighting above the bench area. Position the lamps above the front half of the bench (Figure 21.3). *Outside:* Use mercury vapor (or equivalent) mounting lamps at proper height. Use photocell control.

Convenient Outlets Plan outlets for general-use circuits so there is one 20-ampere duplex outlet for each 4 feet of work bench. Position one outlet at each permanent location for a motor-driven tool when the motor size is ½ hp or less, and at least one extra outlet on each wall of the shop. Place outlets about 48 inches above the floor line. Electrical metallic conduit is suggested for guarding that portion of conductors located within 8 feet of the floor. This helps prevent physical damage to the wiring.

Special Purpose Outlets Provide individual branch circuits for outlets serving motors ½ hp and larger. Design the circuits so one circuit will serve no more than three ⅓-hp motors, two ½-hp motors, or one 1-hp or larger motor. Protect the circuit wiring with conduit as suggested above.

Plan for a 50-ampere, 240-volt, two-wire with ground circuit and outlet for a welder.

Provide an individual circuit for a space heater.

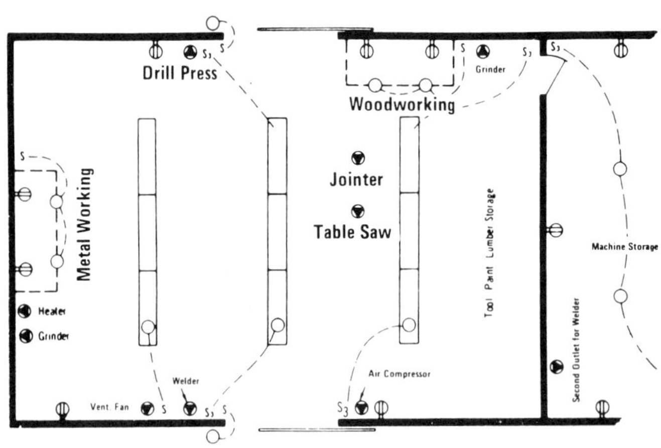

Figure 21.3 Farm shop and adjoining machine storage.

Machine Shed

Lighting Outlets Provide one for every 200 to 300 sq ft with wall-switch control of all outlets.

Convenient Outlets Use one 20-ampere duplex for every 40 feet of wall perimeter, for use with trouble lamp, portable drill, and so on.

Special Purpose Outlets Provide an outlet for an electric welder if machinery repairs are planned for this area. This outlet may be an extension of the branch circuit that serves the welder outlet inside the nearby shop. See Figure 21.3.

Consider placing an outlet for the welder on the yard pole, or somewhere outdoors so large machinery may be repaired without removing parts. An outlet placed just inside the door of the machine shed may also serve the same purpose. An outlet placed out-of-doors must be suitably protected from the weather. Use a type approved for outdoor use.

21-5 SELECTING THE SERVICE EQUIPMENT

Since hardly any two wiring installations are identical, it is difficult to specify any one particular arrangement for the service equipment in circuit breaker assemblies and fused switches; an infinite number of combinations can be devised. However, in any event, it is advisable to provide enough capacity in the service equipment and enough space circuit positions to allow for reasonable future load additions.

Farm Building Services, General

In no case should the service entrance conductors be smaller than No. 6 copper, or equal, with a 60-ampere switch, except for machinery sheds, brooder houses, or other structures where the electrical installation requires not more than two branch circuits; in such cases No. 8 AWG service-entrance conductors may be used (see NEC 230-42 and 230-79).

Most farm buildings with any reasonable load and with allowance for load increases will require at least 100-ampere service. Many well-equipped dairy barns and some poultry houses now require at least 150–200 ampere capacity. Labor cost for installation does not vary greatly from one size of conductor to the next. It is good economy, in the long run, to allow for enough capacity to permit future load additions. Where a service must be increased from, say, 60 amperes to a larger size, there is not much point in changing to 100 amperes even though this might now handle the added load. The cost of changing to 150 or 200 amperes would be very little more.

Section 220-40 of the NEC may be used in determining the size of feeder (or service) supplying an individual farm building. See Table 21.1. If the load consists of a single item such as an irrigation pump or corn dryer, Section 220-40 would not apply. Since this section applies to loads in amperes at 230 volts, it does not cover buildings or loads supplied at 115 volts, nor at voltages above 230/240 volts.

In applying this section of the NEC to a particular situation, first make a list of all items of electrical equipment and their loads. Then determine which of these items will be likely to operate without diversity, that is, which pieces of equipment will be running at the same time to produce the heaviest load.

This will call for good judgment and some knowledge of farm operations. A few examples will illustrate the procedure. In all these examples, assume a reasonable balance of 115 volts on the three-wire service.

EXAMPLE 2 Farm Shop

	Amperes	*Volts*
10 lighting outlets		
6 convenient outlets		
16 outlets at 1½ amperes, = 24 amperes at 115 volts	12	230
Heater, 3kW	13	230
Lathe, 1 hp; 8 amperes × 1.25	10	230
Saw, ¾ hp	6.9	230
Air compressor, ½	4.9	230
Welder	35	230
	81.8	230 volts
	or	
	82 amperes	

If the welder (35 amperes), heater (13 amperes) and half the lights (6 amperes) were in use, we would have a total of 54 amperes without diversity.

Applying Table 21.1, we have:

First 60 amperes at 100 percent demand factor	= 60 amperes
Remaining 22 amperes at 50 percent demand factor	= 11 amperes
Demand	= 71 amperes at 230 volts

Since the loads expected to operate without diversity total less than 60 amperes, the minimum of 60 amperes should be computed at 100 percent demand factor.

From Table 310-16 of the NEC this will require No. 4 type RH copper or No. 2 type RH aluminum conductors in the service entrance.

EXAMPLE 3 Dairy Barn

		Amperes	*Volts*
30 lighting outlets			
10 convenient outlets			
40 outlets at 1½ amperes, 60 amperes = 115 volts		30	230
2⅙-hp ventilating fans at 4.4 amperes = 115 volts			
1 silo unloader, 5 hp; 28 amperes × 1.25		35	230
1 feed conveyor, 5 hp		28	230
Barn cleaner, 5 hp		28	230
Hay elevator, 1 hp		8	230
Hay conveyor, ¾ hp		6.9	230
Milker, 1½ hp		10	230
Milk pump, ½ hp		4.9	230
Tank-truck outlet, 1 hp		8	230
Milk cooler, 3 hp		17	230
Milk house heater, 2 kW		9	230
Water heater, 3 kW	=	13	230
Water pump, ½ hp	=	4.9	230
		204.9 or	230 volts
		205 amperes	

Section 220-21 of the NEC provides that, in adding branch circuit loads to determine the feeder load, the smaller of the two dissimilar loads may be omitted where it is unlikely both loads will be served simultaneously. It is unlikely the milk house heater (9 amperes) and ventilating fan (2.2 amperes) will be running in the summertime when hay dryers are in use, so we can subtract these loads (9 + 2.2 = 11.2) from the total load.

Total load = 205.0 amperes
− 11.2
193.8 or 194 amperes at 230 volts

A barn with all this equipment should have a 200-ampere service anyway to provide for future load additions; however, the application of Section 220-40 will show how much space capacity is available.

First, decide what loads are likely to be drawn at the same time, that is, without diversity. During haying season, the hay elevator and conveyor will probably not be running while the barn cleaner is being operated. The barn cleaner and milker will hardly run at the same time. The milk cooler will probably start before milking is finished. After the milker is shut off and milk house chores are being done, the water heater could come on, and the milk cooler would still be running. In the evening some lights might be on. We can conclude that loads operating without diversity are:

Silo unloader	35 amperes
Feed conveyor	28 amperes
Milk cooler	17 amperes
Lights	5 amperes
Water heater	13 amperes
	98
Loads operating without diversity	98 amperes

Earlier we determined that the net total load in the building is 194 amperes. Subtracting, we have 194 − 98 = 96 amperes of remaining load. Sixty amperes of this can be figured at 50 percent demand factor, leaving 36 amperes at 25 percent demand factor. Thus:

Loads expected to operate without diversity	98 amperes at 230 volts
Next 60 amperes at 50 percent	30 amperes at 230 volts
Remaining 36 amperes at 25 percent	9 amperes at 230 volts
Total computed demand load	137 amperes at 230 volts

With the computed demand load of 137 amperes, the service-entrance conductors must be rated at least at 150 amperes. This would provide spare capacity of 13 amperes for added loads operating with diversity. If a 200-ampere service were installed, spare capacity would be 200 − 137 or 63 amperes. This would be equivalent to two more 5-hp silo unloaders operating without diversity (2 × 28 = 56 amperes) plus a 5-hp bunk feeder (28 × 0.25 = 7 amperes). If all the spare capacity were used for loads operating with diversity, the spare capacity of 63 amperes could serve 63 ÷ 0.25 or 246 amperes of added load.

21-6 SERVICE DROPS FROM METER POLE TO BUILDINGS

The feeder connecting the service-entrance conductors of a building to the meter point may be either overhead or underground. Conductors from the meter pole to each building served from the pole are technically called "service drops" within the definitions of the NEC; however, they perform the same function as feeders and should be designed as feeders.

In selecting conductor sizes for feeders, data needed are: (1) the demand load; (2) the length of feeder and allowable voltage drop; (3) type of conductor; (4) type of insulation.

21-7 CURRENT TRANSFORMER

A "transformer" is a piece of electrical supply equipment used in the transmission and distribution of power to reduce the conductor size in line construction. Electric power, being the product of voltage and current, may result from multiplying a very large current by a very small voltage; that is, in transmitting any given power, if the voltage factor is increased, the current factor will be decreased. Electrical voltage (volts) is the force that moves electrons through conductors to result in current (amperes) flow. As the cross section of the size of the line conductors is determined by the amount of current it will carry, it may be smaller in size if the current is decreased.

Transformers are classified into several general categories: voltage transformers, voltage regulation, metering and protection, and current regulation. It is the current regulation we are going to consider in this section. The current transformer is also called a series transformer because the primary, usually one or more turns, is connected in series with the line.

An ordinary transformer changes the voltage in the primary to a different voltage in its secondary. In a current transformer, the current flowing in its primary is reduced to a much lower current in its secondary. Most current transformers are so designed that when properly installed, the current in the secondary will never be more than 5 amperes. A typical current transformer, shown in Figure 21.4, has the shape of a doughnut, 4 to 6 inches in diameter. It has a single winding: the secondary. The load wire (in which the

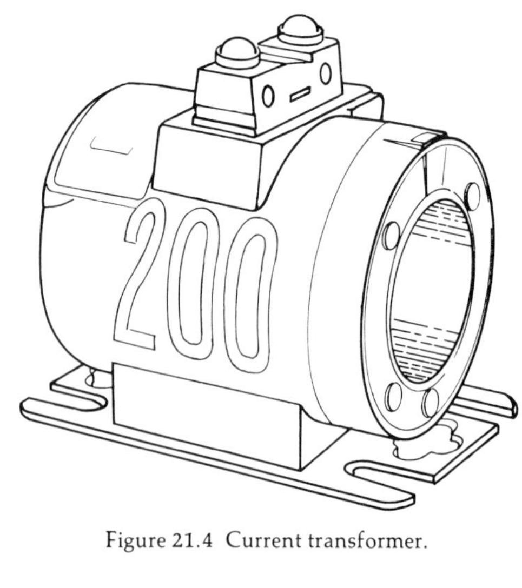

Figure 21.4 Current transformer.

current is to be measured) is run through the hole of the transformer, and becomes the primary. Assuming that the transformer has a 200 : 5 ratio for use with a 200-ampere service, the current in the secondary will be five two-hundredths of the current in the primary. If the current in the primary is 200 amperes, 5 amperes will flow in the secondary. If the service were 400 amperes, the transformer would have a 400 : 5 ratio. The secondary of a current transformer is always rated at 5 amperes, no matter what the ampere rating of the primary may be.

Purpose

If the service at the pole is rated at 200 amperes or more, it means running a very large conductor from the top of the pole down to the meter, and back again to the top. With the price of the conductors, this becomes expensive and clumsy. It is not necessary to run such large conductors down to the meter: a current transformer may be used.

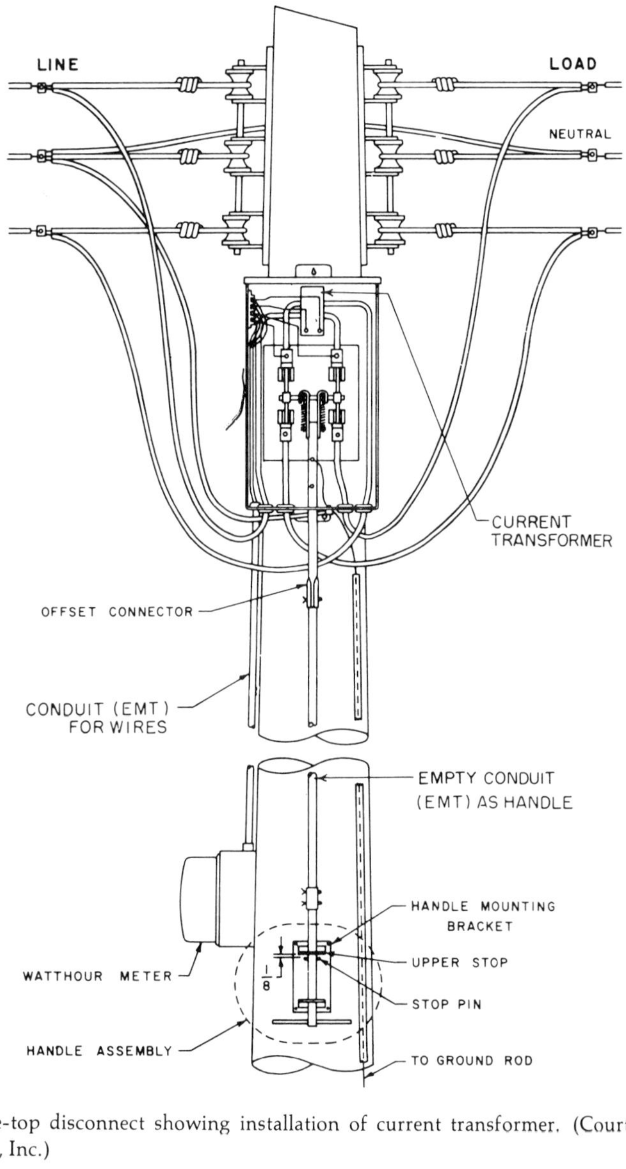

Figure 21.5 Pole-top disconnect showing installation of current transformer. (Courtesy of Midwest Electric Products, Inc.)

Installation

One current transformer is installed at the top of the pole, with the two hot wires running through the hole of the transformer. See Figure 21.5. Four small No. 14 wires run from the pole top to the meter (which must be of the type suitable for use with a current transformer): two from the secondary of the current transformer and two from the hot wires, for the voltage. The meter operates on a total of not more than 5 amperes; however, the dials of the meter will show the actual kilowatt hours consumed.

Caution. The secondary terminals of the current transformer must always be short circuited while any current is flowing in the primary. When connected to a kilowatt hour meter, the meter constitutes a short circuit. Usually when purchased, the transformer will have a short circuiting bar across its secondary terminals; this bar must not be removed until the transformer is connected to the meter.

21-8 EMERGENCY SERVICE

Standby electric power on farms is essential. Power failure may result in serious consequences on some farms as with total confinement buildings, hog operations depending on ventilation; poultry buildings which have electric incubators and brooders; greenhouse operations where the heating system depends on motor-driven circulating pumps and electrically controlled combustion equipment; and with large dairy barns where water cannot be pumped by hand and where labor for milking is not available.

When power failures occur due to icing of lines, high winds, storms or blackouts, standby generation units can serve during such outages.

Standby generating equipment is of two general types, engine-driven and tractor-driven units. The engine-driven units may be further classified into automatic and nonautomatic types. However, all types must be connected to the farm wiring system by means of a double-throw switch (transfer switch) which will disconnect the power supplier's service before the generator is connected to the line. See Figure 21.6. This is to prevent feeding power back into the power supplier's line and thus endangering the lives of linemen. It also prevents damage to the generator by feeding power back into it when normal service is restored.

Engine-Driven Automatic Units

The automatic standby generator is driven by an engine that starts automatically in case of a power outage. These units generally have larger capacity (10 kW or above) than nonautomatic generators. A transfer switch normally energizes the farm wiring system from the power supplier's line. In case of power failure, the engine is automatically started. When it is up to speed, the farm wiring system is transferred to it automatically.

The ampere rating of the transfer control switch selected must be at least equal to that of the incoming service or the sum of the standby circuits when only a part of the normal load is being transferred to the emergency plant. In many cases rewiring only standby circuits is much less expensive than incorporating a much larger, completely automatic unit.

Figure 21.6 Hand-operated transfer switch. (Courtesy of Midwest Electric Products, Inc.)

It is essential that standby equipment be tested frequently, perhaps carrying the load for several hours to make sure it will function automatically in an emergency.

Engine-Driven, Nonautomatic Units

For the average farm it will be equally satisfactory and much less expensive to use a manually started, nonautomatic unit. A "power-off" alarm should be installed to warn the operator of an outage.

The nonautomatic standby generator is driven by a manually started engine. A manually operated double-throw switch must be used to disconnect the power supplier's service when the standby generator is connected to the farm wiring system.

This type of system is suitable for use where automatic operation is not needed and where a tractor is not available to drive the generator. Such an engine should be started frequently to make sure it will run when needed.

With nonautomatic units, an indicator light or separate light on the pole should be installed on the line so the operator will know when the service has been restored.

Selecting the Generator

Electrical characteristics of the generator must be suited to the load. In most cases this will require a 60-cycle unit producing power to successfully operate 120/240-volt utilization equipment. The actual rating of the standby generator may be 110/220, 115/230, 120/240, or 120/208 volts, depending on the system and the rating used.

The capacity of the generator must be sized to carry the maximum load. It should be approximately three times the ampere rating of the largest motor to be started, so as to allow for the starting current of the motor, unless the generator is especially designed for motor-starting duty.

Stating this in another manner, allow 2 kW of generator capacity per brake horsepower of the largest motor to be started. For example: use a 6-kW generator to start a 3-hp motor.

The operator should be sure to follow the manufacturer's requirements to assure engine capacity to run the generator. The horsepower supplied by the engine or tractor should be at least twice the kilowatt rating of the generator: that is, use a 20-hp engine to drive a 10-kW generator.

Some manufacturers have center tap delta systems available that can be used with single-phase, or for three-phase. This type of unit would be satisfactory if the farm were changed from single-phase to three-phase.

The Double-Throw Switch (Transfer Switch)

The capacity of the double-throw switch must be matched to the rating of the conductors supplying the load from the normal source of energy and not to the rating of the generator or to the load supplied when the generator is in operation. The switch is usually of the unfused type, since the generator should have its own set of fuses or circuit breakers, and the service entrance or feeder is already equipped with overcurrent protection. Some inspection authorities and power suppliers require a three-blade, double-throw switch in order to switch the neutral as well as the ungrounded conductors.

A double-throw switch installed on a meter pole or other outdoor location should be of the weatherproof type unless it is installed in an approved weatherproof enclosure (Figure 21.7).

21-9 THREE-PHASE CONVERTERS

Electric motors are used for many jobs on the farm. They are an efficient, compact, and dependable source of power.

Single-phase motors are in general use because farm electric power service is usually single-phase. For some farm operations, however, three-phase motors can be a better choice than single-phase motors. Also they may be the only type readily available in the size needed.

Phase converters make it possible to operate three-phase motors from

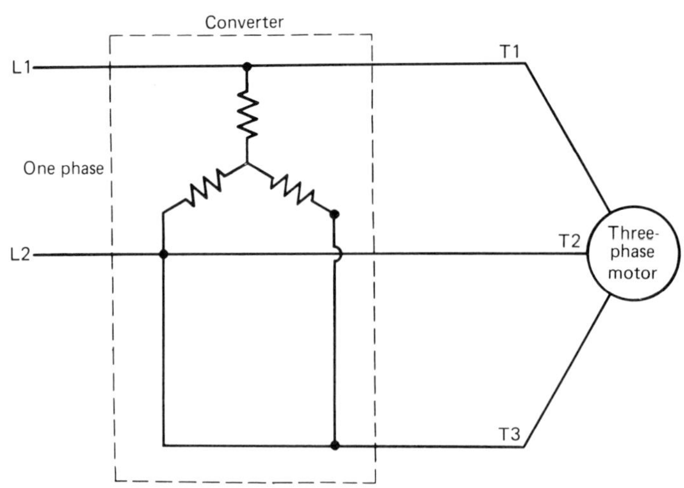

Figure 21.7 Simplified diagram for rotary converter. (Courtesy of RONK Electrical Industries, Inc.)

single-phase power lines when three-phase power is not available. They convert the single-phase line voltage into three-phase voltage.

Proper selection, installation, and protection of both the phase converter and the three-phase motors are essential for satisfactory performance from a converter-motor combination.

Converter-motor combinations are being used to successfully operate many kinds of farm loads. Some examples are corn driers, grain handling systems, irrigation pumps, and animal feeding systems.

Advantages of Three-Phase Motors

Three-phase motors are used instead of single-phase motors for a number of reasons when three-phase power service is available.

In the larger size (above 2 hp), the motors generally are more readily available and less expensive than single-phase motors of the same horsepower rating. Also, they are usually smaller in size, lighter in weight, and offer a greater choice in the type of enclosure and horsepower rating that can be purchased.

Three-phase motors are very simple in construction. No starting windings or starting devices (such as internal centrifugal switches) are required. Power parts generally mean less maintenance and service problems.

The direction of motor rotation is easily reversed with three-phase motors. The electrician simply interchanges any two of the three motor leader connections to reverse the motor.

When to Use Phase Converters

There are three ways to power motor loads:

1. Single-phase motors operating on single-phase power.
2. Phase-converter, three-phase motor combinations operating on single-phase power.
3. Three-phase motors operating on three-phase power where the service is available.

Before installing a phase converter, determine if a converter-motor combination will be the best source of power for the farm operation.

To determine the need, draw up a list of jobs that require electric motors. Include the sizes of motors required, the expected hours of use, and possible future loads. These are all important factors in the decision making.

In some areas, power suppliers often limit the size of motors that may be used on single-phase power lines, because of the high currents that large single-phase motors draw in starting. These high currents reduce the line voltage and may affect service to the other equipment in the operation.

Converter-motor combinations, however, generally draw less starting current than comparable size single-phase motors (however, there is a corresponding decrease in the starting torque). Consequently, power suppliers may permit the use of higher horsepower converter-motor combinations than single-phase motors on their power lines.

The best choice and method of operation depend on:

1. The cost and availability of single-phase motors and three-phase motors of the required size and starting characteristics.
2. Motor size limitations on the single-phase power lines.
3. The availability and cost of three-phase power service.
4. The cost of operation for a converter-motor combination compared with the cost of operation for the other choices. All costs, including the cost of equipment, installation costs, and power rates, should be considered.

Use of a converter-motor combination is often the best choice:

1. When the cost of the line extension to bring three-phase power to the farm is high because of the wire and construction costs for the length of three-phase line required.
2. When the cost per kilowatthour for three-phase service is substantially higher than for single-phase service because of the additional investment the utility company must make to install three-phase service. Over a period of time, the difference in the rates may add up to more than the cost of a converter.
3. When there is a need to use a large motor, but the starting current of a standard single-phase motor of the required size is too high to use on single-phase lines.
4. When a temporary three-phase service is needed until regular three-phase service may become available.
5. When equipment is purchased with a three-phase motor as an integral part of the unit and replacement of the motor is either too difficult or costs more than a converter.
6. When there is a need for a number of motors and the cost of three-phase motors plus the converter is less than the cost of single-phase motors.

Types of Phase Converters

There are two general types of converters—static and rotary. Each type offers advantages for specific kinds of motor loads. However, in this section only the rotary converter will be discussed.

Rotary Converter

A rotary converter consists of a rotating unit similar to a motor and an enclosure containing capacitors. Figure 21.8 shows a simplified diagram for a converter. Two of the three rotating converter terminals are connected directly to the single-phase lines. The third rotating converter terminal is connected to one of the single-phase lines through the capacitor bank. The capacitors provide the rotating magnetic field to start and operate the converter. The generating action of the rotating converter, in combination with the phase shift of the capacitors, produces the third phase voltage to operate the motor.

With the rotary converter, you can operate a number of motors under varying load conditions. The converter is started first (with no load connected). Then while it is running, motors of sizes up to the rating starting horsepower of the converter may be started one at a time. The horsepower

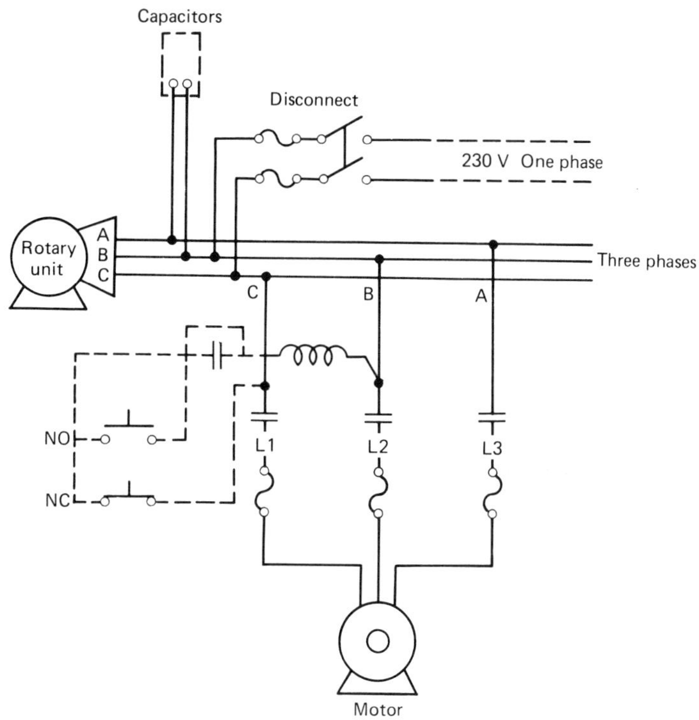

Figure 21.8 Wiring diagram for rotary converter. Showing base unit, disconnect switch, motor starter, remote control, motor, and capacitor unit. (Courtesy of RONK Electrical Industries, Inc.)

rating of the largest motor to be started determines the minimum converter rating needed.

The total horsepower load that can be carried by a rotary converter is determined by its design and the method of switching capacitance when motor loads are connected. Generally this will be from two to four times the horsepower of the largest motor that may be started.

The total horsepower rating indicated by the manufacturer should be with the motors fully loaded and operating at the rated temperature. This rating can be exceeded for short periods if air temperatures are favorable for motor cooling or some motors are not fully loaded.

There is also a minimum size of motor that should be operated alone on a rotary converter to prevent overheating of the motor. The size varies with converters from one-tenth to one-fourth of the converter rating.

In cold weather, capacitors lose capacitance, and the starting torque of the converter will be less. Also, the starting torque of equipment will be greater because of stiff bearings. The converter and bearings must be large enough to provide cold-weather starts. If you oversize the converter, remember that there is a minimum size of motor that should be operated alone on a converter.

Installation

Many of the problems of operation have been due to improper installation, incorrect connections of the wires to the converter, loose connections, or undersizing wiring.

Connection and sizing of the conductors and other installation should be in accordance with the requirements or recommendations of the converter manufacturer, the local utility company supplier, and the NEC and local electrical codes.

The phase converter may be installed outdoors since most units are of drip-proof or totally enclosed construction.

The standard rotary phase converter consists of two basic parts—a base unit and a capacitor panel. The base unit and panel may be installed near the equipment or installed near the service pole in the farm yard.

Wire Size from Converter to Motor

The wiring from the converter to the motor must be of the proper size. If it is too small, there will be excessive voltage drop. This will reduce the motor starting torque, and the motor may not start the load or bring it up to speed. The motor may also draw excessive current and overheat.

To determine the minimum size wire that should be used for 230-volt, three-phase motors, refer to the NEC Article 430, ''Rules governing motor installations.''

1. NEC Article 430-22 provides that branch-circuit conductors supplying a single, continuous duty motor, shall have a current carrying capacity not less than 125 percent of the motor full-load current rating. One reason for the added 25 percent is to allow a heating margin for the high, but short-duration, starting current. Another is to provide for the small percentage of overload which the continuous duty motor is designed to withstand.

EXAMPLE 4

The current value for a single 5-hp, three-phase, 230-volt motor is 19 amperes. Table 430-150 of the NEC shows a 5-hp squirrel cage induction motor to have a running current of 15.2 amperes. The supply conductors must have a carrying capacity of at least 1.25 × 15.2 amperes, or 19 amperes. Assuming that type TW wire is used, the nearest size listed in NEC Table 310-16 is No. 12 which has a carrying capacity of 20 amperes.

2. For multiple-motor load, see NEC 430-24, which states that conductors supplying two or more motors shall have an ampacity equal to the sum of the full-load current rating of all the motors plus 25 percent of the highest rated motor in the group.

EXAMPLE 5

One 10-hp motor and one 3-hp motor will be used on the circuit. The current value of the larger motor (10-hp, 230-volt, induction type) is 28 amperes. Value for the smaller motor is 9.6 amperes. Total current requirement is 28 × 1.25 = 35 + 9.6 = 44.6 amperes. Assuming type THW wire is used, the nearest size listed in NEC Table 310-16 is No. 8 which has a current carrying capacity of 45 amperes. Table 3 of Chapter 9 shows that three No. 8 THW conductors may be installed in a ¾-inch conduit.

Wire Size from Power Source to Converter

The wires from the single-phase power source to the converter must be large enough to prevent excessive voltage drop. These conductors will carry greater current than the conductors to the motor. The single-phase current will be approximately one and one-half to two times the total current of the motors as determined in (1) and (2) of the preceding section. Because single-phase is connected to the converter, Table 430-148 of the NEC must be used.

EXAMPLE 6

The current value for the 10-hp motor was 35 amperes (28 × 1.25). The value for the 3-hp motor was 9.6 amperes. For the wire sizes to the converter, use single-phase Table 430-148 NEC; 10-hp full-load current is 50 amperes + 3-hp full-load current is 17 amperes (50 + 17 = 67 amperes). To determine the wire size required to carry the current of 67 amperes, refer to Table 310-16 of the NEC.

EXAMPLE 7

Copper wire, type TW, direct burial, will be used to carry the current of 67 amperes determined in the above example. That current value is not listed in the table, so the next higher value, 70, must be used. Minimum size copper wire required for 70 amperes is No. 4. For aluminum wire use No. 2 for 75 amperes. Table 3 of Chapter 9 NEC shows that two No. 4 copper TW conductors may be installed in a 1-inch conduit.

Disconnecting Means

Motors, service and branch circuit wiring, and phase-converter should be protected against overload. Too much current causes overheating and damage to equipment.

Every motor larger than ⅛ hp must have a disconnecting means. It shall be a motor-circuit switch rated in horsepower or a circuit breaker. There is an exception to the rule.

Motors rated as 2 hp or less, 300 volts or less, may be disconnected by a general-use switch whose ampere rating is not less than twice full-load current rating. See NEC 430-109. Applying the code rule to a 15-hp example, a motor circuit switch rated at not less than 15 hp must be employed. The switch must be within sight of the device which controls operation of the motor, or else it must be arranged for locking in the open position. See NEC Article 430-102. It must disconnect all ungrounded supply conductors from both motor and controller, and it must have a continuous carrying capacity of at least 115 percent of the nameplate current rating of the motor. See NEC 430-103.

Motor Controller

Magnetic starters are a combination of a magnetic controller, or relay, and overload protection of the motor. Use of one for each motor is highly recommended and may be required by many power suppliers.

It is essential to use a magnetic starter with a rotary converter to provide the means of disconnecting the motor in case of a power outage. Otherwise when the power is restored, the motor will try to start on single-phase power before the converter reaches operating speed and may be damaged.

For maximum protection of the motors, use magnetic starters that have thermal overload protection in each line.

SUMMARY

1. As in other types of installations, farm wiring should provide for expansion of the electrical system.
2. The power supplier decides the location of the service metering equipment.
3. A farm dwelling electrical system must serve a variety of loads.
4. The total load calculations determine the size of the conductors from the service to the separate buildings.
5. It is important that all system grounds and equipment grounds in the various farm buildings be bonded and connected to the well casing or underground water piping system.
6. Emergency electric service facilities are becoming more important to the farm. A standby generator unit can serve the more critical farm electrical loads, should there be an outage.

PROBLEMS

21-1 List three types of service-entrance conductors used on a farm installation.

1. ______________ 2. ______________ 3. ______________

21-2 In the space below, show the method of computing a total farm load. Show largest load first.

21-3 According to the NEC, is it important to have all system grounds and equipment grounds in various buildings bonded and connected to underground water piping.

1. YES ______________ 2. NO ______________

21-4 List at least five reasons why emergency electric service facilities are important to the farm.

1. ____________________

2. ____________________

3. ____________________

4. ____________________

5. ____________________

21-5 Explain the purpose of a transfer switch.

____

____

____

____

21-6 Designing a farm electrical system involves the same basic procedures used for ____________ projects.

21-7 The trend is for utility companies to ____________ their underground conductors.

21-8 Where would a service lateral terminate? ____________________

____

21-9 You are wiring a farm machine shed. How far apart would you space the convenient outlets? ____________________ .

21-10 List some advantages of using three-phase motors. ____________________

____

____

____

chapter 22

STORE BUILDING

Instructional Objectives

1. To understand the importance of good lighting.
2. To become familiar with track systems.
3. To learn how to calculate branch circuits.
4. To identify different types of lighting fixtures and state their application.
5. To understand the purpose of the National Electrical Code.
6. To become familiar with general illumination circuits.
7. To learn how to calculate show window circuits.

Self-Evaluation Questions

Test your prior knowledge of the information in this chapter by answering the following questions. Watch for the answers as you read the chapter. Your final evaluation of whether you understand the material is measured by your ability to answer the questions. When you have completed the chapter, return to this section and answer the questions again.

1. What is the greatest concern to the electrician in modern store construction?
2. In general, what type of lighting fixtures are used in store buildings?
3. How are track systems mounted?
4. What does good lighting do?
5. How many watts of light per square feet are required in a store building?
6. Is it possible to determine the number of lighting circuits by using the watts per square foot?
7. Is show window lighting figured the same as general floor space in a store building?
8. Does the NEC apply to duplex receptacles in show windows? Explain.
9. Are sign circuits restricted to a 20-ampere rating?
10. How are show window lights controlled?

11. According to the NEC what is the minimum wattage for show window lighting?
12. Why does store lighting usually fall into the classification of continuous burning loads?

22-1 GENERAL LIGHTING

The greatest concern to the electrician in modern store construction is the lighting. The requirements for lighting may call for installations ranging from a few outlets for general illumination to a very elaborate and decorative lighting layout. In today's competitive market, there is little choice except to design the lighting system specifically to suit the size structure, the layout, and location of the building, the type of occupancy and merchandise sold, effects desired, and many other factors. This can be accomplished only by an experienced, creative, competent lighting specialist who can select the type and number of fixtures and lay out the arrangements.

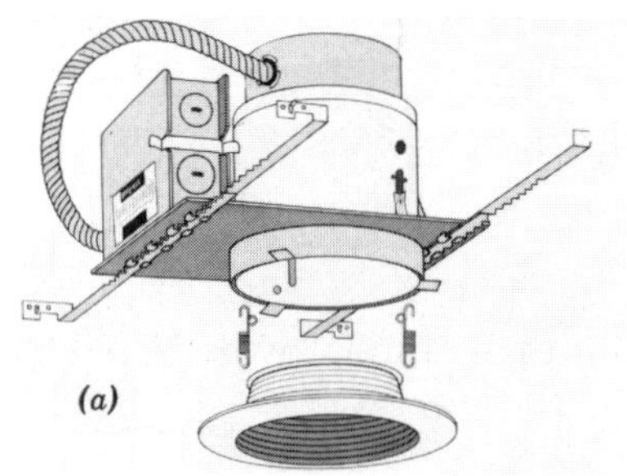

Figure 22.1 Incandescent recessed lighting fixtures. (*a*) Recessed, prewired, round housing. Shows 24-inch adjustable no-notch bar hanger for easy installation to joists. Porcelain socket with copper screw shell adjusts vertically to accommodate 75-watt R-30, 150-watt R-40, or 100-watt A-19 lamps. Once installed, housing is removable from frame and lifts off for easy access to junction box. UL listed for wet locations in bathrooms and outside under overhangs. (Courtesy of Progress Lighting)

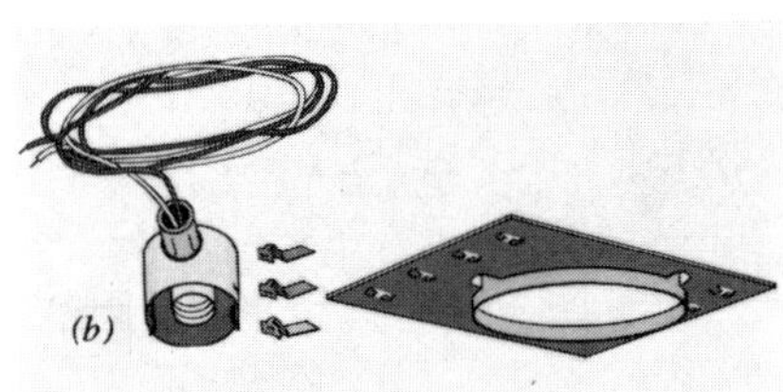

Figure 22.1 Incandescent lighting fixtures. (*b*) Unwired frame in kit suggested for installation in finished, existing ceilings. (Courtesy of Progress Lighting)

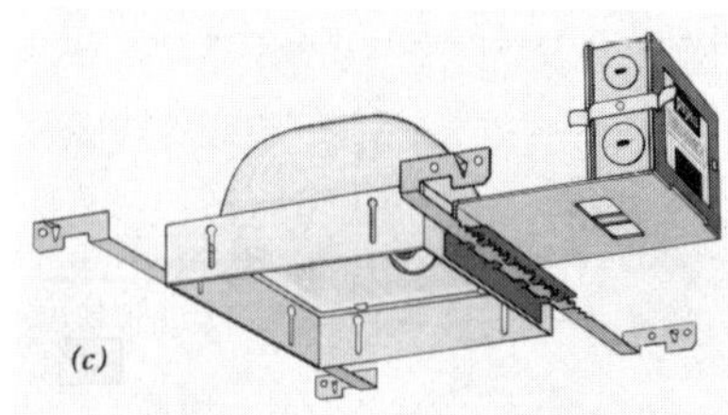

Figure 22.1 Incandescent recessed lighting fixture. (*c*) Square prewired housing. Has removable reflector, allowing access to wiring after housing is installed. Designed to fit standard acoustical tile squares. Is adjustable for easy hanging. Porcelain socket for a 200-watt lamp. (Courtesy of Progress Lighting)

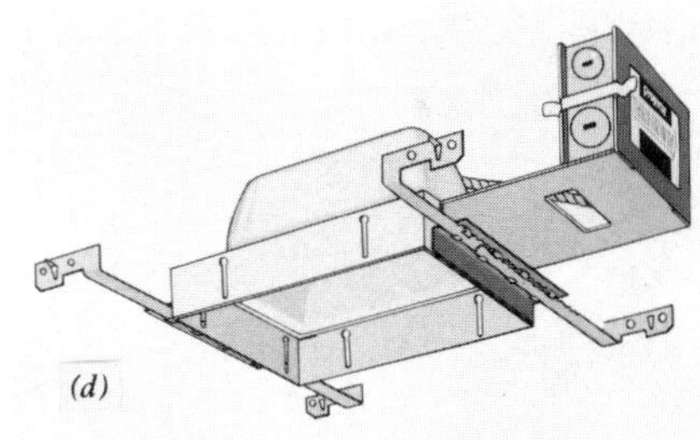

Figure 22.1 Incandescent recessed lighting fixture. (*d*) Rectangle 6½ × 9⅜ inches. UL listed for wet locations such as bathrooms and outdoors under overhangs. Porcelain socket for 100-watt lamp. (Courtesy of Progress Lighting)

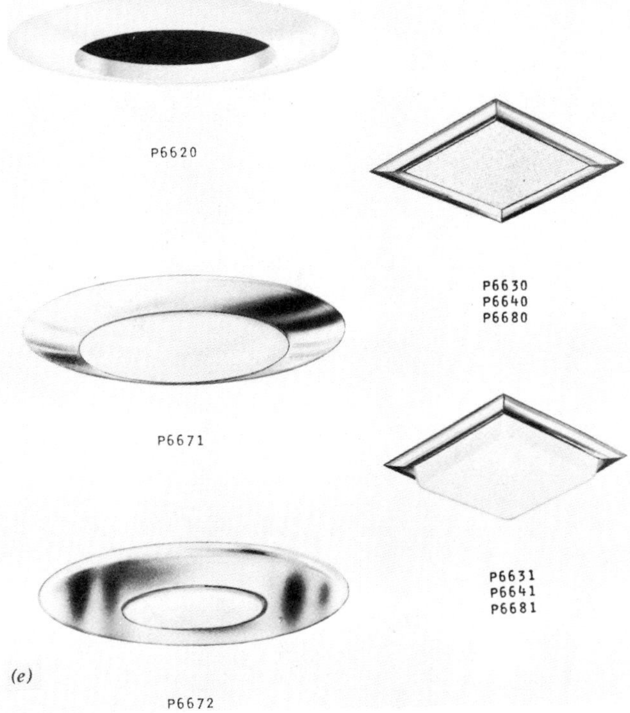

Figure 22.1 Incandescent recessed lighting fixture. (*e*) Typical trim for recessed lighting fixtures. (Courtesy of Progress Lighting)

In general, lighting fixtures will be of the fluorescent type. Fixtures might be suspended from the ceiling, mounted on the ceiling, or recessed in the ceiling; however, many of the modern stores are installing incandescent recessed lighting, using round, square, or rectangular fixtures with interchangeable trims. See Figure 22.1. To really be creative, some designers are installing track systems, that mount vertically, horizontally, or change planes anywhere in a variety of patterns. See Figure 22.2. Type A, R, and PAR incandescent lamps are used with the above recessed and track fixtures.

Good lighting draws attention to the merchandise on display, and encouraging customers to buy what they want, rather than only the things they need.

(a)

Figure 22.2 Show window track lighting. (*a*) Shows various lampholders; notice concealed wiring in stem, minature mounting head. Mounting lock ring secures lamp holder to track. Rotate past 360° horizontally, swivel tilts up to 90° from vertical. (Courtesy of Progress Lighting)

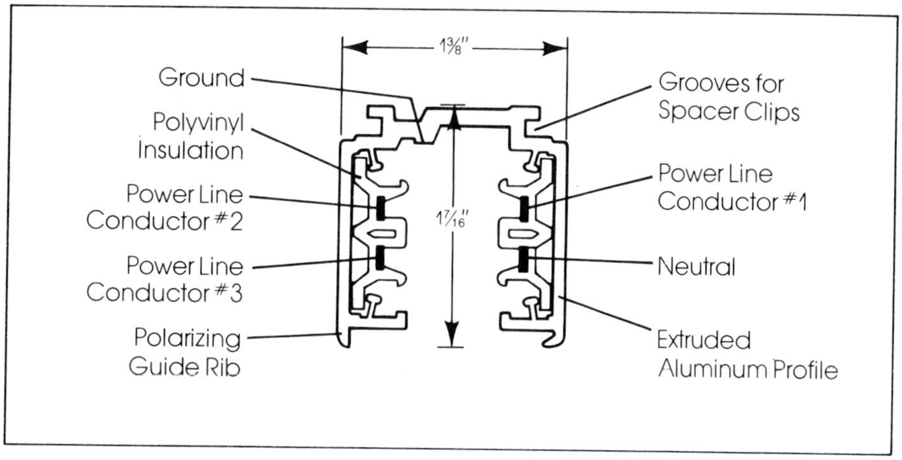

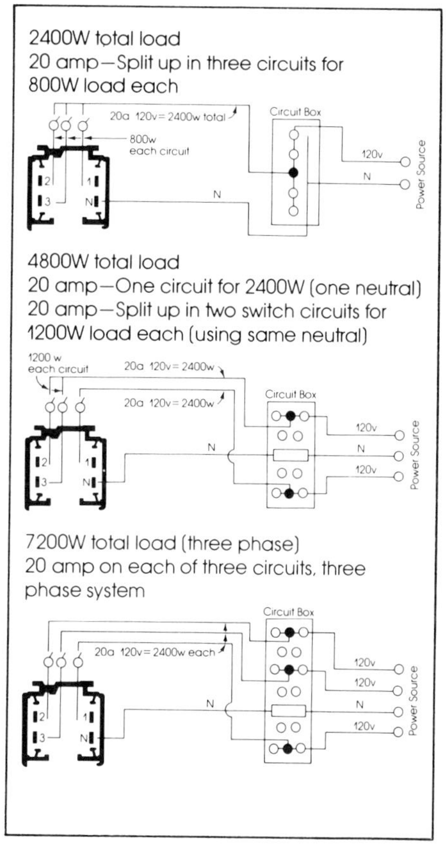

(b)

Figure 22.2 Show window track lighting. (*b*) Shows track section, made in exact 4-foot and 8-foot lengths, may be cut to any length at the job site. Track systems are especially good for high-lighting merchandise, art, meeting rooms, ticket counters, and showrooms where flexible efficient lighting is desired. (Courtesy of Progress Lighting)

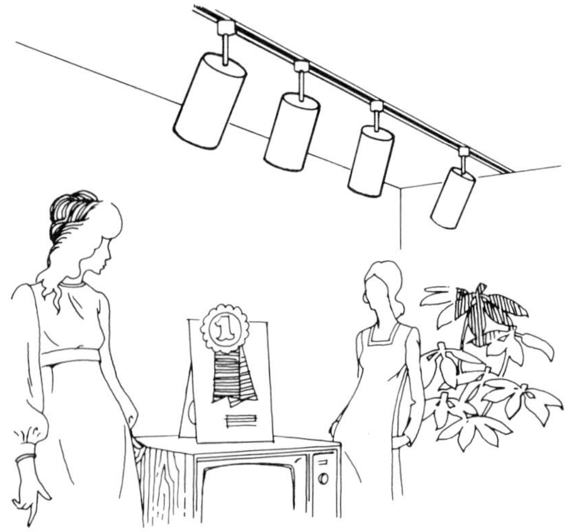

Figure 22.2 (*c*) Typical track lighting installation. (Courtesy of Progress Lighting)

(*c*)

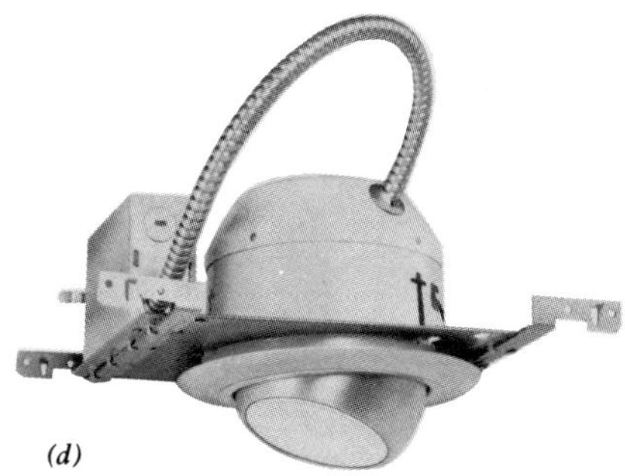

Figure 22.2 (*d*) Typical recessed lighting fixture for show window. (Courtesy of Progress Lighting)

(*d*)

P6976

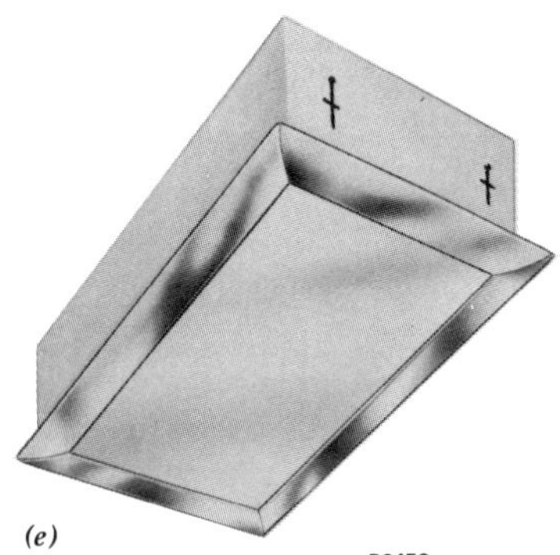

Figure 22.2 (*e*) Recessed lighting fixture for store show window. (Courtesy of Progress Lighting)

(*e*)

P6453
P6455

22-2 NUMBER OF CIRCUITS FOR GENERAL ILLUMINATION

From the dimensions of the store, Figure 22.3, the area measures 60 by 90 feet excluding the show window area, which is roughly 10 feet deep. Its dimensions are noted on the illustration.

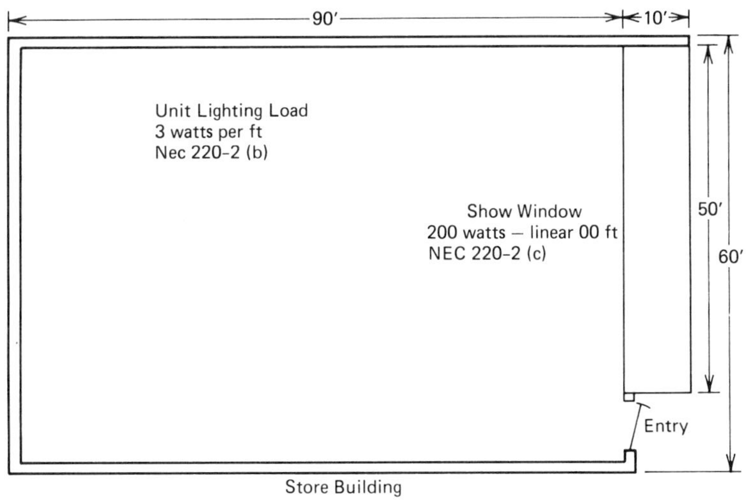

Figure 22.3 Store building floor dimensions.

According to the Table, General Lighting Loads by Occupancies, NEC 220-2(b), stores require 3 watts per sq ft.

The watts load equals the square foot area times the watts per square feet required.

The floor area is equal to:

60 × 90 feet or 5400 sq ft
5400 sq ft × 3 watts = 16,200 watts

Due to a lack of natural light, store lighting usually falls within the classification of continuous burning load. Continuous load supplied by a branch circuit shall not exceed 80 percent of the branch circuit rating output. See NEC 220-2(a).

If the specifications for the project call for 20-ampere circuits to be installed for all lighting, their maximum current is limited to 20 × 80 percent or 16 amperes.

Should the lighting service specify 115/230 volts, then the 16 amperes will transmit power equal to:

16 amperes × 115 volts = 1840 watts

The number of circuits to provide this wattage equals:

16,200 watts (total floor area) ÷ 1840 watts per circuit = 8.8 or 9 circuits

or

16,200 watts ÷ 115 volts = 140.8 amperes ÷ 16 amperes per circuit = 8.8 or 9 circuits

When the specifications do not state the type of branch circuit to be installed, the electrician may decide to use a conventional 15-ampere circuit with No. 14 wire. The example would be thus:

15 amperes × 80 percent = 12 amperes
12 amperes × 115 volts = 1380 watts
16,200 watts ÷ 1380 watts per circuit = 11.7 or 12 circuits

or

16,200 watts ÷ 115 volts = 140.8 amperes ÷ 12 amperes = 11.7 or 12 circuits

When the specifications do not state the exact type of branch circuit to be installed, it might pay to figure both ways, using 15- and 20-ampere circuits. As you can see from the example, it might be less expensive to use the 20-ampere circuit and install three less circuits. Review Chapter 23 for spacing of fixtures.

22-3 SHOW WINDOW LIGHTING

The minimum wattage for show window lighting is set forth in the National Electrical Code.

> NEC 220-2(c) Exception 3. The load can not be less than 200 watts for each linear foot of such window, measured horizontally along its base.

See Figure 22.3. The show window is 50 linear feet along the base of the window, plus 10 linear feet on one side. The minimum wattage is:

50 feet + 10 feet = 60 feet
60 linear feet × 200 watts = 12,000 watts

If 20-ampere circuits are installed, they will each supply:

20 amperes × 115 volts = 2300 watts

The number of show window circuits must be at least:

12,000 watts ÷ 2300 watts per circuit = 5.2 or 6 circuits for the show window lighting

The total load for show window lighting is:

6 circuits × 2300 watts = 13,800 watts

or

12,000 watts ÷ 115 volts = 104.3 amperes ÷ 20 = 5.2 or 6 circuits

For the store measuring 60 × 90 feet, a show window of 50 linear feet as shown by Figure 22.3, would require:

General lighting	9 circuits	(20 amperes)
Show window	6 circuits	(20 amperes)
Total	15 circuits	

However, on a three-wire 115/230-volt, single-phase lighting system, the number of circuits could be reduced by dividing by 230 rather than 115 volts.

EXAMPLE 1

12,000 watts ÷ 230 volts = 52.1 amperes ÷ 20 = 2.6 or 3 circuits

The number of 20-ampere branch circuits required would be two plus or a minimum of three circuits. These would be three 3-wire branch circuits for the show window lighting.

For *emergency lighting*, review Chapter 24.

Duplex Receptacles

Assume there are 100 general-use 20-ampere duplex receptacles installed in the building. The NEC 220-2(c)4 would designate these receptacles as "other outlets," and assign a load rating of 1½ amperes to each outlet. Since the duplex receptacles will be serviced by three-wire branch circuits, the load must be calculated:

50 outlets × 1½ amperes or 75 amperes

which represents power equal to:

75 amperes × 230 volts = 17,250 watts

Because the duplex receptacles are on 20-ampere circuits, each circuit furnishes power equal to:

20 amperes × 115 volts = 2300 watts

The required number of receptacle circuits will be:

17,250 watts ÷ 2300 watts = 7.5 or 8 circuits.

22-4 SIGNS

The sign circuit, which supplies lamps, ballasts, and transformers, or a combination, may be rated at not over 20 amperes, unless it supplies circuits containing electric-discharge lighting transformers exclusively, in which case the maximum is 30 amperes.

Each commercial building and each commercial occupancy with ground floor footage accessible to pedestrians shall be provided at an accessible location outside the occupancy with at least one 20-ampere outlet for a sign or outline lighting.

Each sign installed, other than the portable type, must be controlled by an externally operable switch or circuit breaker to disconnect all ungrounded conductors from the sign. The NEC requires the disconnecting means to be within sight of the sign which it controls.

Switches, flashers, and similar devices controlling the transformer shall be either a type approved for the purpose or have an ampere rating at least double the ampere rating of the transformer it controls. However, for other than motors, a general-use snap switch rated not less than the load it controls may be used on AC circuits to control inductive loads.

Usually the sign is made and installed by a sign company. The electrician has the responsibility to install the circuit for the sign company to make the hookup.

Show window lights and the sign circuit can be controlled manually with a key switch or from a circuit breaker; however, with the push to conserve energy, it would be much better to install a seven-day time clock which will automatically turn the lights and sign on and off as well as skip any particular night the lights and sign are not required.

SUMMARY

1. Good lighting draws attention to the merchandise on display.
2. To really be creative in lighting design, some designers are installing a track system for changing light patterns.
3. Stores require 3 watts of lighting per square foot of floor space.
4. Store lighting usually falls into the classification of continuous burning loads.
5. Minimum wattage for show window lighting is set forth in the NEC at 200 watts per linear foot.

PROBLEMS

22-1 A developer has decided to construct a building 300 by 120 feet. The only natural light will come from the front of the building. How many watts of lighting will be required?

22-2 How many 20-ampere lighting circuits are needed for the above store?

____

22-3 What is the greatest concern to the electrician in modern store building construction? ____________________

22-4 In general, what type of lighting fixtures are installed in a store building? ______________________.

22-5 To be really creative in lighting design, what type of lighting is usually installed? ______________________.

22-6 When a store building has no natural light, this store lighting usually falls into the classification of ____________ burning loads.

22-7 How do you figure the minimum wattage for show window lighting?

____

____

____

____

22-8 What is the maximum load for a sign circuit (in amperes)?

____

22-9 Signs (other than portable type) are controlled by an ____________ ____________ ____________ or ____________ ________.

22-10 Show window lights can be controlled by ____________ ____ ____ or circuit breaker.

chapter 23

LIGHTING—LAMPS AND CONTROLS

Instructional Objectives

1. To identify types of light sources.
2. To understand the operation of a fluorescent lamp.
3. To become more familiar with the advantages of high-intensity discharge lighting.
4. To learn the different methods of lighting calculations.
5. To learn how to estimate maintenance factor.
6. To learn how to use the basic equation for the lumen method of lighting calculation.
7. To understand the need for a ballast in a mercury vapor light.
8. To become familiar with the different types of fluorescent lamps.
9. To learn some disadvantages of the incandescent lamp.
10. To understand the use of a sound control chart.

Self-Evaluation Questions

Test your prior knowledge of the information in this chapter by answering the following questions. Watch for the answers as you read the chapter. Your final evaluation of whether you understand the material is measured by your ability to answer the questions. When you have completed the chapter, return to this section and answer the questions again.

1. What is meant by maintenance factor when calculating lighting problems?
2. What is coefficient of utilization?
3. What is being used to replace mercury vapor lamps?
4. What is a lumen?
5. How are fixtures spaced?
6. What are some of the disadvantages of using incandescent lighting?

7. List the different classes of fluorescent lamps.
8. How do you control fluorescent lighting?
9. What is a high-intensity discharge lamp?
10. What is a high-pressure sodium lamp?

23-1 LIGHTING

There are so many varied types of fixtures and light sources available today that designing a lighting layout allows an architect, lighting engineer, developer, electrical contractor, or electrician to be creative and artistic in laying out a good lighting system.

Commercial building developers are concerned with adequate lighting and artistic qualities as well as energy saving devices. Experience in the lighting industry has taught us that layout, glare, heat, and color have a psychological effect on individuals.

Unfortunately, when developers or builders submit their ideas and design, they will sometimes specify only the footcandle levels for the various areas and the type of lighting to use. In effect, the art of lighting is generally ignored. Economy factors are weighed against all other factors. Usually economy wins over design.

Even though lighting design can be creative, normal design practice will be considered in this section. In office buildings where the general floor space is taken up with desks, then fluorescent lighting is the first consideration.

In areas where the space is "high bay" (high ceiling) the trend is to mercury vapor lighting; however, the most recent development in high-intensity discharge lamps is the high-pressure sodium lamp. This lamp was developed by General Electric and is marketed by them under the trade name "LUCALOX." The outstanding part about this lamp is its high output; in fact the output including ballast losses approaches 100 lumens per watt (lpw). This is double the effectiveness of a color-corrected mercury lamp and is at least 50 percent better than a standard fluorescent. Its small size, high output, and long life make it highly desirable and economical.

Incandescent-type recessed lighting is usually installed in toilets and lobby areas.

23-2 LIGHTING CALCULATIONS

There are two general classes of lighting (excluding floor lighting). One is the closed-room, low-ceiling type of installation. The other is the high-bay, high-ceiling, large-roof-span building, where walls may not exist as far as lighting is concerned. Two different methods of calculation are necessary.

Lumen Method of Calculation

In solving a general lighting problem, six key steps should be taken.

1. Determine the required level of illumination.
2. Select the lighting system and luminaires.

Lighting systems are classified as:

(a) Direct
(b) Semidirect
(c) General diffuse or direct-indirect
(d) Semi-indirect
(e) Indirect

Generally, offices are best lighted by an indirect, direct-indirect, or semi-indirect system. Manufacturing areas usually employ a direct or semidirect system, and merchandising areas may use any system or a combination of systems. Choice as to which of the lighting systems and luminaires best suit a given application will depend upon the seeing tasks to be performed and the characteristics of the area to be illuminated.

3. Determine the coefficient of utilization. The coefficient of utilization is the ratio of the lumens matching the working plane (normally assumed to be a horizontal plane 30 inches above the floor) to the total lumens generated by the lamps. It is a factor that takes into account the efficiency and distribution of the luminaire, its mounting height, the room proportions, and the reflection factors of walls, ceiling, and floor.

The first step in the determining of the coefficient of utilization (*CU*) is the selection of the proper room index. Rooms are classified according to shape into ten groups, each of which is identified by a letter known as the room index. In using a room index table, it should be noted that for direct, semidirect, direct-indirect, and general diffuse equipment, height is the mounting height above the floor, and for semi-indirect and indirect equipment, it is ceiling height.

The next step is the determination of the particular table of coefficients applicable to the luminaire selected. This should be done on the basis of similarity of distribution and efficiency. Care must be exercised in the selection of the reflection factors of ceiling and walls, since they will depreciate from their initial values. The figure selected should represent in-service conditions. However, regardless of what the tables predict, if we use a *CU* that is excessively high there will probably be questions as to its validity. If we use one too low, it will require too many fixtures. A *CU* of 0.6 to 0.7 should be a fair assumption.

4. Estimate the maintenance factor. Footcandles, which will be produced by any lighting installation, are determined by a careful analysis of the conditions under which the system will operate. Up to this point in the problem, careful consideration has been given to the determination of proper values of illumination, the system and luminaire to be used, the dimensions and architecture of the area, the reflection factors of walls and ceiling, and the resulting coefficient of utilization. All of this effort at accuracy is wasted if a haphazard factor for maintenance is applied to attain the final in-service footcandle value.

In the operation of any light system there are three principal variables that affect the amount of light obtained from the system.

1. Loss in light output of the lamp. The average lumen output throughout the life of the lamp is 10 to 25 percent lower than the initial value. The amount of this depreciation depends on the size and type of the lamp.

2. Loss due to accumulated dirt on the reflecting or transmitting surfaces of the luminaire and on the lamps themselves.
3. Loss of reflected light through accumulation of dirt on walls and ceilings; this was considered when selecting the coefficient of utilization.

Maintenance factors covering lamps and luminaires will be suggested for three conditions.

1. *Good Maintenance Factor.* When the atmospheric conditions are good, luminaires are cleaned frequently and lamps are replaced systematically.
2. *Medium Maintenance Factor.* Where less clean atmospheric conditions exist, luminaire cleaning is fair, and lamps are replaced only after burnout.
3. *Poor Maintenance Factor.* When the atmosphere is quite dirty, equipment is poorly maintained.

Careful consideration should be given to the ceiling brightness which may result when high-wattage lamps are used in luminaires with short suspension hangers. This is especially true when indirect luminaires are installed in relatively low-ceilinged areas.

This lumen method of lighting calculation is for rooms. The basic equation is relatively simple:

$$\text{Total lumens} = \frac{\text{fc} \times \text{area}}{CU \times MF}$$

where

MF = maintenance factor
CU = coefficient of utilization
fc = footcandle

$$\text{Lamp lumens} = \frac{\text{fc} \times \text{area}}{CU \times MF}$$

$$\text{Footcandles} = \frac{\text{Lamp lumens} \times CU \times MF}{\text{area}}$$

It is clear that the lamp lumens are fixed, depending on the fixture, and lamps are selected. The footcandles are fixed by specifications. The area is fixed by measurement. The only variables are the coefficient of utilization and the maintenance factor. The maintenance factor can be considered as 75 percent or 0.75 for good conditions with a low of 65 percent or 0.65 for poor conditions. Since one guess is as good as another, in trying to predict maintenance of a nonexistent building, a figure of 70 percent or 0.70 should be a fairly low consistent constant.

Rule of Thumb

A good approximation for power requirements for a lighting design is 5 watts per sq ft for incandescent lighting to obtain 50 fc, and 2 watts per sq ft for fluorescent and mercury vapor to obtain 50 fc. This is not intended to replace the lighting calculations, however, if the lighting calculations do not compare with these rule-of-thumb figures, then recheck the lighting calculations.

Consider a room 20 × 40 feet with fluorescent lighting required at 50 fc.

$$\text{Lamp lumens} = \frac{\text{fc} \times \text{area}}{CU \times MF} = \frac{50 \times 40 \times 20}{0.7 \times 0.6} = \frac{400{,}000}{0.42} = \begin{matrix}95{,}238 \\ \text{lumens}\end{matrix}$$

40 watt rapid start = 3,150 lumens
95,238 ÷ 3,150 = 30.3 lamps

Therefore use 30 or 32 lamps, assuming we have a 4-foot-long fixture and two continuous rows on 10-foot spacing. Delete one fixture to allow 2 feet at each end. This gives 9 fixtures per row. With 2 rows, there are 18 fixtures; and with 2 lamps per fixture, there are 36 lamps. This is two extra fixtures over and above our original calculations. However, to eliminate dark spots at each end, they would be necessary. Now by rule of thumb, check:

1. 800 sq ft at 2 watts per sq ft = 1600 watts.
2. Using 40-watt lamps, 1600 ÷ 40 = 40 lamps.
3. Two lamps per fixture = 20 fixtures.
4. Two rows of fixtures gives 10 fixtures per row.
5. 10 × 4 feet = 40 feet, which is too long, so delete one fixture per row.
6. This gives two-lamp fixtures per row.
7. 18 + 18 = 36 lamps.

We see that we have the same quantity of fixtures with our rule of thumb as we do with our calculations. It may not always work out exactly; however, it is usually a reliable guide or check.

23-3 FIXTURE SPACING

In general, fixtures in office-type rooms and low-bay production assembly rooms are spaced around a 10-foot maximum area. The desired spacing may have to be adapted to available structural conditions. The approach then is to calculate the optimum spacing and then check for mounting conditions and modify accordingly. This will mean installing special mounting supports so as to maintain the desired spacing. See Figure 23.1. For correct spacing the distance from the wall to a fixture is exactly half the fixture-to-fixture distance (Figure 23.2).

One further technique in illumination calculations and fixture spacing should be mastered. Frequently, we are called on to calculate illumination for a very large space such as an office floor. Instead of doing the calculating for the entire floor, it is easier and more meaningful to calculate the number of fixtures required per bay. This can be done by calculating the area covered by a single fixture as follows:

$$\text{Number of fixtures} = \frac{\text{illumination of area}}{\text{lamps per fixture} \times \text{lumens per lamp} \times CU \times MF}$$

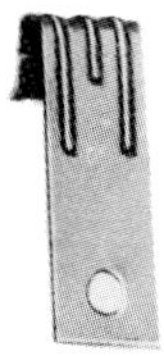

FIG. 1

FIG. 2

FIG. 3

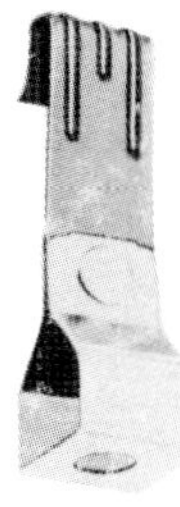

FIG. 4

INSTALLATION INSTRUCTIONS

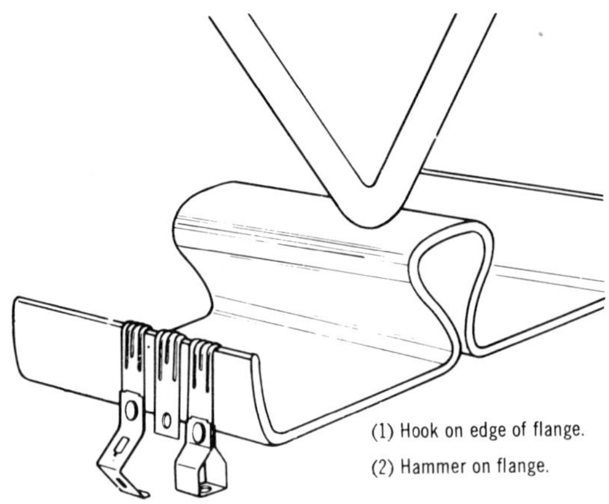

FEATURES

1. Suspend #8, #9 or #12 wire; 3/16", 1/4", or 3/8" plain rods; and 1/4" or 3/8" threaded rod from vertical flanges with thicknesses from 1/16" to 1/4".
2. Requires only hammer to install.
3. Will support 160 pound static load limit.

SIZING CHART

FIG. NO.	CATALOG NUMBER	VERTICAL FLANGE THICKNESS	DESCRIPTION	WEIGHT PER 100 (Lbs.)
1	J-1	1/16" to 5/32"	For wrap wire, "S" hooks, bolting straps	2
1	J-2	5/32" to 1/4"	For wrap wire, "S" hooks, bolting straps	2¼
2	2A-J-1	1/16" to 5/32"	For #8 wire	4¼
2	3A-J-1	1/16" to 5/32"	For 3/16" plain rod	4¼
2	4A-J-1	1/16" to 5/32"	For 1/4" plain rod	4¼
2	6A-J-1	1/16" to 5/32"	For 3/8" plain rod	4¼
2	2A-J-2	5/32" to 1/4"	For #8 wire	4
2	3A-J-2	5/32" to 1/4"	For 3/16" plain rod	4
2	4A-J-2	5/32" to 1/4"	For 1/4" plain rod	4¼
2	6A-J-2	5/32" to 1/4"	For 3/8" plain rod	4½
3	4TI-J-1	1/16" to 5/32"	For 1/4" threaded rod, with thread impression	4¾
3	6TI-J-1	1/16" to 5/32"	For 3/8" threaded rod, with thread impression	4¾
3	4TI-J-2	5/32" to 1/4"	For 1/4" threaded rod, with thread impression	5
3	6TI-J-2	5/32" to 1/4"	For 3/8" threaded rod, with thread impression	5
4	6T-J-1	1/16" to 5/32"	For 1/4" and 3/8" threaded rod, nuts required	5½
4	6T-J-2	5/32" to 1/4"	For 1/4" and 3/8" threaded rod, nuts required	6

Packaged 100 per box.

APPLICATIONS

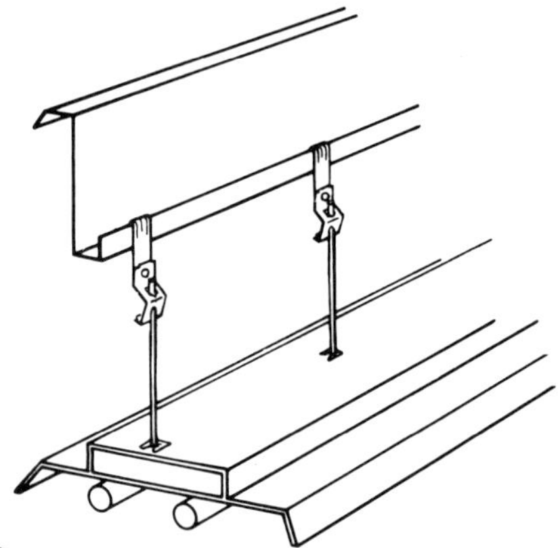

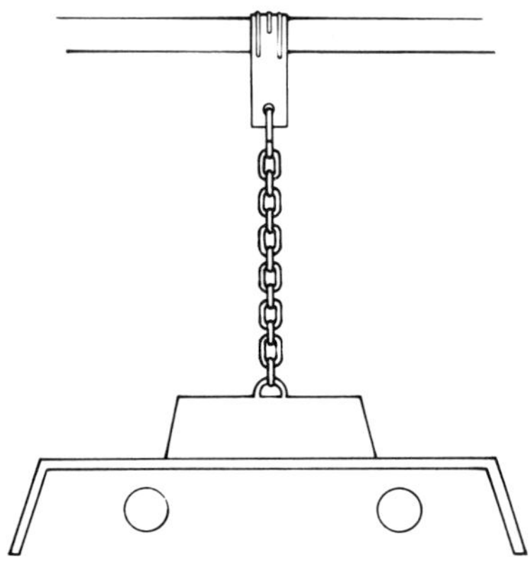

Figure 23.1 Shows the use of vertical flange clamps as special mounting supports for desired fixture spacing. (Courtesy of Erico Products, Inc.)

It follows that the area lighted by a single fixture is

$$\text{Area per fixture} = \frac{\text{lamps per fixture} \times \text{lumens per lamp} \times CU \times MF}{\text{illumination}}$$

EXAMPLE 1

To illustrate, an entire floor of a high rise is to be lighted to an average maintained footcandle of 50. The floor measures 30 × 150 feet and is divided into bays measuring 40 × 25 feet. Using two-lamp recessed fixtures with 40-watt RS lamps, find the number of fixtures required. Assumes an economy grade fixture with a low *CU* of 0.35 and *MF* of 0.7.

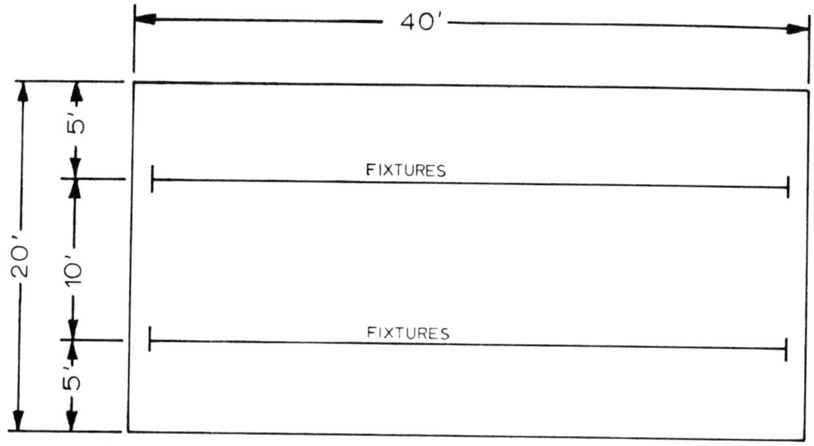

Figure 23.2 Shows proper spacing of fixtures in a commercial building. (Albert Lea Voc Tech)

SOLUTION A

$$\text{Fixtures per bay} = \frac{50 \text{ fc} \times (40 \times 25) \text{ sq ft}}{2 \text{ lamps} \times 3200 \text{ lumens} \times 0.35 \times 0.7} = 31.88$$

This would be either 30 units arranged in 3 continuous rows of 10 the long way or 5 rows of 6 fixtures the short way. In either case, center to center spacing is ± 8 feet.

SOLUTION B

The same result can be obtained by calculating the square feet per fixture.

$$\text{Square feet per fixture} = \frac{2 \text{ lamps} \times 2200 \text{ lumens} \times 0.35 \times 0.7}{50 \text{ fc}} = 31.36$$

Since the fixture is 4 feet long, centerline spacing of fixtures is

$$\text{Spacing} = \frac{31.36}{4} = \pm 8 \text{ feet}$$

$$\text{Fixtures per bay} = \frac{40 \times 25 \text{ feet}}{32 \text{ sq ft per fixture}} = 31$$

$$\text{Actual footcandles} = \frac{30 \text{ fixtures}}{31 \text{ calculated}} \times 50 = 48.5 \text{ fc}$$

We started with 50 fc and the solution shows 48.5 fc. This is close enough.

23-4 TYPES OF LIGHTING

Incandescent Lamp

An incandescent lamp is one in which light is produced by electrically heating a conducting material (the tungsten filament), which is contained in a gas filled bulb. Electrical contact with the base sends current through the lead wires to the filament. The flow of current through the filament sets up a resistance which causes the filament to heat up. Visible light is produced when the filament reaches 572°F or higher (Figure 23.3). The average incandescent filament burns at about 4760°F.

Few people realize that most of the lamps they buy have built-in fuses. If a filament breaks while the lamp is lighted, an arc could form across the break and draw a much heavier current than the lamp was designed for. The fuse is a part of one lead-in wire. See Figure 23.4. Under an abnormal current the fuse will open and protect the lamp and current from excessive current.

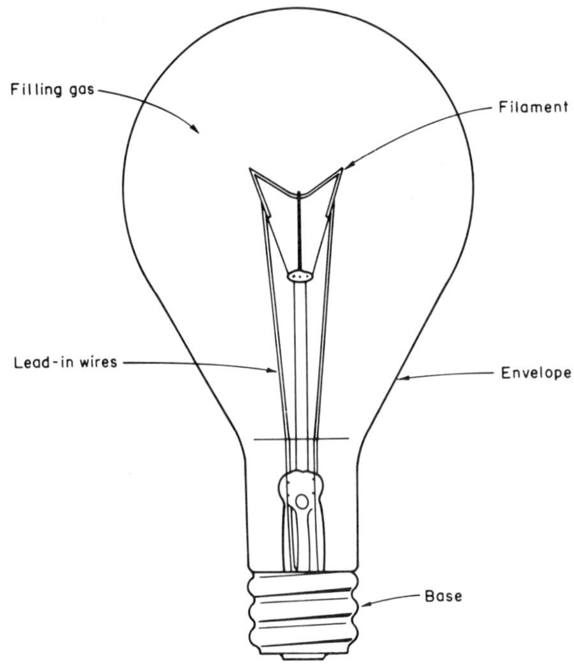

Figure 23.3 Incandescent lamp.

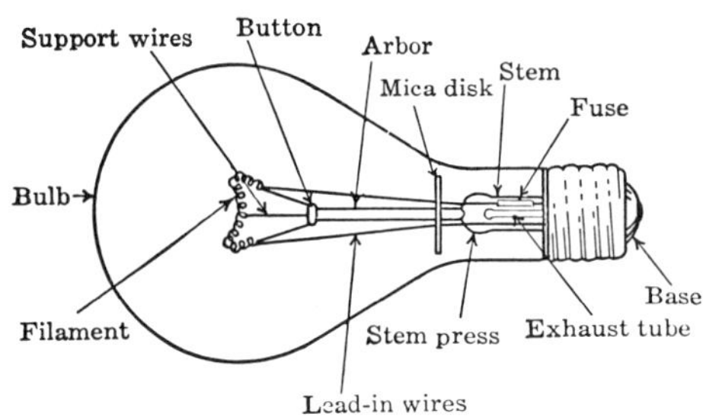

Figure 23.4 Shows principal parts of a typical incandescent lamp. The fuse opens the circuit if a filament should arc across on failure; thus it avoids the sputtering of metal and the cracking of the bulb.

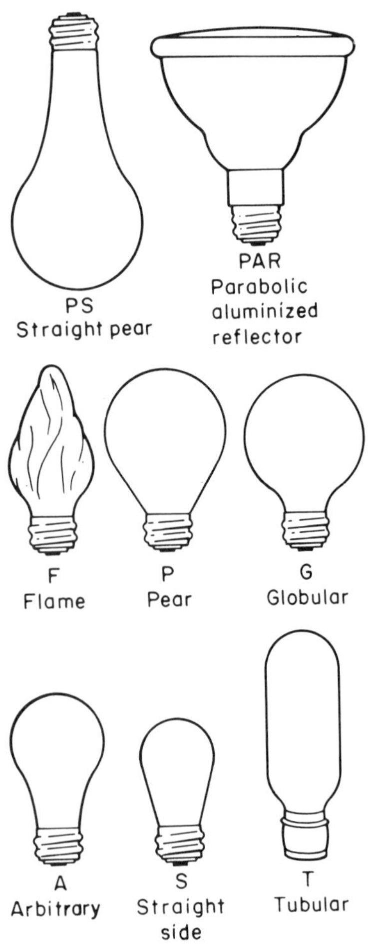

Figure 23.5 Incandescent lamp shapes and sizes.

Incandescent lamps are made in a wide variety of shapes and sizes with different types of bases. See Figure 23.5.

Principal advantages of incandescent lamps are low cost, instant starting and cheap dimming. However, disadvantages outnumber the advantages, since incandescent sources have a relatively short useful life, and the life is very voltage sensitive. At 10 percent undervoltage, life is increased about 350 percent. At 10 percent overvoltage, life is reduced about 75 percent. This means that for a nominal 1000-hour-life lamp, a swing of 10 percent in voltage either way can change lamp life from 3500 to 250 hours. Incandescent lamps are very inefficient producers of light. On the average, less than 10 percent of the wattage goes to produce light; the remainder is heat. Therefore, incandescent sources are a poor choice from an energy conservation point of view. Very few incandescent lamps are used in commercial buildings.

Fluorescent Lamps

A fluorescent lamp is basically a glass tube that is internally coated with phosphor. Sealed within the tube is a fill-gas, and a small amount of mercury. A cathode coated with emission material is located at each tube end.

When the switch is turned on, the cathodes heat up, causing the emission

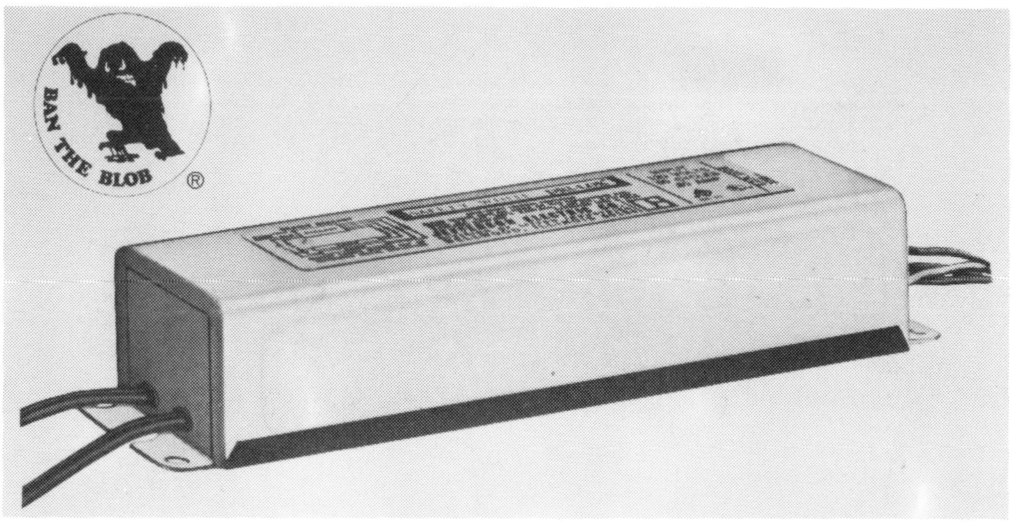

WIRING DIAGRAMS

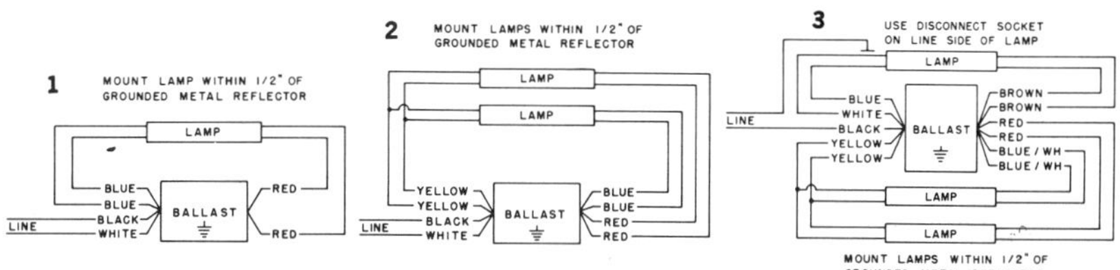

⏚ Note: Ballast case must be grounded to fixture and the fixture must be grounded by the installer for safety and proper starting.

Figure 23.6 Show ballast and wiring diagram for class "P" rapid start, high-power factor, 30–40 watt fixture. (Courtesy of Jefferson Electric)

coating to give off electrons. As the current flows back and forth between the cathodes, these electrons interact with the mercury-argon gas mixture. This creates ultraviolet energy which excites the phosphor and causes it to emit light. By changing the types of phosphors the lamp color and output can be controlled. Like all arc discharge lamps the fluorescent lamp requires a ballast in its circuit (Figure 23.6). The ballast is basically a coil. Its purpose is to limit the current in the circuit. Without the ballast the lamp would draw excessive current, and the fuse or circuit breaker would open.

The fluorescent lamp is in extremely common use, second only to the incandescent lamp. However, like the incandescent lamp, the fluorescent comes in literally hundreds of sizes, wattages, colors, voltages, and specific application designs (Figure 23.7).

Classes of Fluorescent Lamps

1. *Preheat Lamps.* A fluorescent lamp that requires a starting circuit for preheating the cathode. This lamp also requires a ballast and a starter or

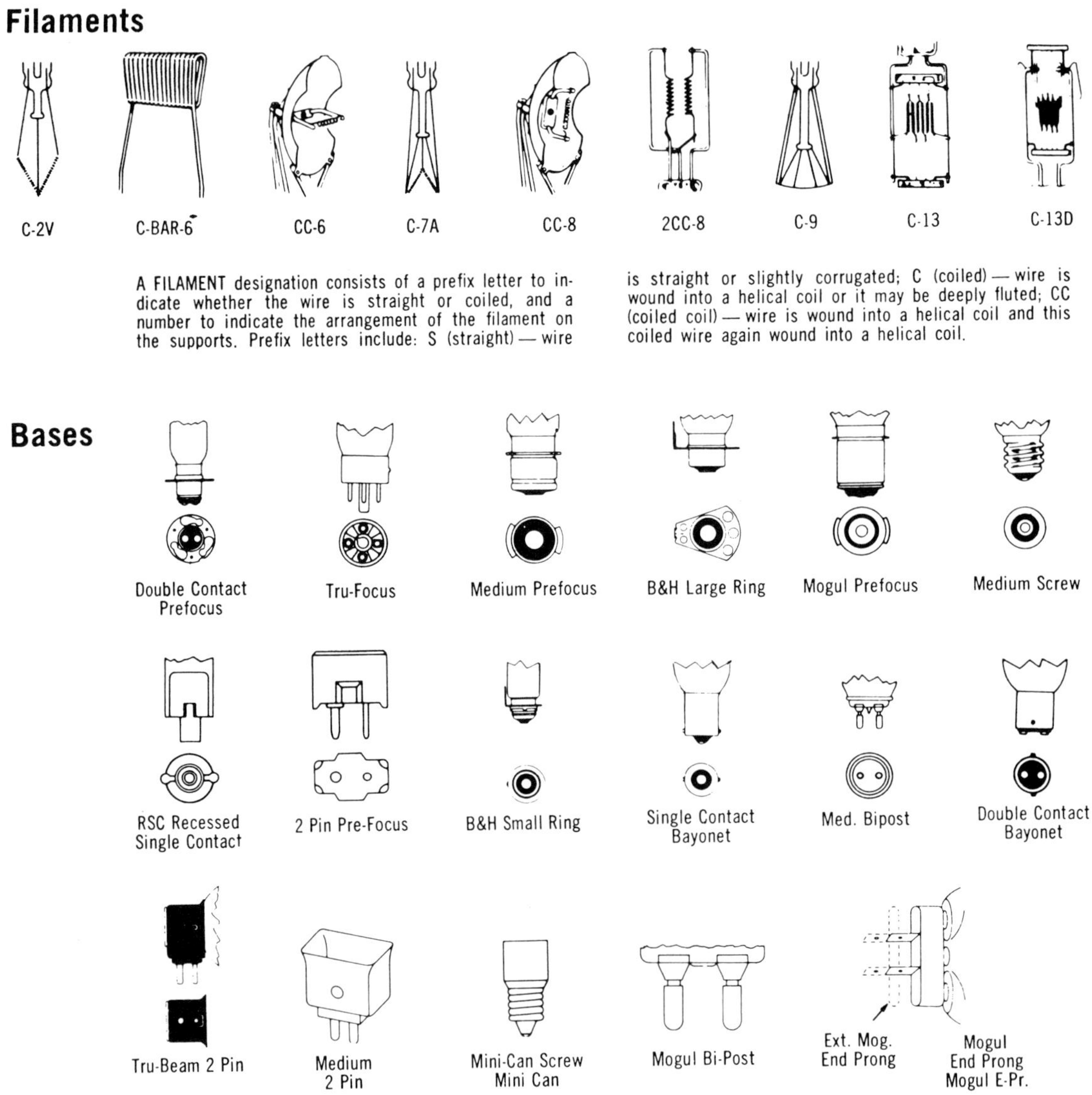

Figure 23.7 Typical fluorescent lamp bases.

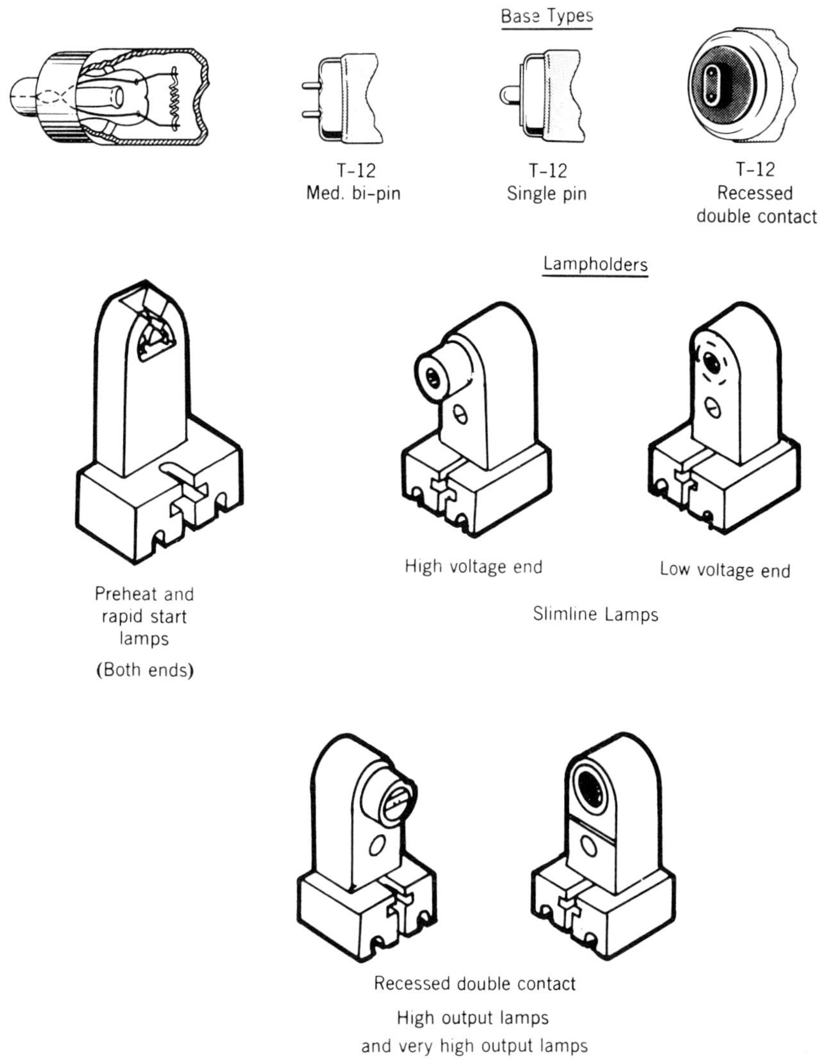

manual starting switch. The preheating process takes a few seconds. All preheat lamps have bipin bases. In large measure preheat lamps have been suspended by the rapid-start and instant-start types.

2. *Instant Start (or Slimline) Lamps.* A fluorescent lamp that does not require an auxiliary starting circuit. A ballast is used, which supplies a very high voltage to strike the lamp. Since the cathodes are started cold, only a single pin on each end of the lamp is required.
3. *Rapid Start Lamps.* A fluorescent that combines the best of the preheat and instant start types. Preheating is supplied by a built-in cathode heating coil in the ballast. No starter is used and the lamp lights almost as quickly as the instant start lamps.
4. *Preheat/Rapid Start Lamps.* A dual purpose fluorescent lamp that may be used in either preheat or rapid start fixtures.
5. *Special Fluorescent Lamp Types.* In addition to the standard tubular fluorescent lamp there are U-shaped lamps, square panel lamps, and tubular types with built-in reflectors and apertures. All these types plus newer developments and their technical data can be found in any current manufacturer's catalog.

23-5 FLUORESCENT DIMMING CONTROL

Fluor-A-Dim is Jefferson's Electric complete system for dimming 40-watt rapid start lamps from full output to an immeasurably low level, with no flicker.

Jefferson's system works through its Control-a-Potentiometer, which controls the auxiliary, which in turn limits the flow of current to the ballast. This then controls each lamp independently. Jefferson circuitry eliminates any interaction between lamps. A defective lamp will not affect the operation of the other lamps.

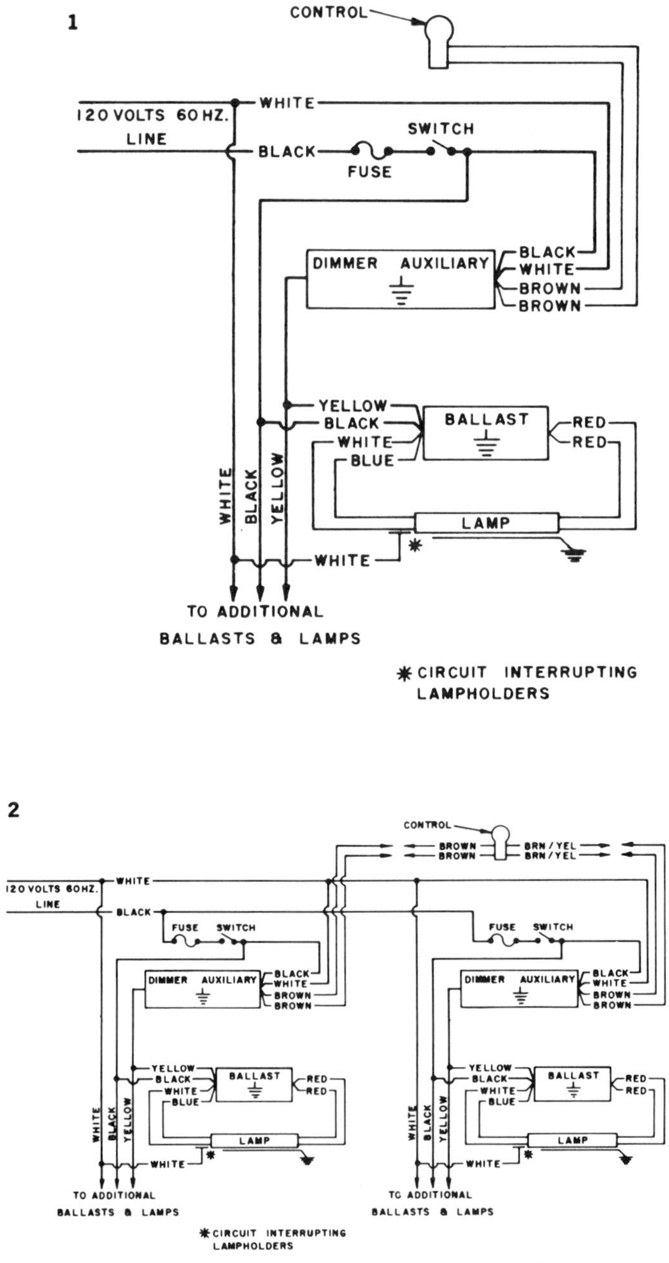

Figure 23.8 Wiring diagram for fluorescent dimming control. (1) Shows single auxiliary 120-volt system. (2) Dual auxiliary 120-volt system. (Courtesy of Jefferson Electric)

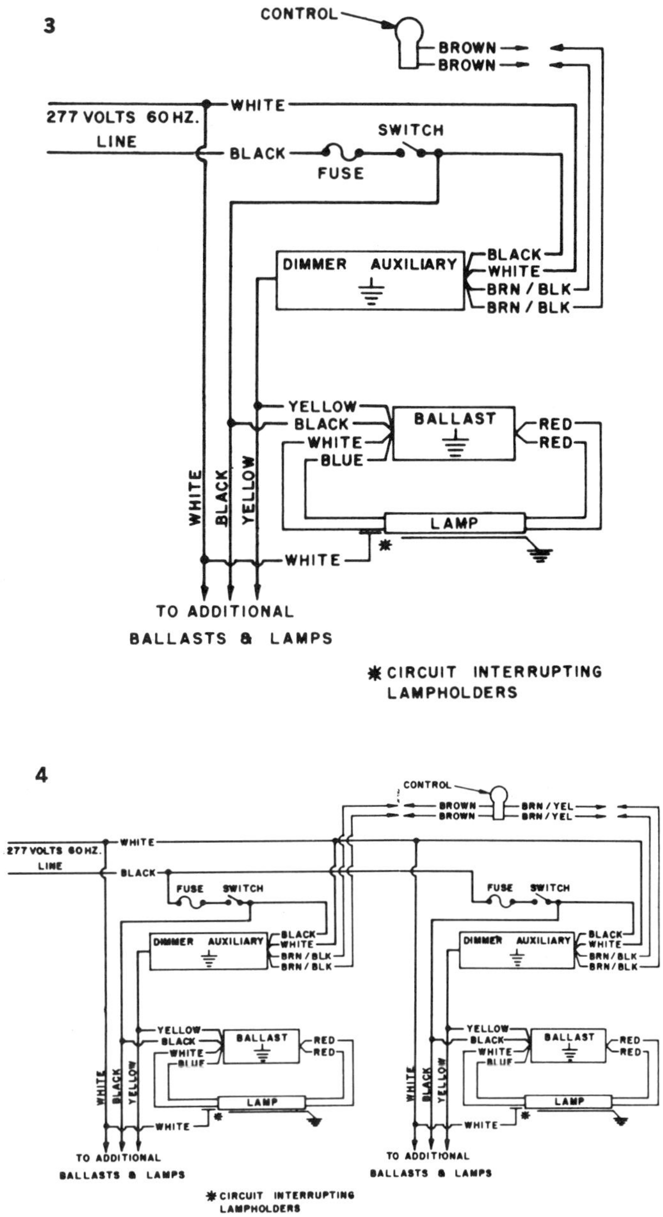

Figure 23.9 Wiring diagram for fluorescent dimming control. (3) Shows single auxiliary 277-volt system. (4) Dual system using 277-volts. (Courtesy of Jefferson Electric)

The system consists of a small control, an auxiliary unit, and special "dimming" ballast. One ballast is needed for each lamp (Figure 23.8). One control and one auxiliary, working as a team, can handle from 4 to 20 lamps on a 120-volt circuit. The 277-volt system, operating in the same manner, will dim from 10 to 50 lamps (Figure 23.9).

A special two-in-one control can be furnished for either system, 120 or 277 volts. This provides for completely independent operational control of two auxiliaries and their respective ballasts and lamps from one location.

Dimmer Installation

The potentiometer and its knob will fit into a standard wall box. The special dimming ballasts (one for each lamp) are mounted in each fixture. While the auxiliary can be remotely mounted, chances are it will be installed in one of the fixtures.

Because of the unique circuitry of the Fluor-A-Dim system, the components cannot be integrated or interchanged with other dimming systems.

High-Intensity Discharge Lamps

The term high-intensity discharge (HID) describes a wide variety of lamps. However, all HID lamps are alike in one way: They produce light from a gaseous discharge arc tube and operate at a pressure above that found in regular fluorescent lamps.

HID lamps can be divided into three major categories: the mercury vapor lamps, the metal halide lamp, and the high-pressure sodium lamp.

Mercury Vapor Lamps

This lamp combines the compact focusable shape of an incandescent with the arc discharge character of the fluorescent. The result is an efficient, long-life source with many applications. Figure 23.10 shows the construction of a mercury lamp.

A mercury differs from a fluorescent in that it is a high-pressure mercury vapor lamp. The fluorescent is a low-pressure lamp.

The operation of the mercury lamp begins with the arc tube. There is a starting electrode beside the main electrode at one end of the tube. Increasing starting voltage from the ballast strikes an arc between these two electrodes, using the argon gas as a current path. The argon ionizes (breaks up) and spreads rapidly throughout the tube. As the ionized argon reaches the main electrode at the opposite end of the tube, the light-producing arc is formed.

A small starting resistor limits the arc current during starting. As a direct result of its high resistance, current flow quickly shifts to the lower-resist-

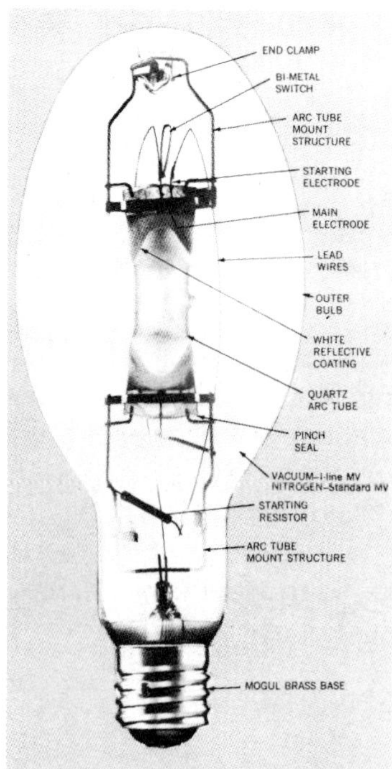

Figure 23.10 Construction of a clear mercury lamp. (Courtesy of General Electric, Lamp Division)

ance, main arc stream. Once the main arc has been established, current flow along its path is approximately 1000 times greater than the flow between the starting electrodes.

Heat from the main arc continues to vaporize the mercury in the arc tube for several minutes after starting. As more and more mercury is vaporized, current flow increases. Only the stabilizing, current-limiting feature of the ballast prevents the lamp from destroying itself.

Light energy is produced by the ionized argon gas particles colliding with the mercury atoms. As electrons in the mercury atom are jarred out of orbit and replaced by electrons from a nearby atom, radiation is given off. The color of light produced depends on which ring of the orbiting electrons has been hit by the colliding particles of argon. High pressures in the arc tube are responsible for deeper penetration than in the fluorescent tube. This results in more visible and less ultraviolet light energy.

If the power supply to a mercury vapor lamp is cut off, the arc will extinguish and not restart for several minutes. This is because sufficient pressure will have built up in the tube during operation, and it will prevent the arc from re-establishing itself immediately. Once the tube has cooled and the pressure lowered, the arc will restart automatically.

Mercury vapor lamps are now in wide use in many schools, stores, banks, and institutions. The advantage over the fluorescents is the shape of the lamp, which permits the use of a small fixture.

The Metal Halide Lamps

The long life, small size, and high output of the mercury lamp led to research to overcome its color problem. The problem was to find metals that could be easily vaporized and still remain chemically stable. This was finally done by adding metals in the form of their salts to the basic mercury arc tube. The result is a metal halide lamp that can produce 50 percent more lumens per watt than a mercury lamp with greatly improved color, making it applicable to almost all indoor use. Its life, however, is shorter. A brief comparison shows the following.

	Mercury	Metal Halide
Color	Poor to good	Good to excellent
Life	20,000 + hours	10,000 + hours
Average efficiency	50 lumens per watt ±	70 lumens per watt ±

Sodium Lamps

The most recent development in high-intensity discharge lamps (HID) is the high-pressure sodium lamp. It is quite different from other HID lamps. It is much simpler in design, because of the tremendous research and development that went into the production of materials used in its construction. See Figure 23.11. The high-pressure sodium lamp is regarded as the most efficient source of white light artificially produced. The 150-, 250-, 400-, and 1000-watt lamps put out approximately 50 percent more light than either mercury vapor or metal halide lamps of the same wattage ratings.

One reason for the success of the high-pressure sodium lamp is the ceramic material used for the arc tube. The ceramic is made of translucent

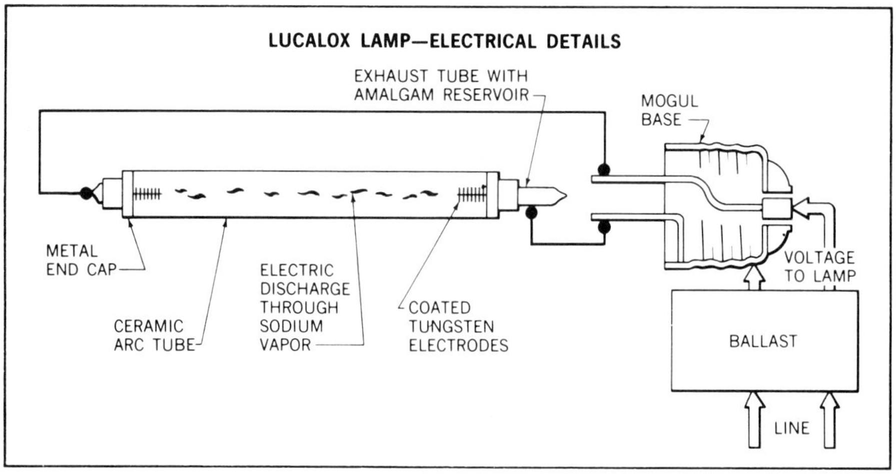

Figure 23.11 Construction details and typical data for high-pressure sodium (HPS) lamp. (Courtesy of General Electric, Lamp Division)

aluminum oxide, developed specially for this lamp by General Electric. The lamps are marketed by General Electric under the trade name "LUCALOX."

These lamps reach full brilliancy in a shorter period of time than mercury or metal halide lamps. Also, they may be restarted without waiting for the tube pressure to drop.

Lucalox lamps are made in two different models for base-up or base-down operation. Each can operate in a near horizontal position. The reason for this is that the excess sodium mixture collects at the coolest point in the arc tube. A special reservoir is fitted to the arc tube at the coolest end of the lamp to collect it. In base-high lamps, the reservoir is placed at the end farthest from the base. In base-down units, the reservoir is placed near the base of the lamp.

High-pressure sodium lamps are used outdoors for parking lots, street lighting, and building floodlighting. Commercial and industrial areas with high ceilings also use them because of their tremendous light output.

23-6 TROUBLE-SHOOTING HINTS

Lamp Starting Problems

Occasionally a field problem will arise involving improper lamp starting within a rapid start fixture. This is most likely with 40-watt, rapid start fixtures. The problem generally is that the lamps start very slowly or not at all. Six common causes are:

1. Low-line voltage.
2. Improper fixture grounding.
3. Insufficient or no filament voltage.
4. Insufficient or no open circuit voltage.
5. Dirty lamps during high humidity operating conditions.
6. Lamps improperly inserted in sockets.

If lamp starting is a problem in your installation, check the fixture grounding, filament voltage (3.4–3.9 volts), and open circuit voltage. If all are normal, the problem, in all probability, is dirty lamps. The lamps should be washed in clean water, drip-dried, and reinstalled.

Short Lamp Life

If the lamp has not given proper length of service as specified by the lamp manufacturer, the following reasons for early failure have to be considered:

1. Improper starting due to insufficient filament voltage.
2. Frequent starting and short operating periods.
3. Improper ballast.
4. Improper voltage supply.
5. Faulty wiring.
6. Defective lamp.
7. Lamps improperly inserted in sockets.

Early lamp failure will be preceded by a dense blackening on either, or both, ends of the lamp. This blackening will extend 2 or 3 inches from the lamp base, and should not be confused with a small dense black spot, which is a mercury deposit and can occur any time during lamp life. Also, it should not be confused with gray bands about 2 inches from either end of lamp, which occur toward the end of normal lamp life.

Lamp Cycling on and off

Ballasts with automatic resetting protectors, such as the Jefferson 600 or "Class P" ballasts - 800, will be removed from the line whenever the ballast exceeds the trip temperature of the protector. The ballast will then cool down and the protector will reset and turn the lights back on. Cycling can be caused by:

1. Line voltage or room temperature too high.
2. Fixture misapplied.
3. Wrong lamps.
4. Short or ground in the fixture.

For some reason the ballast is too hot due to temporary conditions such as nonoperating air conditioning, or high-line voltage. The conditions may be more permanent. For example, a fixture designed to be suspended has been surface mounted, or a recessed fixture has insulation placed on top. To correct, the conditions that are causing the ballast to overheat must be changed so that the ballast runs cooler.

23-7 SOUND LEVEL CONTROL

All electromagnetic devices, such as ballasts, produce sound. However, ballast sound will vary with design, workmanship, and size. Each ballast design has a letter sound rating from A to E, depending upon its decibel rating. A ballast will be considered defective only if it is operating above the sound level stated on the ballast. Sound problems should be solved before the installation is made. The sound chart and other information given in this chapter will help you select the proper ballast for the installation. Ballast sound rating, number of units, fixture design, acoustical treatment, ambient noise level, and location and type of building must be considered for each installation.

Sound Controlled Ballasts

Since a ballast, by its very nature, is an electromagnetic device, operating from an alternating current source, it produces an inherent sound, commonly referred to as "hum." Depending upon the type of ballast involved, the hum varies from an inaudible sound to an audible sound.

While good engineering reduces the sound level—and superior engineering minimizes it—the fact remains that the fluorescent ballast produces sound.

Another recognized fact is that, no matter how superior ballast design may be, the ballast is dependent upon proper application. Therefore, it is most important that, in specifying a particular ballast, several other factors be given very careful consideration, such as: the nature of the installation; the type of fixture; the room acoustics; the sound rating of the ballast.

Characteristics of the Room

1. *Type of Room.* A factory or a busy office, which has a high ambient sound level, does not require the same precautions as a television studio or a church building. Obviously, the elimination of fixture noise is not a requirement in a factory.
2. *Use of the Room.* A school room or an office installation having high ambient sound levels during the day may be quiet in the evening. For this reason, fixture noise that is inaudible during normal room usage may be quite audible during the quiet period. As a result, the ballast-fixture combination that is selected on the basis of average ambient sound level may be the source of a complaint if judged during the abnormally quiet period. Always select the ballast with the best sound rating available for the application. This will help to eliminate complaints that originate during quiet periods.
3. *Acoustics of Room.* Rooms that have sound-absorbent ceilings and perhaps carpeting, drapes, and furniture will help to deaden any ballast hum because sound will be absorbed rather than reflected. It should be remembered that such rooms may have a lower ambient noise level than the typical values shown in the sound control chart (Figure 23.12).

Characteristics of the Fixture

1. *Construction.* Construction of the fixture should be such that it does not amplify sound generated by the ballast. Quiet fixtures are rigid because they are made of heavy gauge metal.
2. *Fittings and Accessories.* Channels, louvers, strips, shields, and reflectors should not transmit or set up sympathetic vibration. These parts should be tightly secured.

HOW TO USE THE SOUND CONTROL CHART

(Chart is Based on Acoustical Ceilings and Average Lighting Levels)

EXAMPLE I: For specifying the ballasts to be used in a "Quiet Office", refer to the Sound Control Chart at right. On the chart, a quiet office is indicated as having an ambient noise level of 20 to 24 decibels. Recommended for this application are ballasts rated A.

EXAMPLE II: For "General Office Areas" having ambient noise levels of 31—36 decibels, a C sound rated ballast is recommended. This will be satisfactory during normal room and average lighting levels. This room will sound noisy during quiet periods and, if this is of concern, then a quieter ballast should be used. A quieter ballast should also be used for higher lighting levels and lack of acoustical ceiling.

TYPICAL APPLICATIONS	TYPICAL AVERAGE NOISE AMBIENTS IN DECIBELS*	SOUND RATING#
STUDY HALL RESIDENCE RADIO— TV BROADCAST STUDIO QUIET OFFICE CHURCH	20 to 24	A
LIBRARY PROFESSIONAL OFFICE EVENING SCHOOL CLASSROOM	25 to 30	B
COMMERCIAL BUILDINGS GENERAL OFFICE AREAS STOCK ROOMS	31 to 36	C
RETAIL STORES NOISY OFFICES	37 to 42	D
LIGHT MANUFACTURING OUTDOOR LIGHTING	43 to 48	E
STREET LIGHTING FACTORY	49 and up	F

FOOTNOTES: *These ambients are typical at the time of normal room usage. Measure ambients with a standard sound meter and 40 db weighing network.
#Where there is a choice of more than one ballast, it is recommended that the quietest one for the application be used.

Figure 23.12 Sound control chart. (Courtesy of Jefferson Electric)

3. *Mounting.* Fixture mounting should properly support the fixture, but should not transmit fixture vibration to the supporting structure. It may be necessary to isolate the fixture with sound absorbent materials. Each ballast should be solidly mounted in the fixture with good ballast-to-fixture contact since this will not only reduce noise but also help to conduct normal operating heat away from the ballast.
4. *Number of Fixtures.* Remember that sources of ballast hum increase as the number of fixtures in a room increases. So the greater the number of fixtures, the more urgent the need for ballasts with superior acoustical ratings.

SUMMARY

1. Mercury vapor lamps produce approximately twice as many lumens per watt as the incandescent type.
2. Mercury vapor lamps operate at a low power factors.
3. The footcandle is an important unit of light.
4. In solving a general lighting problem, six key steps should be taken.
5. The coefficient of utilization is the ratio of the lumens matching the working plane to the total lumens generated by the lamps.
6. A good maintenance factor is considered when the luminaires are cleaned frequently and lamps are replaced systematically.
7. As a general rule, fixtures in office-type rooms are spaced around a 10-foot maximum area.
8. Advantages of incandescent lamps are low cost, instant starting, and cheap dimming.

PROBLEMS

23-1 A classroom 22 × 25 feet is to be lighted to an average maintained footcandle level of 75 fc. Find the number of four-lamp 48-inch fluorescent fixtures required. Assume $CU = 0.38$, $MF = 0.75$.

23-2 How much would it cost to operate two dozen 425-watt high-intensity mercury vapor lamps for a 16-hour working day in a plant where electrical energy costs 2½ cents per kilowatthour?

23-3 Upon what factor does the coefficient of utilization depend?

23-4 In the operation of any light system there are three principal variables that affect the amount of light obtained from the system. Name the three variables.

1.
2.
3.

23-5 Maintenance factors covering lamps and luminaires are suggested for three conditions, poor, medium, and good maintenance. Briefly describe a good maintenance factor.

23-6 A 25-watt incandescent lamp has an efficiency of 10.8 lumens per watt. Determine its light output.

23-7 A 100-watt incandescent lamp yields about 600 lumens, 600 of which fall on a table 5 by 3 feet. What is the average illumination on this table?

23-8 In commercial buildings where the space is "high bay" (high ceilings) the trend is to use ______________ ______________ lighting.

23-9 What is a good maintenance factor? ______________________________

____

____

23-10 In general, fixtures in office-type rooms are spaced around _____ foot maximum area.

chapter 24

EMERGENCY POWER SYSTEMS

Instructional Objectives

1. To understand why standby sources of electrical power are installed.
2. To learn some reason why the National Electrical Code applies to a standby plant.
3. To become familiar with the installation of emergency systems.
4. To list the emergency supply sources.
5. To understand why a transfer switch is required on a standby system.
6. To learn why OSHA has required the installation of EXIT lights and emergency illumination.

Self-Evaluation Questions

Test your prior knowledge of the information in this chapter by answering the following questions. Watch for the answers as you read the chapter. Your final evaluation of whether you understand the material is measured by your ability to answer the questions. When you have completed the chapter, return to this section and answer the questions again.

1. Where are standby sources of electrical power installed?
2. When is a standby source required?
3. Does the NEC require the installation of a standby source?
4. Should a standby source be installed, what does the Code require? List three things required.
5. Where are emergency systems generally installed?
6. What are three sources of supply for an emergency system?
7. How should unit equipment be installed?
8. Before selecting a generator set for a source of emergency power, what consideration must be given?
9. Who usually supervises the installation of a transfer switch? Why?
10. What is the function of an automatic transfer switch?

11. Where do you find the requirements for the installation of EXIT lights?
12. Does OSHA require EXITS to be marked? If yes, how?
13. Is it permissible to supply EXIT or emergency lighting from more than one source? Explain.

24-1 STANDBY POWER GENERATION SYSTEM

Standby sources of electrical power are generally installed in various types of commercial buildings to provide an alternate source of electrical energy to serve in the event of a failure of the normal source of electric service. The cause may be nothing more than a mechanical breakdown. However, power outages are often accompanied by other dangers to our physical well-being. Storms, hurricanes, floods, winds and explosions not only leave downed power lines but a trail of property damage and personal injury as well.

It is during these critical periods that we need the vital public services of our hospitals, police and fire departments, radio-TV stations, and other public agencies. These services are assured if a reliable source of standby power is available to take over in an emergency.

Standby electric power systems are becoming increasingly essential for all public buildings and dwelling units, since each year more of our people depend more and more upon electrical energy.

There is no National Electrical Code requirement that a standby plant must be installed. However, if local, state, or federal law requires that such a plant must be installed, then the requirements of NEC Article 701 apply.

NEC 701-11. Legally required standby power generation systems. (a) Shall be equipped with suitable means for automatically starting the generator set upon failure of the normal electrical service and for automatic transfer and operation of all required electrical functions at full power within 60 seconds of the power failure.

(b) The plant must be provided with an on-premise fuel supply and must be sufficient for at least two hours of operation at full demand.

(c) The plant must be tested at regular intervals.

NEC 701-10. Wiring of the standby systems may occupy the same raceways, boxes, cabinets, and panelboards with other wiring. However, it must be remembered that standby wiring shall not occupy the same raceways, boxes, or cabinets as wiring for emergency systems.

24-2 EMERGENCY SYSTEMS

The provisions of NEC Article 700 apply to the installation, operation, and maintenance of circuit systems and equipment intended to supply illumination and power in the event of failure of the normal supply or in the event of accident to elements of the system supplying power and illumination essential for safety of life and property where such systems are legally required and classed as emergency by municipal, state, federal, or other codes, or by any governmental agency having jurisdiction.

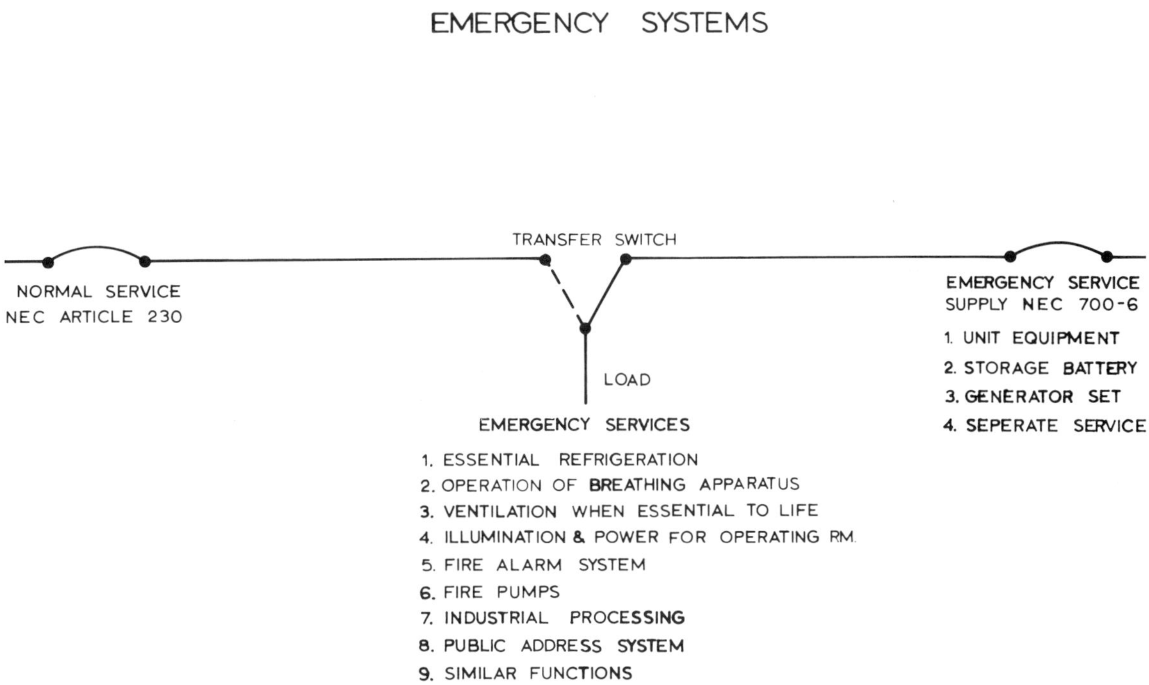

Figure 24.1 Emergency system. Shows normal service to the building, automatic transfer switch and emergency service. Current supply shall be such that in event of failure of the normal supply to or within the building, emergency lighting, or emergency power, or both will be immediately available. (Albert Lea Voc Tech)

24-3 INSTALLATION

Emergency systems are generally installed in places of assembly where artificial illumination is required, such as buildings subject to occupancy by large numbers of persons—hotels, motels, theaters, sports arenas, hospitals, nursing homes, and similar institutions.

Emergency systems may provide power for such systems as essential refrigeration, operation of mechanical breathing apparatus, ventilation when essential to maintain life, illumination and power for hospital operating rooms, fire alarm systems, fire pumps, industrial processing where current interruption would produce serious hazards, public address systems, and similar functions. See Figure 24.1.

24-4 SOURCES OF POWER

The emergency supply systems may be:

1. *Individual Unit Equipment for Emergency Illumination.* The unit may consist of one or more lamps on the unit or supply terminals for remote lamp location. See Figure 24.2. The unit operates by a relaying device so arranged to energize the lamps automatically upon failure of the supply to the unit equipment. Unit equipment shall be permanently fixed in place (i.e., not portable) and shall have all wiring to each unit installed in accordance with the requirements of any of the wiring methods in Chapter 3 of the NEC.

Figure 24.2 (*a*) Single emergency lighting unit. (Courtesy of Teledyne Big Beam)

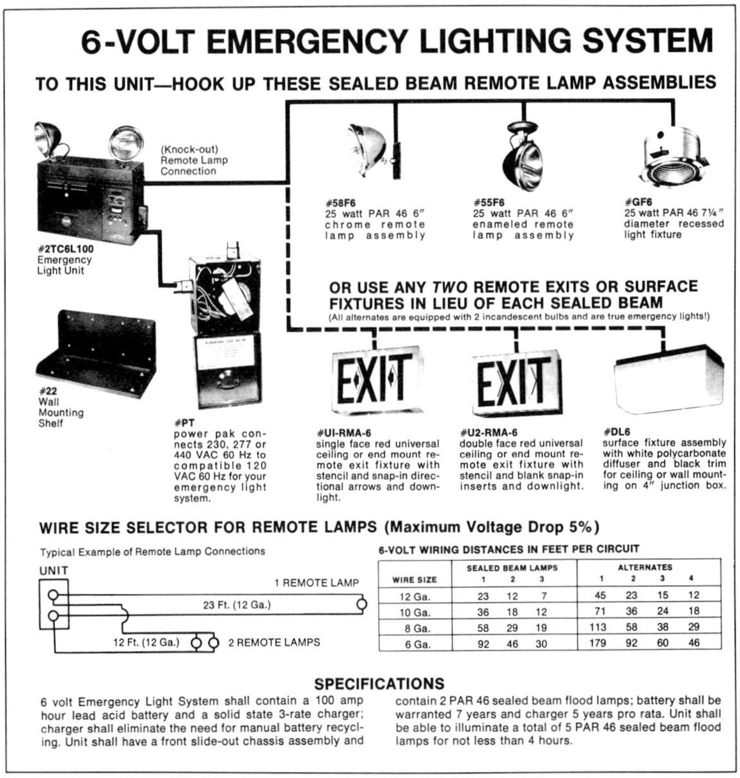

WIRE SIZE	SEALED BEAM LAMPS 1	2	3	ALTERNATES 1	2	3	4
12 Ga.	23	12	7	45	23	15	12
10 Ga.	36	18	12	71	36	24	18
8 Ga.	58	29	19	113	58	38	29
6 Ga.	92	46	30	179	92	60	46

(*b*)

Figure 24.2 (*b*) Shows 6-volt emergency system with typical example of remote lamp connections. (Courtesy of Teledyne Big Beam)

2. *Generator Set.* The most common generating plants are driven by an internal combustion engine as the prime mover. However, before the engine can be selected, careful consideration must be given to selecting the proper type of fuel. Fuel options are gasoline, LP gas, natural gas, and diesel. There are specific advantages and disadvantages of each, however. Several general points that may outrule all others should be considered.
   (a) Availability of the various fuels in a particular location.
   (b) Local regulations governing the storage of gasoline and gaseous

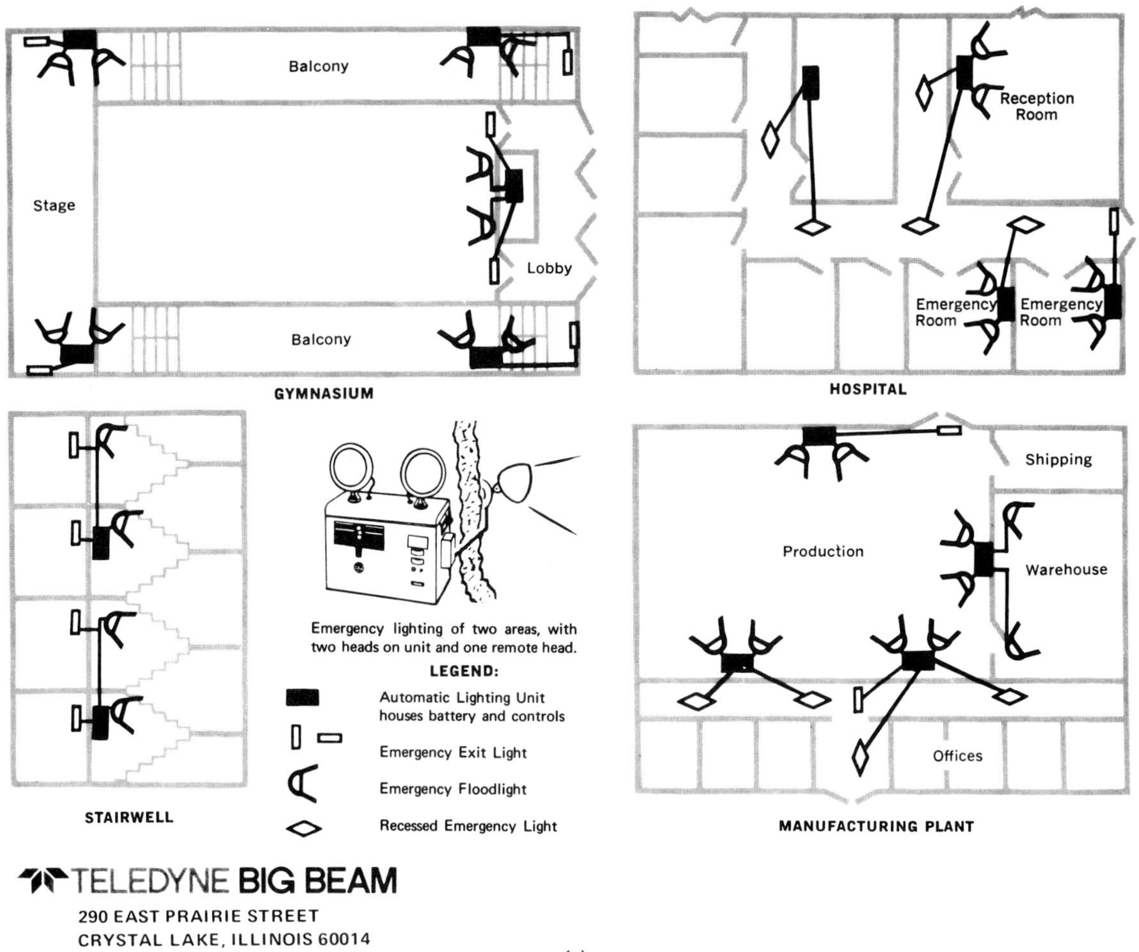

(c)

Figure 24.2 (*c*) Emergency illumination, shows location of individual unit equipment. (Courtesy of Teledyne Big Beam)

fuels. Enough on-site fuel supply shall be provided to operate the prime mover at full demand load for at least 2 hours, if the prime mover is solely dependent upon a public utility natural gas system for its supply. Means shall be provided for automatically transferring from one fuel supply to another where dual fuel supplies are used. Figure 24.3 shows a properly installed plant.

3. *Storage Batttery.* Storage batteries may be used for emergency lighting and emergency power if they are capable of supplying and maintaining the total load of the circuits supply at not less than 87½ percent of the system voltage for a period of not less than 1½ hours. However, batteries shall be designed and constructed to meet the requirements of the emergency service. Batteries must be compatible with the charger for the installation.

24-5 LOAD TRANSFER SWITCH

Electric utilities require load transfer switches and usually supervise their installation, since power company lines on which utility personnel work are affected.

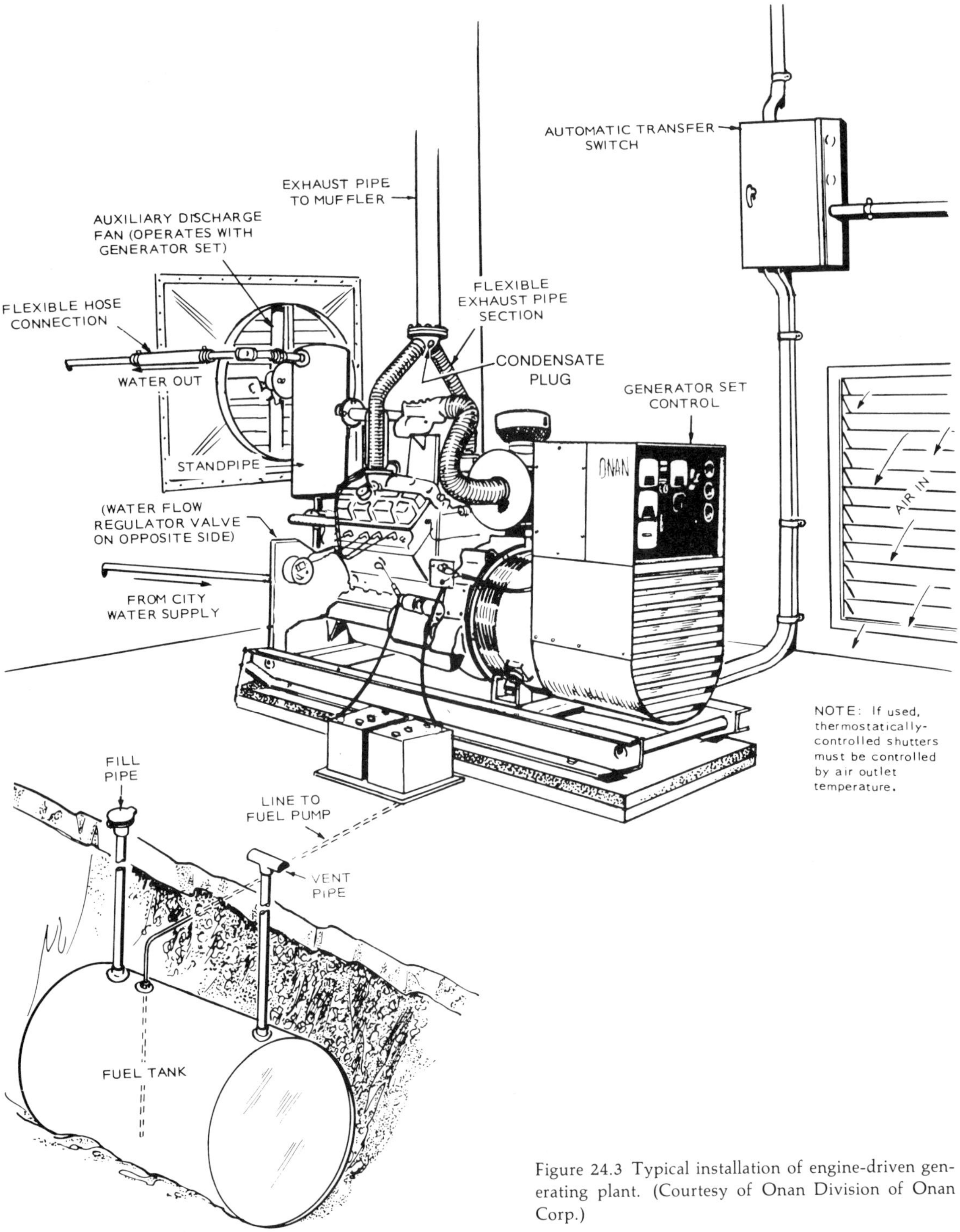

Figure 24.3 Typical installation of engine-driven generating plant. (Courtesy of Onan Division of Onan Corp.)

There are two types of transfer switches—manual and automatic. The manual transfer switch must be operated by the operator on duty either at the plant or a remote station. The plant is started by manual switch, and the load is transferred from one source to another by a hand-operated double-throw switch. See Figure 24.4.

An automatic transfer switch starts the plant and transfers the load automatically, not requiring the attention of an operator. With automatic transfer switches the power outage can be limited to less than 10 seconds.

Figure 24.4 Automatic transfer switch. (Courtesy of Onan Division of Onan Corp.)

An automatic load transfer switch is a necessity where uninterrupted power is of prime importance, such as in a hospital, where health or safety is at stake.

24-6 EXIT SIGNS

Every apprentice electrician and contractor should give full attention to OSHA regulations on exit signs. All of OSHA rules are presented in Subpart E of the Occupational Safety and Health Standards Part 1910, Federal Register page 23531. These requirements on location and lighting of exit signs apply to all places of employment in all new buildings and also in all existing buildings.

Prior to OSHA, there were no universal requirements that all commercial buildings or all places of employment have exit signs. The NEC does not require them. And although the NFPA Life Safety Code does cover rules on emergency lighting and exit signs, that code was only enforced where state and local government bodies required it.

Section 1910.35 of OSHA makes clear that every building must have a means of egress—a continuous, unobstructed way for occupants to get out of a building in case of fire or other emergency. Exits from all parts of a building or structure must be provided at all times the building is occupied.

The law says:

> Every exit shall be clearly visible or the route to reach it shall be conspicuously indicated in such a manner that every occupant of every building or structure who is physically and mentally capable will readily know the direction of escape from any point, and each path of escape, in its entirety, shall be so arranged or marked that the way to a place of safety outside is unmistakable.

Then it says:

Exit Markings

1. Exits shall be marked by a readily visible sign. Access to exits shall be marked by readily visible signs in all cases where the exit or way to reach it is not immediately visible to the occupants.

2. Any door, passage, or stairway which is neither an exit nor a way of exit access, and which is so located or arranged as to be likely to be mistaken for an exit, shall be identified by a sign reading "Not An Exit" or similar designation, or shall be identified by a sign indicating its actual use, such as "To Basement," "Storeroom," "Linen Closet," etc.
3. Every required sign designating an exit or way of exit access shall be so located and of such size, color, and design as to be readily visible.

It is well for the apprentice to note that marking must be supplied for the way to the exit as well as for the exit door itself.

4. Sign reading "EXIT," or similar designation, with an arrow indicating the direction, shall be placed in every location where the direction of travel to reach the nearest exit is not immediately apparent.
5. Every exit sign shall be suitably illuminated by a reliable light source giving a value of not less than 5 footcandles on the illuminated surface.

From the above, note that an exit sign does not have to be internally illuminated, although it may be. And this lighting is required on "every exit sign" which means exit signs over doors and exit signs indicating the direction of travel. See Figure 24.5.

Because the OSHA regulations on exit signs do not require emergency power for lighting of such signs, the light units that illuminate exit signs may be supplied from regular (nonemergency) circuits. It is not necessary to supply such circuits from a tap ahead of the service main or from batteries or an emergency generator.

As far as OSHA is concerned, Article 700 of NEC on emergency systems does not apply to circuits for exit sign lighting.

Installation of Exit Signs When exit signs are installed by the electrician, they shall be installed according to Article 700 of NEC.

[NEC 700-18 states "Current supply shall be such that in the event of failure of normal supply to or within the building, emergency lighting, emergency power, or both, will be immediately available."]

Figure 24.5 Exit signs. (*a*) Wall mounted. (*b*) Ceiling mounted. (Courtesy of Progress Lighting) (*c*) Location of signs. (Courtesy of Lightcraft of California)

(c)

Consideration shall be given to the type of service to be rendered, whether of short time duration, as for EXIT lights of a theater, or of long duration, as for supplying emergency power and lighting due to a long period of current failure from trouble either inside or outside the building as in the case of a hospital.

Loads on Emergency Branch Circuits NEC 700-15. No appliance and no lamps, other than those specified as required for emergency use shall be supplied by the emergency lighting circuits.

Emergency Illumination NEC 700-16. Emergency illumination shall include all required EXIT lights and all other lights specified as necessary to provide sufficient illumination. Emergency lighting systems shall be so designed and installed that the failure of any individual element, such as the burning out of a light bulb, cannot leave any space in total darkness.

Circuits for Emergency Lighting NEC 700-18. Branch circuits intended to supply emergency lighting shall be so installed as to provide service immediately when the normal supply for lighting is interrupted. Such installations shall be provided by either one of the following:

1. An emergency lighting supply.
2. Two or more separate and complete systems with independent power supply, each system providing sufficient current for emergency lighting purposes.

This is the only location where the two lighting systems would be permitted to be together in the same enclosure. In this way no space would be left in the dark or be without an EXIT light to show the way out of the building.

Independent Wiring NEC 700-9. Emergency circuit wiring shall be kept entirely independent of all other wiring and equipment. Emergency circuit wiring shall not enter the same raceway, cable, box, or cabinet with other wiring. However,

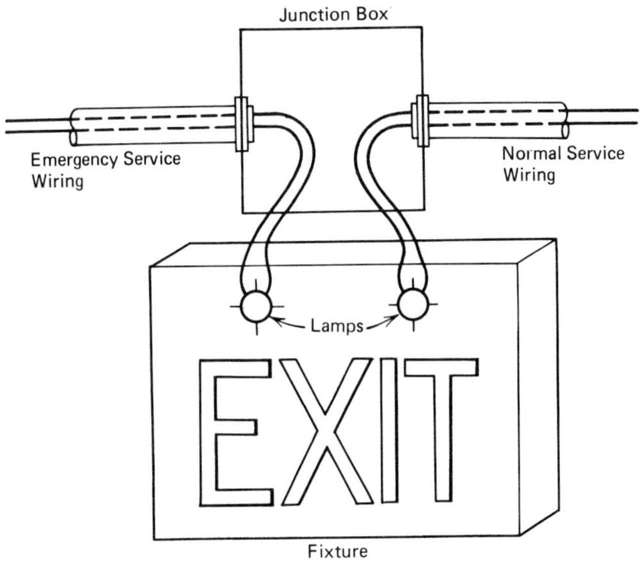

Figure 24.6 Wiring for exit or emergency lighting. Lighting fixtures supplied from two sources are permitted in the same fixture [NEC 700-17].

the exception states that EXIT and emergency lighting fixtures supplied from two sources shall be permitted within the same fixture. Normal wiring and emergency wiring shall be permitted in a common junction box attached to EXIT or emergency lighting fixtures supplied from two sources. [See Figure 24.6.]

SUMMARY

1. Standby sources of electrical power are generally installed in various types of commercial buildings to provide an alternate source of electrical energy.
2. The National Electrical Code does not require the installation of a standby plant.
3. Emergency systems are generally installed in places of assembly where artificial illumination is required.
4. Storage batteries may be used for emergency lighting and emergency power if they meet the requirements.
5. An automatic transfer switch starts the plant and transfers the load automatically.
6. When exit signs are installed by an electrician, they must be installed according to the National Electrical Code.
7. No appliances and no lamps, other than those specified as required for emergency use shall be supplied by the emergency lighting circuits.
8. Emergency circuits shall be kept entirely independent of all other wiring and equipment.

PROBLEMS

24-1 List six causes for a power failure.

1. ______________ 4. ______________
2. ______________ 5. ______________
3. ______________ 6. ______________

24-2 If state, federal, or local laws require the installation of a standby power generator system, what are at least five requirements of the National Electrical Code?

1. ______________ 4. ______________
2. ______________ 5. ______________
3. ______________

24-3 Emergency systems are generally installed in places of assembly where artificial illumination is required. List six types of commercial buildings where an emergency system would be installed.

1. ______________ 4. ______________
2. ______________ 5. ______________
3. ______________ 6. ______________

24-4 You are to install an internal combustion engine as a prime mover for a generating plant. You must make a decision as to the type of fuel used. What two general points should be considered?

1. ______________

2. ______________

24-5 Exit and emergency lighting fixtures may be supplied by two sources in the same fixture. Draw a wiring diagram showing an exit light being supplied by two sources.

24-6 When exit signs are installed by the electrician, they should follow the requirements of what Article of the NEC?

Article ______________________

24-7 Standby sources of electrical power are generally installed in various types of commercial buildngs to provide ______________________

____

____

24-8 Standby power wiring shall or shall not (circle one) occupy the same raceway as wiring for emergency systems.

24-9 Storage batteries may or may not (circle one) be used for emergency lighting.

24-10 Who should supervise the installation of the transfer switch?

____

____

____

____

PART 7

MOBILE HOME AND RECREATIONAL VEHICLE PARKS

chapter 25

MOBILE HOMES AND PARKS

Instructional Objectives

1. To identify a mobile home park.
2. To understand the need for feeder and service-entrance conductor demand factors.
3. To learn how to calculate a service for a mobile home park.
4. To become aware of the National Electrical Code as applied to mobile home parks.
5. To learn how to install service-entrance equipment.

Self-Evaluation Questions

Test your prior knowledge of the information in this chapter by answering the following questions. Watch for the answers as you read the chapter. Your final evaluation of whether you understand the material is measured by your ability to answer the questions. When you have completed the chapter, return to this section and answer the questions again.

1. What is a mobile home park?
2. What is the voltage of the service equipment for a mobile home park?
3. What demand factor is permitted when calculating a load on feeders and services?
4. Where is the service equipment for a mobile home installed?
5. Does the NEC apply to mobile homes?
6. Are receptacles located outside a mobile home protected by GFCI?
7. Is it necessary to bond the cabinet of the service equipment?
8. If a mobile home is served by a cord, what size breaker must be used for the receptacle?
9. Is it permissible to install the electric service on the side of a mobile home?
10. On what basis is a park electrical wiring system calculated?

25-1 MOBILE HOME DEFINED

The National Electrical Code in Article 550-2 defines a mobile home as "A factory assembled structure or structures equipped with the necessary service connections and made so as to be readily movable as a unit or units on their own running gear and designed to be used as a dwelling unit(s) without a permanent foundation." The phrase "without a permanent foundation" indicates that the support system is constructed with the intent that the mobile home will be moved from time to time at the convenience of the owner. When a mobile home is placed on a foundation, it indicates a permanent location.

The electrician will be much more concerned with the wiring of the mobile home park so little will be discussed here relative to the actual wiring of the factory constructed home.

25-2 MOBILE HOME PARK

A mobile home park is a parcel of land used for the accommodation of occupied mobile homes.

A mobile home lot is a space within a designated portion of a mobile home park for the accommodation of one mobile home and its accessory building or buildings, for the exclusive use of the occupant (Figure 25.1). An accessory building could be any one of the following: awning, cabana, carport, fence, storage cabinet, or porch established for the use of the occupant of the mobile home lot.

The park usually contains a park office for the owner or manager, a facility for laundry equipment, a recreation room, and many times a swimming pool.

25-3 PARK ELECTRICAL WIRING SYSTEM

The service equipment for the entire mobile home park must be single-phase, 115/230 volts. The details will vary with the size of the park. In a small park, the service equipment might be located in the park office or in the laundry room, with distribution to groups of individual lots. In larger parks, where the park service exceeds 240 volts, transformers and secondary distribution panelboards shall be treated as services, that is, each transformer serves a feeder for a number of homes. An example would be in a city, where one transformer serves a number of houses.

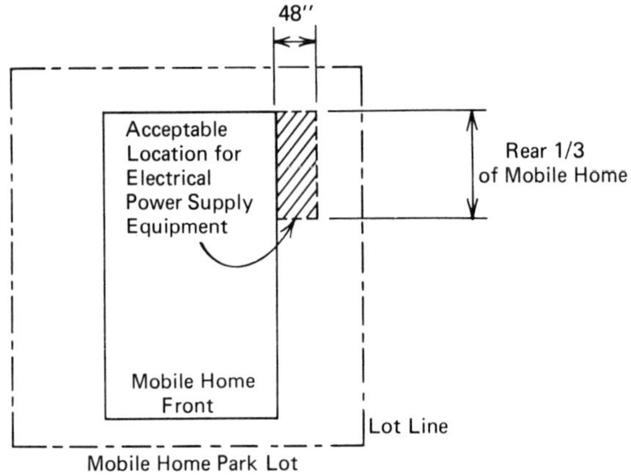

Figure 25.1 Mobile home lot, showing acceptable location of electrical power equipment.

TABLE 25.1 Demand Factors for Feeders and Service-Entrance Conductors (NEC Table 550-22)

Number of Mobile Home Lots	Percent Demand Factor
1	100
2	55
3	44
4	39
5	33
6	29
7–9	28
10–12	27
13–15	26
16–21	25
22–40	24
41–60	23
61 and over	22

25-4 CALCULATING THE LOAD

Park electrical wiring systems shall be calculated on the basis of not less than 16,000 watts at 110/230 volts per each mobile home service (NEC 550-22). The demand factors that are set forth in Table 550-22 of the NEC (reproduced here at Table 25.1), shall be permitted in calculating load on feeders and services.

Notice in Table 550-22 that the demand factor for one mobile home is 100 percent; however, the demand factor drops off rapidly as the number of homes increases. For example there are 20 lots calculated at a 25 percent demand factor, the minimum ampacity of the conductors in the feeder for the 20 homes would be:

$$\frac{(16000)\,(20)\,(0.25)}{230} = \frac{80,000}{230} = 347 \text{ amperes}$$

Similar calculations show that the minimum ampacity for a feeder for 10 lots would be:

$$\frac{(16000)\,(10)\,(0.27)}{230} = \frac{43200}{230} = 187 \text{ amperes}$$

It should be remembered that a somewhat larger service would be needed when considering the park office, laundry, or a park swimming pool.

25-5 MOBILE HOME SERVICE EQUIPMENT

Unlike a house on a permanent foundation, the service equipment for a mobile home cannot be installed in or on the mobile home. It is installed at a distance of not more than 30 feet from the point of entry of the feeder assembly into the home. An individual service equipment might contain only a 50-ampere three-pole, four-wire 125/250-volt receptacle into which the cord from the home is plugged. Because it is protected by a two-pole 50-ampere circuit breaker (Figure 25.2), it could contain more receptacles and

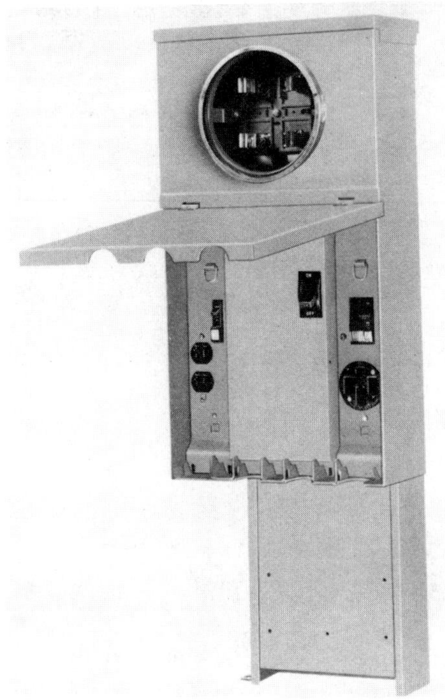

Figure 25.2 Mobile home meter pedestal for underground distribution of electrical power to individual mobile home site. (Courtesy of Square D Company)

Figure 25.3 Ground-fault circuit protection. (Courtesy of Pass & Seymour)

breakers because the NEC permits additional receptacles for connection of electrical equipment located outside the mobile home. All of these receptacles shall be protected by approved ground-fault circuit protection. See Figure 25.3.

Note that while the NEC states that the service equipment must be rated at a minimum of 100 amperes (NEC 550-23) (it must have terminals suitable for wire with at least 100 ampacity), any receptacle and the breaker protecting it shall not exceed a 50-ampere rating. However, larger breakers may be installed in the service if the home is not served by cords but by direct and permanent connection to terminals in the service equipment.

It must be kept in mind, that each service equipment must have a grounding breaker for the neutral wire of the cord, the grounded wire of the smaller receptacles, and the green grounding wire of the cord. The cabinet must be bonded and a ground wire run from the neutral bus to an underground water system or if not available to a driven ground rod.

SUMMARY

1. A mobile home park is a parcel of land used for the accommodation of occupied mobile homes.
2. The service equipment for the entire mobile home park must be single-phase, 115/230 volts.
3. The park electrical wiring system shall be calculated on the basis of not less than 16,000 watts at 110/230 volts per each mobile home service.
4. The service equipment for a mobile home cannot be installed in or on the mobile home.
5. All receptacles outside the mobile home must be protected by ground-fault circuit protection.

PROBLEMS

25-1 Draw a sketch of a mobile home park lot. Show the location of the mobile home on the lot, and mark the acceptable location for the electrical power supply equipment.

25-2 Compute the minimum ampacity for the service feeders of a mobile home park expecting to park 50 mobile homes. Show the demand factor.

25-3 A person has a small parcel of land. The plans are to develop the land and park three mobile homes. What demand factor would you use?

25-4 For the above question, calculate the service-entrance conductors. How many amperes?

25-5 For the following number of mobile homes parked in a park, list the percent demand factor.

14 homes ______________ percent

1 home ______________ percent

16 homes ______________ percent

20 homes ______________ percent

23 homes ______________ percent

50 homes ______________ percent

75 homes ______________ percent

100 homes ______________ percent

25-6 How does the NEC define a mobile home?

____

____

____

____

25-7 What does the phrase "without a permanent foundation" mean?

____

____

____

____

25-8 The service equipment for the entire mobile home park must be
__________, and __________ volts.

25-9 What is an accessory building? Name several. __________
____
____
____

25-10 In larger parks, where the park service exceeds 240 volts, __________ and __________ distribution panelboards shall be treated as service.

chapter 26

RECREATIONAL VEHICLE PARKS

Instructional Objectives

1. To become familiar with NEC relating to recreational vehicle parks.
2. To understand the need for using a demand factor when calculating service-entrance conductors.
3. To learn to lay out and draw a recreational vehicle park.
4. To learn how to calculate a service.
5. To become familiar with the vehicle park secondary electrical distribution system.

Self-Evaluation Questions

Test your prior knowledge of the information in this chapter by answering the following questions. Watch for the answers as you read the chapter. Your final evaluation of whether you understand the material is measured by your ability to answer the questions. When you have completed the chapter, return to this section and answer the questions again.

1. Are all recreational vehicles equipped for 115-volt service?
2. What percentage of all recreational vehicle sites supplied with electrical supply shall be equipped with a 30-ampere 125-volt receptacle?
3. Are ground-fault circuit-interrupter type receptacles required in recreational vehicle parks?
4. How many watts are figured per site, using only 20-ampere supply facilities?
5. What is the special class of service equipment called?
6. On what side of the vehicle is the service cord attached?
7. Who has the final authority, relative to inspection?
8. What is the permanent demand factor for a park having eleven vehicle sites?

26-1 RECREATIONAL VEHICLE

The National Electrical Code Article 551-2 defines a recreational vehicle as a unit that is self-propelled or is mounted on or pulled by another vehicle and is primarily designed as temporary living quarters for camping, travel, or recreational use. It is normally identified as a travel trailer, camping trailer, truck trailer, truck camper, or motor home.

The travel trailer is mounted on wheels, has a living area less than 220 sq ft (excluding wardrobes, closets, cabinets, kitchen units, fixtures, etc.), is of such size and weight so that a special highway use permit is not required, and is designed as temporary living quarters while camping or traveling.

NEC Article 551-13(e) Power-supply assembly shall be located within 15 feet of the rear, on the left (driver or road) side, or at the rear, left center of vehicle, within 18 inches of the outside wall.

NEC Article 551-6(a) Equipment and material indicated for connection to system rated 115 volts. Two-wire with ground, or 115–230 volts, three-wire with ground, shall be approved and installed as per 551, Part A.

26-2 RECREATIONAL VEHICLE PARK AND LOTS

A recreational vehicle park is defined as a plot of land upon which two or more vehicles sites are located, established, or maintained for occupancy by recreational vehicles of the general public as temporary living quarters for recreation or vacation purposes. A lot is a plot of ground within the recreational vehicle park intended for the accommodation of either a recreational vehicle, tent, or other individual camping unit on a temporary basis.

Some vehicles are equipped with wiring for use at 115 volts, having no other electric power except while in a park. Such vehicles are equipped with a permanently attached cord that must be a minimum of 23 feet and a maximum of 28 feet long. The cord must be of a size to carry the load of the vehicle (NEC 551-13).

The requirements for recreational vehicle sites and parks are similar to those for mobiles homes; however, there are some differences.

The equipment at each lot site is a "power outlet" meeting special requirements, from which the vehicle is supplied. The service equipment is of a special class called "Recreational Vehicle Service Equipment," (Figure 26.1).

The recreational vehicle park secondary electrical distribution system to recreational vehicle sites is derived from a single-phase, 120/240-volt, three-wire system, and like the mobile home park is usually installed in the park office or utility house, with feeders to groups of power outlets for individual vehicles (Figure 26.2).

The NEC rules for the layout and installation of electrical equipment in a recreational vehicle park; however, the local inspector has the final authority. The Code specifies, "A minimum of 75 percent of all recreational vehicle sites with electrical supply shall be equipped with a 30-ampere, 125-volt receptacle conforming to the configuration shown in NEC 551-13(c)."

All 15–20 ampere 125-volt receptacles for connection of electrical equipment outside the vehicle shall be 125-volt, 15–20 ampere, single-phase

Figure 26.1 Individual recreational vehicle site service-entrance equipment for supplying electrical power underground. (Courtesy of Midwest Electric Products, Inc.)

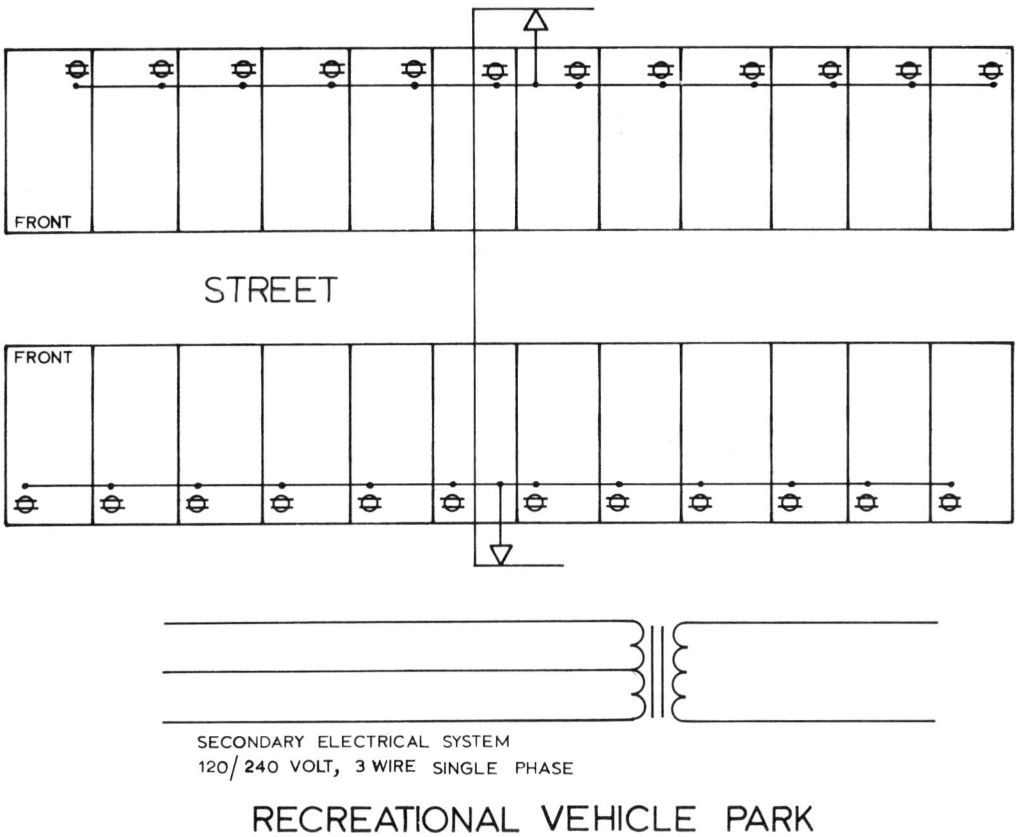

Figure 26.2 Recreational vehicle park, showing supplying electrical power underground to individual campsites. All 15- and 20-ampere 125-volt receptacles have approved ground-fault protection for the connection of electric equipment outside the vehicle, and are on 125-volt, 15- and 20-ampere single-phase ground-fault circuit interrupters.

receptacles of approved ground-fault circuit-interrupter type. The remainder of all recreational vehicle sites with electrical supply shall be equipped with one or more of the receptacle configurations conforming to Section 551-52 of the NEC.

It is recommended that the newer parks include a 50-ampere 120/240-volt receptacle in addition to the 20- and 30-ampere type to accommodate larger luxury motor homes that require a larger load.

26-3 CACULATIONS FOR THE SERVICE

The demand factor for an electrical service and feeders shall be calculated on the basis of not less than 3600 watts per site equipped with both 20-ampere and 30-ampere supply facilities and 2400 watts per site equipped with only 20-ampere supply facilities. See Table 551-44 (Table 26.1) for the minimum allowable demand factors that shall be permitted in calculating the load for services and feeders.

EXAMPLE 1

A recreational vehicle park has 24 sites. NEC 551-42 states that 75 percent are on a minimum basis when the electrical supply is equipped with 30-ampere 125-volt receptacles. (1) 75 percent or $0.75 \times 24 = 18$ sites to be equipped with 30-ampere facilities; and (2) $24 - 18 = 6$ sites equipped with 20-ampere facilities. (3) Total wattage would be $(3600 \times 18) + (2400 \times 6) = 64{,}800 + 14{,}400 = 79{,}200$ watts. (4) Table 551-44 (Table 26.1) shows 24 sites have a demand factor of 25 percent; therefore the demand for watts would be $79{,}200 \times 0.25 = 19{,}800$ watts

Remember the service is supplied at 120/240 volts so the minimum size of the service equipment for the park would be 83 amperes. You would want to install a larger service conductor to take care of additional loads in the park.

TABLE 26.1 Demand Factors for Feeders and Service-Entrance Conductors for Park Sites (NEC Table 551-44)

Number of Recreational Vehicle Sites	Demand Factor Percent
1	100
2	100
3	70
4	55
5	44
6	39
7–9	33
10–12	29
13–15	28
16–18	27
19–21	26
22–40	25
41–100	24
101 and over	23

SUMMARY

1. The National Electrical Code rules for the layout and installation of electrical equipment in a recreational vehicle park.
2. The demand factor for an electrical service shall be calculated on the basis of not less than 3600 watts per site equipped with both 20- and 30-ampere supply facilities.
3. All 15–20 ampere 125-volt receptacles for use outside the vehicle shall be 125-volts, 15–20 ampere, single-phase receptacles of approved ground-fault circuit-interrupter type.
4. Some recreational vehicles are equipped with wiring for use at 115 volt, and have no other electric power except while in a park.
5. The service equipment is of a special class called "Recreational Vehicle Service Equipment."

PROBLEMS

26-1 Draw a recreational vehicle park, showing 14 sites, and including outside GFCI receptacles.

26-2 Calculate the service feeders for the above park.

26-3 Calculate the service load for a park having 20 sites.

26-4 List the percent demand factor for the following:

a. 4 sites ________________ percent

b. 8 sites ________________ percent

c. 17 sites ________________ percent

d. 150 sites ________________ percent

26-5 Draw a sketch of a recreational vehicle, showing where the service cord would be attached.

26-6 Define a recreational vehicle according to the NEC. ____________

____

____

26-7 Where is the power supply located? ____________________ .

26-8 What is the service equipment called? ____________________ .

26-9 All 15–20 ampere receptacles for connection of electrical equipment outside the vehicle shall be ____________ volts.

26-10 It is recommended that newer parks include a _____ ampere receptacle to accommodate larger mobile motor homes.

PART 8

ESTIMATING

chapter 27

ESTIMATING ELECTRICAL WIRING

Instructional Objectives

1. To understand the importance of basic estimating.
2. To make you aware of the estimating errors.
3. To become familiar with material and labor costs.
4. To learn how to figure overhead costs.
5. To understand project and selling price.

Self-Evaluation Questions

Test your prior knowledge of the information in this chapter by answering the following questions. Watch for the answers as you read the chapter. Your final evaluation of whether you understand the material is measured by your ability to answer these questions. When you have completed the chapter, return to this section and answer the questions again.

1. What are the qualifications for an estimator?
2. Where does an estimator get the information for the job?
3. Name several common estimating errors.
4. How should materials be purchased to make a profit?
5. What factors bear on whether or not to bid a job?
6. How do you figure selling price?
7. How does unit pricing apply?
8. Is the loadcenter or panelboard figured with the service?
9. What is included in direct job expense?
10. What is meant by overhead costs?

27- 1 DEFINITION OF ESTIMATING

Estimating is the process of forecasting *future costs of labor, material,* and *job expense* by envisioning from plans the conditions under which the installation will be made, then counting or measuring and listing the items of materials required to make the installation in accordance with specifications, and assigning a standard labor unit to each item; and finally factoring them as required to conform with the conditions of the job at hand so that a summary of prices of material, labor, job expense, overhead, and profit may result in a successful bid.

27-2 BASIC CONSIDERATIONS

Estimates may be classified into several types, depending upon their purpose and the degree of accuracy required.

Most normally used methods will fall into the following categories:

1. *Detailed Unit Quantity Estimate.* This method consists of finding the costs of a work unit involving material, labor, and equipment. These costs are multiplied by the number of units on the job.
2. *Detailed Total Quantity Estimate.* The total quantity of material,labor, and equipment costs are found and then added together for the total estimate.
3. *Quantity Survey.* This method consists of a complete estimate of all the quantities of material required for a given project. The labor units are added to each material item, material is priced, and then overhead and profit are added for a selling price.
4. *Square-Foot and Cubic-Foot Methods.* These are crude methods used to check other types of estimates.
5. *Progress Estimates.* These are periodic checks on work progress to determine the amount of payment due the contractor at the time the estimate is made.

27-3 QUALIFICATIONS OF AN ESTIMATOR

Since the estimate is made before the actual work is started, an estimator's ability to visualize the phases of construction is of utmost importance.

The following are other necessary traits:

1. An orderly mind
2. Ability to concentrate
3. Patience
4. Neatness
5. Reasonable amount of mathematical ability
6. Experience in the electrical field
7. Knowledge of the details of construction
8. Ability to collect, classify, and file data

9. Knowledge of labor, equipment, and construction materials
10. Good judgment in selecting workers, materials, and equipment

To arrive at a reasonable price for a given project, estimators must rely on not only what they can obtain from the specs and drawings, but also on their knowledge of electrical construction details.

27-4 ESTIMATING ERRORS

Serious mistakes in estimating can result in loss of contracts or, worse, in contracts that can be fulfilled only at a loss to the contractor.

Common estimating errors can be reduced to a negligible amount if a few simple precautions are taken.

1. The use of proper forms will greatly reduce or eliminate errors and omissions, regardless of the size of the project. Repeated use of the same forms and work sheets will result in a more accurate, more detailed estimate made with greater confidence than if a different hit-and-miss method is used for each job. See Section 27.6.
2. Arithmetical errors—especially improper location of the decimal point—haunt all estimators. The best method of detecting mathematical errors is to have a second person go over them. If this is not possible, they should be checked by the estimator using a slightly different procedure—summing a column from bottom to top instead of top to bottom, for example.

   There are many other possible errors:

   (a) omitting overhead and profit
   (b) allowing for waste of materials
   (c) neglecting job factors, such as weather conditions, performance of general contractor, labor strikes, types of structure, etc.
   (d) neglecting transportation costs, storage costs, etc.
   (e) changing prices of labor and materials
   (f) errors in copying items
   (g) omitting rent or purchase of special tools or equipment required on a project that is unique
   (h) not knowing the correct overhead factor to apply to a certain size project.

There are other sources of error, of course, but those listed above should be sufficient to provide a basis for evaluating the accuracy of existing procedures.

27-5 SUCCESSFUL BIDDING

The ultimate objective of every estimator is to achieve the best possible success ratio in bidding. An excellent ratio is one job awarded out of every four bids. However, a more realistic ratio is one out of seven. Taking everything into consideration it is probably possible to obtain 40 percent of the jobs bid, but that's with optimum conditions!

Four steps influence the success ratio in competitive bidding:

1. Taking the necessary measures to wisely select the jobs bid.
2. Carefully preparing electrical plans for takeoff.
3. Selecting the tools that will give the best results for takeoff.
4. Selecting estimating forms that are complete and inter-related.

Estimating is a disciplined, step-by-step procedure required to achieve the highest percentage of bidding success, to ascertain that nothing is omitted, and to ensure that all of the work done is accurate and complete.

Job Selection

Getting the work for the electrical contractor can be very time consuming, and time means money.

The cost-per-workerhour successfully bid can be substantially lowered by carefully selecting the jobs that are bid. At the very best, this is a difficult thing to do, but the rewards are great.

Several factors have an important bearing on whether or not a job should be bid:

1. Number and caliber of competing contractors.
2. Familiarity with the general contractors who are bidding.
3. Your tooling and equipment capabilities.
4. Your familiarity with the type of construction involved.
5. Relation of the size of the job to your total capabilities.
6. Need for the job to meet your workerhour budget.
7. How the timing of the job fits into your work schedule.

A consistent analysis of these conditions on a point value basis will contribute a great deal toward providing the basis for an intelligent decision.

Function	Points
Competition of electrical contractor	20
Relationship with general contractor	20
Job familiarity from previous experience	10
Size of job relative to capability	20
Need for job in manhour budget	20
Capacity of manpower to handle job	10
	100

Profit and Loss

Profitable operation of a contracting business demands the ability to apply honestly the following:

Materials + labor + direct job expense + overhead = cost
Cost + profit = selling price

Because of the highly competitive field of residential electrical contracting, the contractor must be extremely careful about applying accurate costs and a reasonable profit in calculating the selling price for a given project.

If he overestimates his costs, he doesn't get the job.

If he underestimates the job, he gets the job but with too low a profit or with a loss.

It doesn't take too many low profit or loss inflicting jobs to force a contractor out of business.

Other operations involved in profit and loss that should be considered involve the following:

1. Buy materials in large quantities where possible.
2. Select items of material that will save installation time. For example, a certain type of outlet box may require less time to install than another. If the saving in labor is more than the additional cost of the outlet box, then it will probably pay to use the time saving outlet box.
3. Keep all tools in good working condition.
4. Look into the possibility of building "custom" tools to help reduce labor.
5. Don't purchase unnecessary tools, but don't be stingy on tools that can make your firm additional profits.
6. Use an estimating method specifically designed for residential work where accurate estimates can be turned out in the shortest possible time.

Unit Pricing

Wiring installations in most commercial and industrial buildings vary so greatly that a definite standard of unit pricing is not practical; however, for the more standard operations of residential work, unit pricing provides a simple, fast, and reasonably accurate method of estimating—an absolute necessity in this highly competitive field.

In general, unit pricing is the process of combining items that are commonly used together in a given section of an electrical system, and then using this price as a unit when the particular group or section comes up in the estimating process.

For example, a conventional duplex wall receptacle installed in a residence will normally consist of the following items:

1 2¾-inch deep switch box with NM clamps	0.65
2 16 penny nails to secure box to wood stud	
1 ground clip	0.40
1 duplex receptacle	1.65
1 duplex receptacle plate	0.40
23 feet 122 with ground NM cable at 0.06	1.40
	$4.50

It might be found that only 21 feet of NM might be better than 23 feet. Use your own judgment. Make the adjustment to suit your own requirements. Once the material price has been calculated, the labor units will have to be applied for all the materials, including the installation of the NM cable.

For now, let's assume that the installation will be in new work where all partitions are open, there is adequate working space, and average working conditions.

It has been found that the average electrician in residential wiring will average 0.55 workerhour per duplex receptacle (wall outlet). If the electrician gets $10.00 per hour, the labor on the outlet will then cost $10.00 × 0.55 or $5.50. This, plus the material cost of $4.50 gives a total base cost of $10.00 per outlet.

To the above figure will have to be added your direct job expense, overhead, and profit. These figures will vary from contractor to contractor; however, if yours happens to total $2.55 the total unit price per duplex receptacle will be $10.00 + $2.55 or $12.55.

Obviously, a few of the items indicated in the material list could change; however, the price change will be insignificant in relation to the whole job. Therefore, it can be assumed that $12.55 would be the average selling price of a duplex wall receptacle on the average residential project.

In old work, where the NM cable must be fished in concealed partitions, the labor may go up to 1.30 workerhours per outlet, making it necessary to raise the total selling price to $19.75 per outlet.

1.30 workerhours × $10.00 = $13.00 + $4.50 material + $2.25 profit = $19.75

From the above, we can easily see that if a conventional new residence were to contain, say, 90 duplex receptacles, the total installation cost should be close to 90 × $12.55 or $1129.50. By the same token, a house with 43 outlets should cost 43 × $12.55 or $539.65 at a prevailing labor rate of $10.00 per hour.

To check this unit price, an actual residence containing 43 outlets will be used. Allowing time for unloading the necessary materials from the truck, marking the outlets locations, nailing on the boxes, drilling the holes, pulling in the wire, and installing the wire to the outlet box we have the following:

Materials (Price to Your Area)		
43	2¾-inch switch box with NM clamps	$27.95
700 feet	12/2 with grd. NM cable at 0.06	42.00
120 feet	12/3 with grd. NM cable at 0.10	12.00
	100 staples	2.75
2	74B wire connectors	0.10
50	crimp connectors	0.75
43	duplex receptacles at 1.65	70.95
50	ground clips	2.00
		$175.70

Labor (workerhours)	
Unloading truck	0.05
Layout and installing boxes	5.50
Drilling holes through plate under box	1.00
Drilling, pulling, securing, stripping	7.00
Installing duplex receptacles and plates	9.00
	23.00

The installation averaged 23 ÷ 43 or 0.54 workerhour per outlet. The material costs averaged $175.70 ÷ 43 or $4.08, a little less than estimated.

Notice that the actual materials used were not exactly those listed in the unit price schedule. Still the variation made little difference in the overall cost of the project.

To obtain a unit price for a given portion of a residential wiring installation, modifications may have to be made from time to time to keep the prices accurate. Typical reasons for modification could include:

1. a change in material price
2. a change in the contractor's overhead
3. hiring a group of new electricians with not too much experience in residential wiring

The intelligent contractor will compare actual material and labor cost after each job has been completed to keep a close watch on his own unit prices as well as to improve his estimating skill and management ability. If there is a problem, the contractor can correct it before too much damage is done.

Other units will include incandescent lighting outlets (wall, ceiling, etc.) single-pole, three-way, and four-way switches, weatherproof receptacles, bathroom fans, heaters, ground-fault receptacles, range outlet, and circuit wiring.

Unit Pricing for Electrical Services

To make a unit price for electrical services, first list the items that seldom vary. These will include:

1	2-inch weatherhead	$6.00
2	watertight (compression) connectors at 5.50	11.00
1	ground clamp for 1-inch water pipe	1.75
1	2-inch EMT connector	3.00
4	fasteners for meter	0.60
	fasteners for inside load center	0.50
10 feet	2-inch EMT conduit	3.35
		$26.20

Some estimators include the loadcenter or panelboard with the nonvarying items, since a home requiring a 200-ampere service will normally require between 36 and 40 spaces. The difference in price is only a few dollars, so they include the cost of the panel with main breaker (about $81.00) and breakers; that is, two GFCI circuit breakers for bath and outdoor circuits $45.00 each and 38 breakers calculated at $2.45 each for a total panel cost of $253.10. This added to the other items brings the total cost to $264.10 + $26.20 or $290.30. This price will remain constant for the majority of 200-ampere services, as will the labor requirements for these items. The variables include the service-entrance conductors (which will vary in length from job to job), the number of straps and fasteners for the amount of conduit, and the length of the ground wire from the service equipment to ground.

Similar unit prices are possible for nearly every section of a residential wiring system and for a variety of wiring methods.

Unit prices are readily applicable to such items as:

1. switches
2. special-purpose circuits
3. signal wiring (fire and intruder)

There are some items that are just not suited for unit pricing. One category is residential lighting fixtures. Because of the wide variation in size, type, quality, and price, their cost and installation must be included in the estimate as separate items.

Another point to remember about unit prices is that most of them are calculated on an average complete installation.

Keep in mind that most of the units are priced for a complete residential wiring project having 30 or more total outlets, a complete service entrance, sometimes recessed fixtures, and electric heaters.

When the number of outlets are less than 30—such as in a remodeling project—the estimate should reflect an increase in each unit price used or a contingency factor should be added to the overall bid according to the judgment of the estimator.

Material Cost and Pricing

Standard materials commonly used in residential electrical systems include nearly all types of outlet boxes, wire and cable, fittings, safety switches, service-entrance equipment, and lighting fixtures. Very seldom do any of these items have to be custom built; therefore the prices for them are readily available without a special quotation from suppliers on each job.

In some instances, the electrical contractor may be able to obtain a better discount by buying large quantities of materials for several residential projects at one time; then a special quotation from electrical suppliers is warranted. When such a situation occurs, it is recommended that the contractor not change his normal unit pricing of materials (unless it means losing a very important job). Rather, the savings should be put into the contractor's business either to help cover some of the firm's operating expenses or to increase the firm's yearly profits.

To help keep abreast of the constantly changing prices of electrical materials, most electrical contractors—particularly those involved primarily in residential and small commercial projects—subscribe to one of the price services that publishes current prices of practically all standard electrical products. Most of the price service firms offer an extremely rapid, accurate reporting service to keep the contractor up to date with the constantly changing prices of electrical construction materials.

Regardless how you choose to go about it, you must know the exact price of your materials to stay in business. In fact most of the contractors who go out of business each year—either going broke or finding it unprofitable to continue—do so because they do not know their costs accurately enough to bid work at a profit. Your competitors' bid cannot be used as a basis for bidding your work. The one you pattern your bids after may be taking work at a loss and could cause both of you to go out of business.

Direct Job Expense

To operate a successful contracting business, a thorough understanding of direct job expense is necessary, since these costs must be recovered from the job bids or contracts.

Direct job expenses are all of the costs in addition to material and labor which have to be met directly because they would not exist if the job had not been undertaken. All other expenses would fall under overhead, which will be discussed later.

Items that should always be considered as direct job expenses in estimating residential electrical construction are:

1. bid bonds (if required)
2. blueprint payments
3. inspection fees
4. building permits
5. special license
6. labor adders for payroll
   - (a) social security
   - (b) workmen's compensation
   - (c) unemployment insurance
   - (d) employee's liability
   - (e) other dues and assessments paid in the form of a percentage of the production labor payroll

In addition there are items that lie in the fringe area between direct job costs and overhead expenses. When the exact amount of costs for these items can be allocated to a particular job, they should become direct job expenses. However, because of the difficulty in doing so, they are most often charged against overhead. These items include:

1. layout and job supervision
2. freight and postage for materials shipped
3. estimating charges
4. interest on borrowed money
5. supplemental drawings
6. job truck
7. depreciation and consumption of tools

To illustrate, let's assume that a truck is used for two full days to carry workers and materials to a project. The expense of operating the truck definitely can be charged to direct job expense, as can a certain percentage of the depreciation calculated for the truck. However, if the truck was used to service many jobs in one day—like delivering materials to, say, six different projects—it would be difficult to charge the truck expenses for the day to the various jobs on which it was used.

Another example would be the use of tools. On small jobs where tools and equipment are used only for a relatively short period of time on each, it is probably impractical to charge an item of tool depreciation directly as a direct job expense.

A prorata amount of the tool cost could be charged and used in the estimate to help recover the original cost of the tool. One of the best ways of doing this is to establish a monthly cost or rental charge per tool—calculated on the basis of its useful life—then further break this cost down into hours. It should be much easier to determine how much direct job expense to charge each job.

An electric drill may cost the contractor $100.00. If the average life of a ½-inch drill motor used on residential construction is approximately 18 months, the cost of using each drill will be $100 ÷ 18 or $5.55 per month or about 25 cents per working day (22 working days per month). If two drills were used on a project for two days, it would be reasonable to charge about 2 × $0.25 × 2 days or about $1.00 cost against the job for direct job expenses. This sounds like small change, but the net effect of such items as stepladders, trucks, and electric generators used on a project can cut down the firm's overhead expenses if they are charged to each individual project on which they are used. Remember, someone has to pay for the tools, and it might as well be your customers.

In some instances of construction, it may become necessary for workers to spend the night at out-of-town locations. These workers must be compensated for room and board in addition to the travel expenses. The estimator should try to foresee such instances and include an amount in the total contract to cover the costs incurred (under direct job expenses).

Where a company truck is used to transport electricians to or from out-of-town jobs, any expense incurred should definitely be charged as a direct job expense, not to overhead.

You may be thinking by now that if you included in your estimate all the necessary items of direct job expense, you would never get a contract. However, look at it this way. Anytime you fail to include any item of direct job expense that is going to be paid directly or indirectly, you are giving your customer money—the contractor's or your money. You wouldn't think of grabbing $50.00 out of the cash drawer and throwing it out the window, but the effect is the same when you don't figure all of your direct job costs and include them in your bid.

Overhead Costs

In general, overhead expenses are items that cannot be charged to a particular job or project but must be paid to remain in business.

The contractor's (estimator's) first problem is to determine what his overhead is, and then find some practical method of distributing this amount on a fair basis to the various jobs performed over a period of a month, three months, or a year. From this, we can see that the most practical method would be to add a certain percentage of the yearly overhead to each contract.

The following items are usually considered as overhead expenses, because a definite amount of money is needed by all electrical contractors, regardless of their size.

1. *Advertising.* Any sales promotion—whether devices or services—used to obtain or increase business would fall under this heading. Such items would include:
   (a) newspaper ads
   (b) radio or TV commercials
   (c) classified phone directory listings
   (d) business cards
   (e) direct mail brochures
   (f) circulars
   (g) signs
   (h) truck advertising
   (i) letterheads may also be listed—could be office expense

   The average residential electrical contractor normally spends about 1 percent of his gross sales for advertising. Larger contractors spend much less.
2. *Association Dues and Subscriptions.* Business clubs, the chamber of commerce, electrical associations, and similar dues would be considered under this heading. Also the subscription to trade journals—if necessary for the operation of the business—should be charged under this heading and will therefore be valid as a tax exemption.
3. *Automobiles and Trucks.* Items like state inspections, licenses, registration fees, gas, oil, minor repairs, and similar costs of automobiles and trucks used in the normal operation of business should be included here.
4. *Bad Debts.* This account should be debited each month for uncollectable or doubtful accounts. The electrical contractor should make an allowance each year for such bad debts and a proportionate part should be included in each job estimate.
5. *Collections.* The cost of collecting past-due accounts should be charged to this account—that is, legal fees, commissions paid for collections, etc.
6. *Charitable Contributions.* All donations to charitable or educational organizations as defined by the federal (and/or state) income-tax regulations should be included under this heading.
7. *Defective, Lost, Obsolete, or Stolen Goods.* The actual value falling under any or all of these categories is chargeable to overhead and may be deducted from income taxes.
8. *Delivery Costs.* The total expense of all freight, express, and postage that cannot be charged to a given job as direct job expense belongs here.
9. *Depreciation.* This account may be debited with the amount of depreciation on buildings, autos and trucks, equipment and heavy tools, office furniture, and office equipment, based on the estimated useful life of each.
10. *Insurance.* The amount paid for insurance is usually broken down into several parts (general insurance, workmen's compensation, etc.) However, insurance fees should be charged as overhead (unless a percentage of them can be charged as direct job expense). Some types would include fire, theft, liability, and the like.

11. *Interest.* The amount of interest paid or accrued on outstanding interest-bearing obligations of the business, with the exception of interest that may be charged to a particular job, should be charged here.
12. *Legal and Accounting Fees.* As the names imply, all accountant fees or CPA fees, as well as those of an attorney, will fall under this heading.
13. *Office Expense.* This includes the cost of such supplies as pencils, pens, staples, printed forms, and similar items. It does not include office equipment (file cabinets, adding machines, typewriter, etc.).
14. *Rent.* The total amount paid for rental or lease of buildings, land, storage space, etc., used in the operation of the business which cannot be charged to one particular job would be included here.
15. *Salaries and Commissions.* This includes salaries and commissions not readily chargeable to jobs. A secretary's salary would normally fall under this heading, but if a secretary was, say, typing written specifications for a particular project, his or her time would be charged directly to the job.
16. *Taxes.* Any city, county, state, federal, and other taxes not charged directly to a job should be included under this division. Many contractors break this category down into two parts—general taxes and payroll taxes.
17. *Telephone.* Charge this account with all telephone expenses (including answering services) that cannot be charged directly to a job.
18. *Miscellaneous.* All expenses allowed by the IRS not otherwise provided for in the above categories are included here.

The first step in determining the amount of applicable overhead for each job is to determine a base cost including material cost, labor cost, direct cost, and so forth, to which the overhead percentage will have to be estimated, based on good judgment and an examination of the records or experiences of other contractors. Once the contractor has his own records, however, he can determine his own overhead more accurately. This determination should be based on several years' operations rather than only a few months.

The estimator should determine whether a single rate of overhead can be equitably applied to all jobs or whether because of both large and small work being performed it will be necessary to develop more than one rate.

The next step is to determine the total normal overhead expense of the business consistent with the total base rate developed previously.

The final step is to apportion the total general overhead cost to the overhead cost totals for large and small jobs, respectively, on the basis of percentages used in segregating the rate base for each type of work. The rate base applicable to that type of work is applied accordingly.

Profit and Selling Price

Although many electrical contractors derive some pleasure from operating a successful business, few, if any, would continue their operations if a reasonable profit could not be made. This is the main reason for being in business—to make money.

The amount of profit that a firm makes over a given period depends upon a lot of factors:

1. The amount of work performed.
2. The amount charged for the work performed.
3. Correct job management.
4. Buying materials at the right price.
5. One of the most important is the ability to determine the correct selling price for the greater part of the work performed.

You might miss a few, but the majority of your bids must be performed within a certain predetermined budget if you and the firm are to make a profit.

The amount of profit (percentage-wise) that a contracting firm should realize depends mainly upon the gross amount of work the firm will perform during a year's time.

For example, if a contractor could make $25,000 a year working for another contractor, it stands to reason that he should at least aim for $40,000 annually, considering the headaches of running a business.

Often, the first year or two will be low on profits; but after that time, the contractor should be putting more money into his own pocket than he could make working for someone else.

A contracting firm that grosses say, $250,000 each year may get by charging between 10 and 15 percent profit on its work, while a firm doing half this amount of work will probably have to count on a 25 percent profit. If you're doing only $50,000 volume (about the minimum for even the smallest contracting firms), you will have to charge around 40 percent profit for most of your work.

The first question that probably comes to mind is how, then, can the smaller contractors compete with the larger firms if they have to charge twice as much profit on each job? The answer is simple. Usually, the larger a contracting firm, the less efficient it is. A large firm's overhead is usually much more (percentage-wise) than the smaller firms. Therefore, what the smaller firm has to charge in the way of profits can usually be made up by charging less overhead and by performing the work more efficiently.

The calculation for determining the cost of construction is:

Materials + labor + direct job expense = prime cost

To determine the selling price:

Cost + anticipated profit = selling price

It may seem very simple to run an electrical contracting business, but the truth is that there are many complications. Items such as constantly fluctuating material prices and labor problems are prime examples. A further complication arises from the fact that the average contractor has difficulty in determining his overhead expenses accurately and more often than not underestimates them. Since the costs are higher than anticipated, the profits are naturally lower.

The business is further complicated by intense competition in construction, which makes it compulsory to come up with a correct price—an almost exact price.

To illustrate, if a contractor overestimates his costs, he does not get the job. If he underestimates the job, he may get it but with either too low a profit or a loss.

To obtain the amount of profit that you want out of the business, you must be able to buy materials at the best possible cost, have electricians trained, and practice good job management. These are the basic steps. Then you must know almost exactly what your materials are going to cost for any given job; you must apply the correct labor units to these figures; and finally you must add all direct job expenses to come up with a prime cost for the project.

By making careful records of the business over periods of time, you should know almost exactly what your current overhead is and what percentage is required to be added to the prime cost of each job. This figure varies; however, most contractors keep their overhead at about 25 percent of the prime cost of their gross income.

When the overhead is applied to the prime cost, you come up with a price that should cause you to break even. That is, if you take the job after calculating all material costs, labor, direct job expense, and overhead, you should be able to do the job without losing money. You will not of course, make any profit, as this has yet to be added.

Now comes the profit and selling price that can be determined only by you. Few contractors who specialize in residential and small commercial projects can get by on less than 20 percent profit, because their gross sales are seldom over $100,000 annually. Many charge 25 percent or 30 percent profit, depending on the type of work and how much the builders or general contractors are willing to pay.

When you decide upon your own figure, this should be added to the prime cost. The result (hopefully) is the selling price that will yield the profit that you have anticipated.

Keep records of each job, giving a cost breakdown on the various stages of construction. If you are making a mistake in your estimates, you should be able to detect the problem immediately and rectify it. Remember losses go into overhead.

27-6 ESTIMATING AND CONTRACTING FORMS

Many calculations and records must be made during the process of estimating. Such calculations are best performed when a systematic pattern is followed, using appropriate forms. In general, all forms have spaces for the name of the project, the date, the names of the estimator, foreman, and so forth, and other standard data. Forms discussed here were developed by the Minnesota Electrical Association in Minneapolis. See Figure 27.1.

GENERAL WIRING PROPOSAL & CONTRACT

No. ______________ Date ______________

Submitted By	To ______________ (Purchaser)
______________	______________ (Address)
	Wiring to be done on the following premises:
______________	Address ______________
ELECTRICAL CONTRACTOR	Owned By ______________

The undersigned hereby agrees to wire the premises named above, furnishing all labor and materials required to complete the electrical work as specified herein,

(A) For Sum of: ______________ Dollars ($ ______________)

(B) On Time and Material Basis ______________

payable as follows: ______________

____

and subject to the following conditions:

1. **This proposal shall be a contract between the parties hereto provided the purchaser signs acceptance within ______________ days from date.**

2. **The materials and workmanship furnished under this proposal shall comply with the rules and regulations set forth in the National Electrical Code, and all State and Local regulations governing such work. The price quoted includes required insurance, permits and inspection fees.**

3. **Any changes or additions in this proposal or the work installed hereunder shall be made in writing, setting forth the details of such changes and the cost thereof, and signed by both parties.**

4. **The contractor shall not be held responsible or liable for any loss, damage or delay caused by fire, strikes, civil or military authority, insurrection or riot, or by any other cause beyond his control.**

5. **If the purchaser disposes of the property by sale or otherwise, the full unpaid amount of this contract shall be due and payable at once.**

6. **All equipment and devices installed as a part of this proposal shall be guaranteed for a period of one year from date of completion of this installation, except as otherwise noted, or in accordance with manufacturers warranty. Contractors liability shall be limited to the replacement of defective parts.**

SPECIFICATIONS

The Electric wiring and equipment to be installed under this Proposal shall be in accordance with:

C. Plans and Specifications furnished and prepared by ______________

____

with the following changes or exceptions: ______________

____

____

____

D. Details set forth on the following pages of this proposal ______________

____

____

ACCEPTANCE

The foregoing proposal is hereby accepted and the work ordered to be done subject to terms of payment and conditions stated above.

____
Purchaser

Signed By ______________
Official Capacity

Dated ______________

Work to Start ______________

To Be Completed ______________

RESPECTFULLY SUBMITTED:

____
Electrical Contractor

Signed By ______________

Form S-2 Minnesota Electrical Ass'n

(a)

Figure 27.1 (*a*) General Wiring Proposal & Contract. Good for small and medium sized commercial wiring installations, schools, churches, creameries, and so forth, for which specifications, proposal, and contracts are usually made up by the electrical contractor. (Courtesy of Minnesota Electrical Association, Inc.)

HOUSE WIRING SPECIFICATIONS & PROPOSAL

No.______________________ Date______________________

SUBMITTED BY	TO______________________ (Purchaser)
______________________	______________________ (Address)
______________________	______________________ (Telephone No.)
Electrical Contractor	Wiring to be done on the following premises: ______________________

The undersigned contractor hereby agrees to wire the premises named above, furnishing all labor and materials required to complete the electrical work as specified herein, for the sum of:

______________________ Dollars ($______________)

payable as follows______________________

____

and subject to the following conditions:

1. This proposal shall be a contract between the parties hereto, provided the purchaser signs acceptance within______________ days from date.

2. The materials and workmanship furnished under this proposal shall comply with the rules and regulations set forth in the National Electrical Code, and all State and Local regulations governing such work. The price quoted includes customary insurance, permits and inspection fees.

3. Any changes or additions in this proposal or the work installed hereunder shall be made in writing, setting forth the details of such changes and the cost thereof, and signed by both parties.

4. The contractor shall not be held responsible or liable for any loss, damage or delay caused by fire, strikes, civil or military authority, insurrection or riot, or by any other cause beyond his control.

5. If the purchaser disposes of the property by sale or otherwise fails to have the building complete, the full unpaid amount is payable at once.

6. All equipment and devices installed as a part of this proposal shall be guaranteed for a period of one year from date of completion of this installation. Contractors liability shall be limited to the replacement of defective parts.

7. No fixtures, lamps, chimes or other equipment are included in this proposal unless specified below.

8. This Proposal covers complete finished job (including switches, receptacles, plates, etc.)

9. The following detailed specifications shall apply to this job:

(a) Outlets are to be installed as per schedule on reverse side of this proposal.
Make of switches and receptacles ______________________ Color Plates______________

(b) Wiring System ______________________

(c) Service Entrance (Size wires and materials) ______________ Main Switch______________ AMPS

(d) No. and Size of Branch Circuits ______________________
Fuse Holders (Cutouts) ______________ Circuit Breakers ______________

(e) Special wiring or equipment included in this proposals:

____

____

____

____

PURCHASER'S ACCEPTANCE

The proposal as outlined above and on the reverse side of this sheet is accepted. My signature here is the order for this work.

____
Purchaser

Dated______________________

Respectfully submitted,

____

ELECTRICAL CONTRACTOR

By ______________________

Minnesota Electrical Ass'n — Form S-1

(b)

Figure 27.1 (*b*) Electrical Wiring Proposal. Proposal is signed by contractor and when signed by the customer, becomes a contract to start work. (Courtesy of Minnesota Electrical Association, Inc.)

1. *Service-Entrance and Service-Entrance Neutral Sizing Forms.* These are designed for use with the methods described in Chapter 9 of the NEC to size the service-entrance conductors. Data to be entered in the blanks follow the NEC procedure step by step.
2. *Estimate Sheet.* This form includes columns for description, material, and labor which permit orderly recording of costs. A separate sheet should be used for each individual unit (wall switches, receptacles, etc.) when they are used to obtain unit prices. See Figure 27.2.
3. *Bid and Estimating Summary.* Material and labor costs from each estimate sheet may be copied on to this form to get all totals on one sheet. The second and third parts of this form provide an orderly procedure for calculating labor job factor, nonproductive labor, and job expense. Close attention to each of the items listed will help avoid some of the common estimating errors. See Figure 27.3.
4. *House Wiring Summary.* This is a handy checklist of some 57 items encountered in residential work. It includes columns to be filled for showing quantity, unit price, material cost, labor unit, and labor cost for each item. It can be used at a later date to check the actual installation cost against the bid. See Figure 27.4.
5. *Job Record Insert.* This form may act as a proposal for small projects if work to be done is described. Upon completion, a space for the customer's signature is provided for approval; this can be a great aid in collecting the bill at a later date. The inside of the folder has spaces for material used, extra material used, special material, and so forth. The back of the form has room to list any extra work that is to be done as well as a space for the customer's signature to approve this extra work. Upon completion of the project, the form should be inserted in the job folder. See Figure 27.5.
6. *Job File.* This file is used for filing estimate sheets, job cost records, invoices, and other paper pertaining to the particular job. The printed form allows for complete details of the job, and the entire folder will provide a neat and orderly job file system that can be stored in any standard drawer or filed in any standard filing cabinet. The job files may be arranged either alphabetically or numerically, with easy access to filing information on the front.
7. *Job Record Envelope.* This envelope may be used like the job file; however, it is more suited for residential jobs. In addition to spaces for the job name, location, and so forth, there is space to describe the project and the work to be done as well as miscellaneous notes and the billing dates. It further enables the contractor to see at a glance the total cost of the project, and if all necessary agencies have been notified. The back of the folder has spaces to list all labor (including the electrician's name and a summary of the material cost). See Figure 27.6.

ESTIMATE SHEET

JOB ____________________ PAGE ______

____________________ OF ______ PAGES

ESTIMATED BY ______ CHECKED BY ______ DATE ______

DESCRIPTION	MATERIAL				LABOR		
	Quantity	Unit Price	Per	Amount	Unit	Per	Amount
		TOTAL					

MISCELLANEOUS		RECAPITULATION	
		Material Cost	
		Hours Labor @	
		Hours Labor @	
		Direct Job Expense	
		Total Prime Cost	
		Overhead Expense	
		TOTAL COST	
		Profit	
		Selling Price	
Form E-3 MINNESOTA ELECTRICAL ASS'N		**BID SUMMITTED**	

(*a*)

Figure 27.2 (*a*) Estimate Sheet. This form is used for precise estimating in making bids. Separate sheets are suggested for each floor or section of the job. (Courtesy of Minnesota Electrical Association, Inc.)

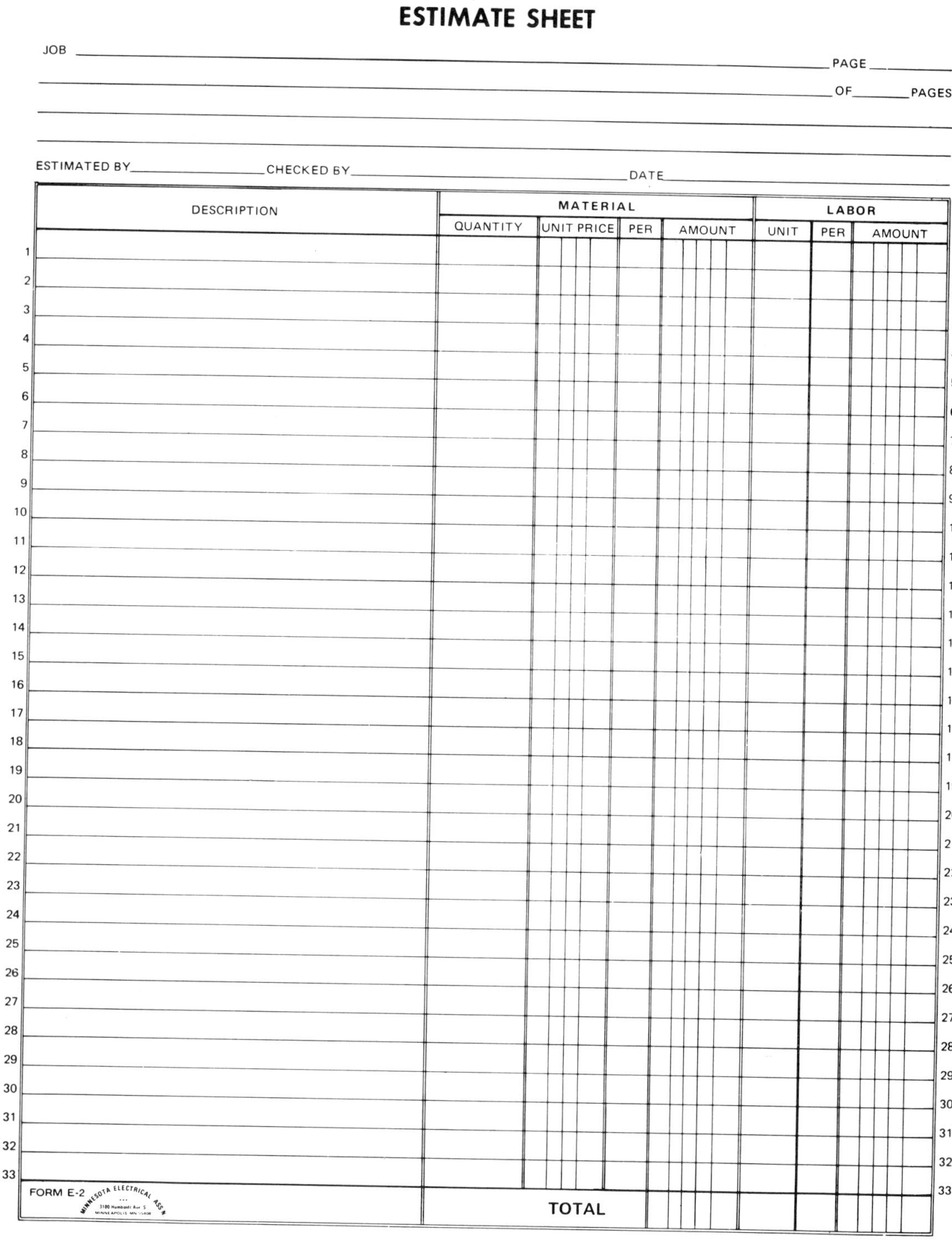

ESTIMATE SHEET

JOB ______________________________ PAGE ______
______________________________ OF ______ PAGES
____
____

ESTIMATED BY __________ CHECKED BY __________ DATE __________

	DESCRIPTION	MATERIAL				LABOR			
		QUANTITY	UNIT PRICE	PER	AMOUNT	UNIT	PER	AMOUNT	
1									1
2									2
3									3
4									4
5									5
6									6
7									7
8									8
9									9
10									10
11									11
12									12
13									13
14									14
15									15
16									16
17									17
18									18
19									19
20									20
21									21
22									22
23									23
24									24
25									25
26									26
27									27
28									28
29									29
30									30
31									31
32									32
33									33
	FORM E-2 MINNESOTA ELECTRICAL ASS'N.	TOTAL							

(b)

Figure 27.2 (*b*) This form is commonly used for small jobs that require a limited number of items. (Courtesy of Minnesota Electrical Association, Inc.)

BID SUMMARY SHEET

JOB ______ SHEET NO. ____ OF ____ SHEETS

ESTIMATED BY ______ CHECKED BY ______ DATE ______

SHEET NO.	DIVISION	MATERIAL-DOLLARS	LABOR-HOURS

NON-PRODUCTIVE LABOR	HOURS
Handling Material	
Superintendent	
Traveling Time and Lost Time	
Job Clerk	
TOTAL (A)	

JOB EXPENSE	DOLLARS
Tools, Scaffolds	
Pro Rata Charges	
Insurance, Public Liability, Etc.	
Cutting, Patching, Painting	
Watchman	
Telephone	
Drawings	
Inspection and Permit Fees	
License	
Storage	
Freight, Express and Cartage	
Transportation	
Board ____ Men ____ Weeks At ____	
TOTAL (B)	

	MATERIAL-DOLLARS	LABOR-HOURS
Miscellaneous Material and Labor		
Non-Productive Labor		(A)
TOTALS-MATERIAL (C) & LABOR (D)		
(D) ____ Hours Labor @ ____		
____ Hours Labor @ ____		
____ Hours Labor @ ____		
Taxes: Soc. Sec. ____ Unemp. ____		
Workmen's Compensation Insurance		
LABOR COST GROSS TOTAL		
Job Expense (B)		
Material Cost (C)		
TOTAL PRIME COST		
____ % Overhead		
TOTAL NET COST		
____ % Profit		
Selling Price Without Bond		
Bond		
Selling Price With Bond		
PRICE QUOTED		

Form E-1

To Avoid Errors, Check List on Reverse Side   Let's Upgrade Our Electrical Industry

Figure 27.3 Bid Summary Sheet. Provides space to list each estimate sheet and put in the total of material and labor. (Courtesy of Minnesota Electrical Association, Inc.)

HOUSE WIRING SUMMARY

JOB ______________________________ ESTIMATED BY ______________________ PAGE ____________

______________________________ CHECKED BY ______________________ OF __________ PAGES

DESCRIPTION	OUTLETS – MATERIAL				LABOR		
	QUANTITY	UNIT PRICE	PER	AMOUNT	UNIT	PER	AMOUNT
Ceiling or Wall Light Outlet							
Recessed Box with Outlet Box Att.							
Single Pole Switch							
Three Way Switch							
Four Way Switch							
Weatherproof Switch, Add for Each							
Dimmer Switch							
Door Switch							
Pilot Light on Switch, Add for Each							
Duplex Receptacle 2 Wire							
Duplex Receptacle 3 Wire Grd.							
Duplex Receptacle Split Wired or Switched							
Duplex Receptacle Weatherproof with G.F.I. Recept. or Brkr.							
Floor Receptacle							
Clock Hanger Receptacle							
T.V. or Radio Receptacle							
Telephone Outlet, Box & Plate							
Disposal							
Kitchen Exhaust Fan (or Hood)							
Dishwasher							
Electric Range - with Recept.							
Electric Oven							
Electric Cook Top							
Electric Clothes Dryer with Recept.							
Electric Water Heater							
Electric Water Pump 110 V. ☐ 220 V. ☐							
Furnace Oil or Gas Burner							
Thermostat							
Furnace Blower w/switch							

(a)

Figure 27.4 (*a*) and (*b*). House Wiring Summary. A simple complete outline on one sheet for a full summary of equipment and wiring for a house wiring job. Later can be used for a check sheet after installation. (Courtesy of Minnesota Electrical Association, Inc.)

Electric Heat Units Recessed																		
Electric Heat Units Baseboard																		
Electric Heat Units Wall Mount																		
Electric L.V. Thermostats																		
Electric Highvoltage Thermostats																		
Receptacle Insert Sections																		
Bath Room Heater																		
Bath Room Exhaust Fan																		
Receptacle 20 Amp. 120 V. Grd.																		
Receptacle 20 Amp. 240 V. Grd.																		
Receptacle 30 Amp. Air Cond., etc.																		
Low Voltage Switch - Single Relay																		
Relay & 1 Flush Switch																		
Added Flush Switch																		
Relay & 1 Surface Switch																		
Added Surface Switch																		
Master Switch																		
Gang Box Relay																		
Relay, Install & Connected																		
Flush Switch, Install																		
Surface Switch, Install																		
Master Switch Install																		
Transformer Mounted & Connect.																		
Electronic Air Cleaner or Humidifer																		
Bell Wiring w/Transformer																		
Smoke Detectors																		
Service: 60 Amp. – Circuit																		
100 Amp. – Circuit																		
150 Amp. – Circuit																		
200 Amp. – Circuit																		
Fuse Centers or Panels																		
Electric Permit/Inspection Fee																		
Temporary Service Set Up																		
Underground Service																		
Form E-20	TOTAL																	

(b)

JOB SHEET

PERMIT NUMBER ____________ JOB NUMBER ____________

NAME ____________ ADDRESS ____________

JOB AT ____________ KIND OF JOB ____________

START ________ FINISH ________ RECAP ________ INVOICE ________ CUSTOMER LEDGER ________

ORIGINAL PRICE			BILLING PRICE		
EXTRAS			TOTAL COST (See Below)		
TOTAL BILLING PRICE OF JOB			NET PROFIT OR LOSS		

DIRECT JOB EXPENSE		
PERMIT AND INSPECTION		
COMPENSATION INSURANCE		
FINANCING		
COMMISSIONS		
CAR FARE		
DRAYAGE		
TOTAL JOB EXPENSE		

COST SUMMARY				
ITEM	Estimated		Actual	
MATERIAL (See Other Side)				
LABOR (________ Hours)				
DIRECT JOB EXPENSE				
PRIME COST				
OVERHEAD EXPENSE				
TOTAL COST OF JOB				
Remarks				

LABOR COST

Dates	Workman	Hours	Rate	Cost	Rate	Charge
			TOTAL			

TOOLS AND EQUIPMENT

Quantity	Item	Workman	Returned	Used	Charge	Remarks

Figure 27.5 Job Sheet. Shows a complete cost record for the average job. (Courtesy of Minnesota Electrical Association, Inc.)

JOB FILE

Job ______________________ Job. No. ______________________

Location ______________________ Date ______________________ 19____

Name	Address	Telephone Number
Architect:		
Engineer:		
Owner:		

Bid to: ☐ Architect ☐ Engineer ☐ Owner Bid to be in by ______________________ 19____ Submit ________ copies of bid

Form E-19 BP284H

Figure 27.6 Job File envelope. For filing estimate sheets, job cost records, invoices, and other papers pertaining to one job. (Courtesy of Minnesota Electrical Association, Inc.)

27-7 TAKEOFF

Just as all elements of the wheel are equally important but must be dependent on the hub, so are all functions of the estimate equally important but most dependent on the takeoff.

By definition, the takeoff is the process of counting and measuring quantities of material and equipment and recording them in a manner that indicates the relationship between the actual conditions of installation and those of the standards for subsequent factoring of labor units where required. See Figure 27.7.

The accuracy of the estimate is largely governed by the accuracy of the takeoff.

The effectiveness of job management depends on the accuracy of the bill of material, which in turn is dependent on the accuracy of the takeoff. Therefore, to get the job, it is essential to have an accurate estimate, and then it is highly desirable to manage the job so that it is within the limits of the estimate.

LARGE TAKE-OFF & LISTING SHEET

JOB ____________________ ESTIMATED BY ________ PAGE ______ OF ______ PAGES

____________________ CHECKED BY ________ DATE ____________

TOTALS													

(a)

Figure 27.7 (*a*) Large Takeoff & Listing Sheet. Use this form to list various materials as they are measured or counted from plans, using one cross line for each circuit or each room as desired. (Courtesy of Minnesota Electrical Association, Inc.)

SMALL TAKE-OFF & LISTING SHEET

JOB ______________________________ PAGE ________

______________________________ OF ________ PAGES

ESTIMATED BY ____________ CHECKED BY ____________ DATE ____________

Form E-8

(b)

Figure 27.7 (*b*) Small Takeoff & Listing Sheet. As the length of conduit runs, conduit, wire, outlet boxes, switches, and so forth, are measured and counted from the plans, they must be listed first to obtain total quantities of each item or type material before being transferred to regular estimate or price sheets. (Courtesy of Minnesota Electrical Association, Inc.)

To ease the recording of quantities and prevention of oversights and omissions, use proper estimating forms for each system shown on the plans, as follows:

1. Service, metering, and grounding
2. Lighting fixture schedule
3. Distribution equipment schedule
4. Raceway equipment schedule
5. Feeder and busway schedule
6. Branch circuit schedule
7. Outlet detail and summary
8. Power system
9. Telephone system
10. Special systems

Each of these systems requires conduit, wire, boxes, and equipment in varying types and quantities, but for the sake of continuity and accuracy, it is essential to take off each system separately and record items on individual estimating forms designed for that particular system.

Another important factor in takeoff for both speed and accuracy is the type of equipment used to count and measure items of material and equipment shown on the plans.

SUMMARY

1. Estimating is the process of forecasting future costs of labor, material, and job expense.
2. Estimates may be classified into several types, depending upon their purpose and the degree of accuracy required.
3. An estimator's ability to visualize the different phases of construction is of utmost importance.
4. Mistakes in estimating can result in loss of contracts.
5. Several factors have an important bearing on whether or not a job should be bid.
6. Material + labor + direct job expense + overhead = cost.
7. Cost + profit = selling price.
8. Overhead expenses are items that cannot be changed to a particular job or project but must be paid to remain in business.
9. The amount of profit a firm makes depends on many factors.
10. Just as all elements of the wheel are equally important, but must depend on the hub, so are all functions of the estimate equally important but most dependent on the takeoff.

PROBLEMS

27-1 Write the formula for figuring prime cost.

27-2 Write the formula for figuring the selling price.

27-3 List at least six qualifications of an estimator.

1. ______________________ 4. ______________________

2. ______________________ 5. ______________________

3. ______________________ 6. ______________________

27-4 List eight possible errors made by estimators.

1. ______________________
2. ______________________
3. ______________________
4. ______________________
5. ______________________
6. ______________________
7. ______________________
8. ______________________

27-5 What five labor factors are figured in as direct job expenses?

1. ______________________
2. ______________________
3. ______________________
4. ______________________
5. ______________________

27-6 An electric drill costs the contractor $155.00. The average life of this drill is estimated to be 12 months. How much tool cost must be charged off per month?

27-7 What happens if the contractor overestimates the selling cost of the bid?

____

____

____

____

27-8 What is material unit pricing? ______________________

____

27-9 What is the meaning of "direct job expense"?

____

____

____

____

27-10 You have been asked to contribute $100.00 to a charity. What would you charge this off to (circle one)?

1. Direct job expense
2. Unit pricing
3. Overhead

27-11 A contractor bids five jobs and gets one. The cost of the unsuccessful bids (circle one):

1. is zero.
2. must be included in overhead.
3. is forgotten.
4. is absorbed by the architect.

References

National Fire Protection Association, *National Electrical Code,* New York: McGraw-Hill, 1977.

W. J. McGuinness and B. Stein, *Building Technology: Mechanical and Electrical Systems,* New York: Wiley, 1977.

H. P. Richter, *Practical Electrical Wiring,* Tenth Edition, New York: McGraw-Hill, 1976.

R. C. Mullin, *Electrical Wiring Residential,* New York: Delmar, 1975.

W. N. Alerich, *Electrical Construction Wiring,* American Technical Society, Chicago, 1971.

Sears, Roebuck and Co., "Simplified Electrical Wiring," 1969.

American Heating and Air Conditioning Wholesalers Association, *Fundamentals of Heating,* Second Edition. Columbus, 1971.

Energy Utilization Systems, "Energy Consumption and Life-Cycle Costs of Space Conditioning Systems," Pittsburgh, 1976.

ALA Research Corporation, "Solar Dwelling Design Concepts," Washington, D.C., 1976.

Glossary

AC. Alternating current

AWG. American Wire Gauge, the standard wire size measuring system in the United States.

Access. Opening in a building through which equipment can be carried in or out, or removable panels in equipment for servicing parts.

Air changes. The number of times the air is changed per hour in a room.

Air conditioning. A method of filtering air and regulating its humidity and temperature in buildings.

Air density. The weight of air, pounds per cubic foot.

Alternating current. An electric current that reverses its direction of flow at regular intervals.

American wire gauge. Gauge used for designating the sizes of solid copper wire used in the United States.

Ampacity. A wire's ability to carry current safely, without undue heating. The term formerly used to describe this characteristic was current-carrying capacity of the wire.

Ampere. The practical unit that indicates the rate of flow of electricity through a circuit.

Annunciator. Electric signal equipment having a number of pushbuttons located at different places which are wired to an electromagnetic in the annunciator box. When pressure is on any pushbutton, it causes a signal to be displayed showing what button was operated.

Appliance circuit. A branch circuit that supplies outlets specifically intended for appliances.

Appliance outlet. An outlet connected to an appliance circuit. It may be a single or duplex receptacle or an outlet box intended for direct connection to an appliance.

Architectural-electrical plan. Architectural plan on which electrical work is shown.

Auxiliary circuit. Another circuit besides the main circuit; often a control circuit.

Auxiliary resistance heating. Electric resistance heaters that supplement the heat from a heat pump or solar unit.

BX. Trade name for NEC type AC flexible armored cable.

Baseboard. Hot water heater or electric heater along the wall.

Bonding. To assume a low-impedance path to ground should a fault occur on any of the service-entrance conductors.

Box connector. An attachment used for fastening the ends of cable to a box.

Branch circuit. That part of the wiring system between the final set of fuses protecting it and the place where the lighting fixtures or drop cords are attached.

British Thermal Unit. Quantity of heat.

Bus duct. An assembly of heavy bars of copper or aluminum that acts as a conductor of large capacity.

Cable. A conductor composed of a number of wires twisted together.

Cable grip. A clamp that grips the cable when it is being pulled into place.

Carrying capacity. The amount of current a wire can carry without overheating.

Cartridge fuse. A fuse enclosed in an insulating tube in order to confine the arc or vapor when the fuse blows.

Circuit. The path taken by an electrical current in flowing through a conductor from one terminal of the source of supply to the other.

Circuit breaker. A device used to open a circuit automatically.

Circular mil. The area of a circle one-thousandth of an inch in diameter; area in circular mils = diameter, in mils, squared (multiplied by itself).

Closed circuit. A complete electric circuit through which current will flow when voltage is applied.

Common neutral. A neutral conductor that is common to, or serves, more than one circuit.

Conductivity. The ability of a substance to carry an electric current.

Conduit. A pipe or tube, made of metal or other material, in which electrical conductors or wires are placed.

Conduit box. An iron, steel, or nonmetallic box located between the ends of the conduit where the wires or cables are spliced.

Conduit bushing. A short threaded sleeve fastened to the end of the conduit inside the outlet box; inside of sleeve is rounded out on one end to prevent injury to the wires.

Conduit coupling. A short metal tube threaded on the inside and used to fasten two pieces of conduit end to end.

Conduit elbow. A short piece of conduit bent to an angle, usually to 45 or 90 degrees.

Conduit rigid. A mild steel tubing used to enclose electric wires.

Conduit wiring. Electric wires placed inside conduit.

Connected load. The sum of all loads on a circuit.

Contactor. A device used to open and close an electrical circuit rapidly and often.

Continuous rating. The output at which a machine can operate continuously without overheating or exceeding a certain temperature.

Contract documents. Legal papers that include the contract, the working drawings, and the specification.

Contractor. One who does a certain job for a sum of money agreed upon before the work is started.

Control switch. A small switch used to open and close a circuit which operates a motor or an electromagnetic coil; this motor or electromagnetic coil is used to operate or control some electric machine.

Convector. A heating element that warms the air passing over it which, in turn, rises to warm the space by convection.

Convenience outlet. A duplex receptacle connected to a general-purpose branch circuit, not intended for any specific item of electrical equipment.

Copper. A metal used for electrical conductors because it has less resistance than any other metal except silver.

Cord. Two insulated flexible wires or cables twisted or held together with a covering of rubber, tape, or braid.

Current. The electric flow in a electric circuit, expressed in amperes (amps).

DP. Double pole.

DPS. Double pole snap switch.

DPST. Double pole single throw.

DPDT. Double pole double throw.

Dead end. The end of a wire to which no electrical connection is made; the end used for supporting the wire; the part of a coil or winding that is not in use.

Dead wire. A wire in which there is no electric current or voltage.

Degree day. The number of Fahrenheit degrees that the average outdoor temperature over a 24-hour period is less than 65°F.

Demand. Amount of electric current needed from a circuit or generator.

Demand factor. Ratio of the maximum amount of power required by a system to the total connected load of the system.

Dimmer. A resistance coil connected in series with a lamp to reduce the amount of current flowing through it, consequently dimming or reducing the light.

Disconnect. To remove an electrical device from a circuit, or to unfasten a wire, making part or all of the circuit inoperative.

Distribution. Division of current between the branches of an electrical circuit.

Distribution box. Small metal box in a conduit installation, giving accessibility for connecting branch circuits.

Distribution lines. The main feed line of a circuit to which branch circuits are connected.

Distribution panel. Insulated board from which connections are made between the main feed lines and branch lines.

Distribution system. The whole circuit and all of its branches which supply electricity to consumers.

Duct. A space in an underground conduit to hold a cable or conductor.

Electrical codes. Rules and regulations for the installation and operation of electrical devices and currents.

Electronic air cleaner. A filter somewhat more efficient than the usual filter, particularly for the removal of small suspended particles.

Entrance cable. Fiber-insulated cable of two large-size insulated wires covered with a bare spiral wound sheathed that serves as the neutral conductor.

Entrance head. A weatherproof housing for supporting the entrance cable.

Fault. A short circuit either line to line or line to ground.

Feeder. Line supplying all the branch circuits with the main supply of current.

Fish wire. Flat, narrow, flexible steel wire used to pull conductors through lengths of conduit.

Fixture. Device for holding electric lamps which is wired inside and is attached to the wall or ceiling.

Fixture wire. Insulated, stranded wire used for wiring fixtures.

Flexible cable. Cable consisting of insulated, stranded, or woven conductors.

Flexible conduit. Nonrigid conduit made of fabric or metal strip wound spirally.

Flush receptacle. Type of lamp socket, the top of which is flush with the wall into which the socket is recessed.

Fossil fuels. Oil, gas, and coal.

Four-way switch. Switch that controls the current in four conductors by making or breaking four separate contacts.

Four-wire, three-phase system. Distribution system having a three-phase star connection, one lead being taken from the end of each winding and the fourth from the point where they are all connected together.

Fuse. Safety device to prevent overloading a current; it consists of a short length of conducting metal which melts at a certain heat and thereby breaks the circuit.

Fuse block. Insulated block designed to hold fuses.

Fuse clip. Spring holder for a cartridge-type fuse.

Fuse plug. Fuse mounted in a screw plug, which is screwed in the fuse block like a lamp in a socket.

GFCI. Ground-fault circuit interrupter—a device that senses ground faults and reacts by opening the circuit.

Gang. One wiring device position in a box.

Ganged switches. A group of switches arranged next to each other in ganged outlet boxes.

General-purpose branch circuit. One that supplies a number of outlets for general lighting and convenience receptacles.

Greenfield. Trade name for flexible steel conduit.

Ground circuit. Part of an electric circuit in which the ground serves as a path for the current.

Ground clamp. Clamp on a pipe or other metal conductor connected to the ground for attaching a conductor of an electrical circuit.

Ground electrode. A piece of metal physically connected to ground. Can be pipe, rod, mat, pad, or structural member.

Ground fault. An unintentional connection to ground.

Ground plate. Metal plate buried in moist earth to make a good ground contact for an electrical circuit.

Ground wire. Conductor connecting an electrical device or circuit to the ground.

Grounded neutral wire. The neutral wire connected to the ground in a three-way distribution system.

Heat pump. An all-electric heating/cooling device that takes energy for heating from outdoor air.

Home-run. Cables that run between the service distribution panel and the first outlet in the branch circuit.

Hot conductor. A conductor or wire which is carrying a current or voltage.

Humming. A noise caused by the rapid magnetizing and demagnetizing of the iron core of a transformer, motor, or generator.

Illumination. The directing of light from its source to where it can be used to the best advantage.

Individual branch circuit. One that supplies only a single piece of electrical equipment.

Infiltration. Cold air that leaks in. Expressed as air changes per hour.

Insulate. To place insulation around conductors or conducting parts.

Insulating compound. An insulating wax which is melted and poured around electrical conductors in order to insulate them from other objects.

Insulating joint. A thread or coupling in which the two parts are insulated from each other.

Insulator. A device used to insulate electric conductors.

Interior wiring. Wiring placed on the inside of buildings.

Junction box. A box in a street distribution system where one main is connected to another main; also a box where a circuit is connected to a main.

Key switch. A switch for turning on and off electric circuits which are operated by means of a special key.

Kilowatt. One thousand watts.

Knife switch. A switch having a thin blade that makes contact between two flat surfaces or short blades to complete the circuit.

Labeled. Materials and equipment having a label that indicates compliance with standards or tests to determine suitable usage.

Lamp. A device used to produce light.

Lamp base. The metal part of a lamp which makes contact with the socket.

Lamp dimmer. An adjustable resistance connected to a lamp circuit in order to reduce the voltage and the brightness of the lamps.

Lamp socket. A receptacle into which the base of the lamp is inserted, which makes connection from the lamp to the circuit.

Leads. Short lengths of insulated wires that conduct current to and from a device.

Lighting fixture. An ornamental device which is fastened to the outlet box in the ceiling and which has sockets for the lamps.

Lightning arrester. A device that allows the lightning to pass to the ground, thus protecting electrical machines.

Line side. The side of a device electrically closest to the source of current.

Line voltage thermostat. A thermostat that is connected directly to the line. Full line voltage is fed through it to the controlled heater.

Live. A circuit carrying a current or having a voltage on it.

Load. The work required to be done by a machine. The current flowing through a circuit.

Load side. The side of a device electrically furthest from the current source.

Lumen. Unit of light flux.

Luminaire. An ornamental electric lighting fixture.

MCM. Thousand circular mils used to describe large wire sizes.

Main. The circuit from which all other smaller circuits are taken.

Main feeder. A feeder supplying power from the generating station to the main.

Master control. Control of all the outlets from one point.

Metal conduit. Iron or steel pipe in which electric wires and cables are installed.

Meter base. A device intended to hold the kilowatthour meter.

Mutlipole. Connects to more than one pole, such as a two-pole circuit breaker.

NEC. National Electric Code; often called.

Negative. The point toward which current flows in an external electrical circuit; opposite of positive.

Neon. An inert gas used in electric lamps.

Neutral. Not positive or negative, although it may act as positive to one circuit and negative to another.

Neutral conductor. A middle conductor of a three-wire direct-current or single-phase circuit.

Neutral terminal. A terminal that may be positive to one circuit and negative to another circuit.

Open circuit. A break in a circuit; not having a complete path or circuit.

Outlet box. A steel box placed at the end of conduit where electric wires are joined to one another and to the fixtures.

Overcurrent device. A device such as a fuse or a circuit breaker designed to protect a circuit against excessive current by opening the circuit.

Overhead. Electric light wires carried outdoors on poles.

Overload. Carrying a greater load than the machine or device is designed to carry.

Overload capacity. The amount of load beyond a rated load that a machine will carry for a short time without dangerously overheating.

Panelboard. A box containing a group of overcurrent devices intended to supply branch circuits.

Panel box. The box in which switches and fuses for branch circuits are located.

Panel schedule. A schedule appearing on the electrical drawing detailing the equipment contained in the panel.

Parallel. Two lines that are equally distant at all points. Connecting machines or devices so that the current flows through each one separately from one line wire to another line wire.

Parallel circuit. A multiple circuit; a connection where the current divides and parts flow through each device connected to them.

Pilot lamp. A small lamp used on switchboards to indicate when a circuit switch or device has operated.

Pole. An electrical connection point. In a panel, the point of connection. On a device, the terminal that connects to the power.

Positive. The point in a circuit from which the current flows; opposite of negative.

Primary. That which is attached to a source of power, as distinguished from the secondary.

Pull box. A metal cabinet inserted into a conduit run for the purpose of providing a cable pulling point.

Raceways. Metal moulding or conduit that has a thinner wall than standard rigid conduit used in exposed wiring.

Radiant cables. Electric cables embedded in the ceiling for heating.

Range hood. Hood over a stove to collect odor-laden air that is to be exhausted.

Receptacle. A device placed in an outlet box to which the wires in the conduit are fastened, enabling quick electrical connection to be made by pushing an attachment plug into it.

Recessed. A converter cabinet that extends partially or fully into a pocket in the wall.

Riser diagram. Electrical block-type diagram showing connection of major items of equipment. It is also applied to signal equipment connections, as a fire-alarm riser diagram. Generally applied to multistory building.

Romex. One of several trade names for type NM nonmetallic-sheathed cable.

R-value. The resistance rating of thermal insulation.

Service connections. The wiring from the distributing mains to the building.

Service drop. The overhead service wires that serve a building.

Service entrance. The place where the service wires are run into a building.

Service lateral. The underground service conductors.

Service switch. The main switch that connects all the lamps or motors in a building to the service wires.

Service wires. The wires that connect the wiring in a building to the outside supply wires.

Short circuit. An accidental connection of low resistance joining two sides of a circuit, through which nearly all the current will flow.

Single pole. Connects to a single hot line.

Single-pole switch. A switch that opens and closes only one side of a circuit.

Stranded wires. Wires or cables composed of a number of smaller wires twisted or braided together.

Switch. A device for closing or opening of a circuit.

Switch glade. The movable part of a switch.

Switch plate. A small plate placed on the plastered wall to cover a switch.

Symbol. A letter, abbreviation, or sign that stands for a certain unit or thing.

Tap. A wire connected some distance from the end of the main wire or conductor.

Terminal. A connecting device placed at the end of a wire, appliance, or machine to enable a connection to be made to it.

Terminal lug. A lug crimped on to the end of a cable so it can be bolted to another terminal.

Thermal transfer. Moving heat into or out of occupied space.

Three-way switch. A switch with three terminals by which a circuit can be completed through any one of two paths.

Three-wire circuit. A circuit using a neutral wire in which the voltage between outside wires is twice that between neutral and each side.

Time switch. A switch controlled by a clock that opens and closes a circuit at the desired time.

Toggle-type switch. A device with a projecting lever whose movement causes contact to be made with a snap action.

Transducer. A device that will convert one form of energy into another form.

Transformer. A device used to change alternating current from one voltage to another. It consists of two electrical circuits joined together by a magnetic circuit formed in an iron core.

Travelers. Hot black or red conductors in a three-wire circuit controlled by a three-way switch.

Tungsten filament. A filament made from tungsten (a very hard metal with a high melting point) and used in a lamp bulb.

Two pole. A switch that opens or closes both sides of a circuit or two circuits at one time.

Unbalanced load. A distribution system where there is a greater load on one phase or side than on the other.

W. Symbol for watt.

Watt. The unit of electric power.

Wire gauge. A method of measuring the diameter of different wires.

Wire nut. Trade name for small, solderless, twist-on branch circuit conductor connector.

Wireway. Term generally used to mean a surface raceway.

Wiring connector. A device for joining one wire to another.

Wiring device. Receptacle, switch, pilot light, dimmer switch, or any device that is wired in a branch circuit and fits into a 4-inch outlet box.

Wiring diagram. Diagram showing actual wiring, with numbered terminals. All wiring is shown.

Wiring symbols. Small signs placed on a wiring diagram to indicate different devices and connections.

U coefficient. The rate of heat transmission.

Ventilation. Controlled air.

Voltage. The electric pressure in an electric circuit, expressed in volts.

Zone. Section of a heating and/or cooling system separately controlled.

INDEX